Springer Series on
SIGNALS AND COMMUNICATION TECHNOLOGY

SIGNALS AND COMMUNICATION TECHNOLOGY

Foundations and Applications of Sensor Management
A.O. Hero III, D. Castañón, D. Cochran, and K. Kastella (Eds.)
ISBN 978-0-387-27892-6

Human Factors and Voice Interactive Systems, Second Edition
D. Gardner-Bonneau and H. Blanchard
ISBN 978-0-387-25482-1

Wireless Communications: 2007 CNIT Thyrrenian Symposium
S. Pupolin
ISBN 978-0-387-73824-6

Adaptive Nonlinear System Identification: The Volterra and Wiener Model Approaches
T. Ogunfunmi
ISBN 978-0-387-26328-1

Wireless Network Security
Y. Xiao, X. Shen, and D.Z. Du (Eds.)
ISBN 978-0-387-28040-0

Satellite Communications and Navigation Systems
E. Del Re and M. Ruggieri
ISBN: 0-387-47522-2

Wireless Ad Hoc and Sensor Networks
A Cross-Layer Design Perspective
R. Jurdak
ISBN 0-387-39022-7

Cryptographic Algorithms on Reconfigurable Hardware
F. Rodriguez-Henriquez, N.A. Saqib, A. Díaz Pérez, and C.K. Koc
ISBN 0-387-33956-6

Multimedia Database Retrieval
A Human-Centered Approach
P. Muneesawang and L. Guan
ISBN 0-387-25627-X

Broadband Fixed Wireless Access
A System Perspective
M. Engels and F. Petre
ISBN 0-387-33956-6

Distributed Cooperative Laboratories
Networking, Instrumentation, and Measurements
F. Davoli, S. Palazzo and S. Zappatore (Eds.)
ISBN 0-387-29811-8

The Variational Bayes Method in Signal Processing
V. Šmídl and A. Quinn
ISBN 3-540-28819-8

Topics in Acoustic Echo and Noise Control
Selected Methods for the Cancellation of Acoustical Echoes, the Reduction of Background Noise, and Speech Processing
E. Hänsler and G. Schmidt (Eds.)
ISBN 3-540-33212-x

EM Modeling of Antennas and RF Components for Wireless Communication Systems
F. Gustrau, D. Manteuffel
ISBN 3-540-28614-4

Interactive Video
Methods and Applications
R. I Hammoud (Ed.)
ISBN 3-540-33214-6

ContinuousTime Signals
Y. Shmaliy
ISBN 1-4020-4817-3

Voice and Speech Quality Perception
Assessment and Evaluation
U. Jekosch
ISBN 3-540-24095-0

Advanced ManMachine Interaction
Fundamentals and Implementation
K.-F. Kraiss
ISBN 3-540-30618-8

Orthogonal Frequency Division Multiplexing for Wireless Communications
Y. (Geoffrey) Li and G.L. Stüber (Eds.)
ISBN 0-387-29095-8

Circuits and Systems Based on Delta Modulation
Linear, Nonlinear and Mixed Mode Processing
D.G. Zrilic   ISBN 3-540-23751-8

Functional Structures in Networks
AMLn—A Language for Model Driven Development of Telecom Systems
T. Muth   ISBN 3-540-22545-5

RadioWave Propagation for Telecommunication Applications
H. Sizun   ISBN 3-540-40758-8

Electronic Noise and Interfering Signals
Principles and Applications
G. Vasilescu   ISBN 3-540-40741-3

DVB
The Family of International Standards for Digital Video Broadcasting, 2nd ed.
U. Reimers   ISBN 3-540-43545-X

(continued after index)

Edited by
Alfred O. Hero III
David A. Castañón
Douglas Cochran
Keith Kastella

Foundations and Applications of Sensor Management

ditors:

Alfred O. Hero III
University of Michigan
Ann Arbor, MI
USA

David Castañón
Boston University
Boston, MA
USA

Douglas Cochran
Arizona State University
Tempe, AZ
USA

Keith Kastella
General Dynamics Michigan Research and
  Development Center
Ypsilanti, MI
USA

ISBN 978-0-387-27892-6   e-ISBN 978-0-387-49819-5

Library of Congress Control Number: 2007933504

© 2008 Springer Science+Business Media, LLC
All rights reserved. This work may not be translated or copied in whole or in part without the written permission of the publisher (Springer Science+Business Media, LLC, 233 Spring Street, New York, NY 10013, USA), except for brief excerpts in connection with reviews or scholarly analysis. Use in connection with any form of information storage and retrieval, electronic adaptation, computer software, or by similar or dissimilar methodology now known or hereafter developed is forbidden. The use in this publication of trade names, trademarks, service marks and similar terms, even if they are not identified as such, is not to be taken as an expression of opinion as to whether or not they are subject to proprietary rights.

Printed on acid-free paper.

9 8 7 6 5 4 3 2 1

springer.com

Preface

This book grew out of a two day workshop that was held in May 2005 and was funded by the U.S. Defense Advanced Projects Research Agency (DARPA) and the U.S. National Science Foundation (NSF). The express purpose of this workshop was to gather together key contributors to the field of active sensing and sensor management to discuss the state-of-the-art in research, the main mathematical approaches to design and performance approximation, the problems solved and the problems remaining. At the end of the workshop the participants had generated an outline and agreed on writing assignments.

The intended audience for this book are graduate students, engineers and scientists in the fields of signal processing, control, and applied mathematics. Readers would benefit from a rudimentary background in statistical signal processing or stochastic control but the book is largely self contained. Appendices cover background material in information theory, Markov processes, and stopping times. A symbol index and a subject index are also included to facilitate the reader's navigation through the book.

Thus the book lies somewhere between a coherent textbook and a loose collection of papers typical of many recent edited collections on emerging topics in engineering. Like an edited collection, the chapters were written by some of the principal architects of recent advances in sensor management and active sensing. However, authors and editors attempted to adopt a common notation, cross reference other chapters, provide index terms, and adhere to an outline established at the NSF workshop. We hope the reader will find that the book has benefited from this extra planning and coordination.

Alfred Hero, David Castañón, Douglas Cochran, and Keith Kastella

Ann Arbor, Boston, Tempe, Ypsilanti
July 2007

Acknowledgments

The editors wish to acknowledge the U.S. Defense Advanced Research Projects Agency (DARPA) and the U.S. National Science Foundation (NSF) for supporting a workshop, NSF grant number CCF 0524865, at which the structure of this volume was initially worked out.

Contents

Preface	v
Acknowledgments	vii
Contributing Authors	xv
Symbol Index	xvii

1
Overview of Book — 1
Alfred O. Hero III, David A. Castañón, Douglas Cochran, Keith Kastella

1.	Introduction	1
2.	Scope of Book	2
3.	Book Organization	3

2
Stochastic Control Theory for Sensor Management — 7
David A. Castañón, Lawrence Carin

1.	Introduction	7
2.	Markov Decision Problems	10
3.	Partially Observed Markov Decision Problems	19
4.	Approximate Dynamic Programming	26
5.	Example	27
6.	Conclusion	32

3
Information Theoretic Approaches to Sensor Management
Alfred O. Hero III, Christopher M. Kreucher, Doron Blatt

1.	Introduction	33
2.	Background	35
3.	Information-Optimal Policy Search	40
4.	Information Gain Via Classification Reduction	43
5.	A Near Universal Proxy	44
6.	Information Theoretic Sensor Management for Multi-target Tracking	47
7.	Terrain Classification in Hyperspectral Satellite Imagery	53
8.	Conclusion and Perspectives	57

4
Joint Multi-target Particle Filtering
Christopher M. Kreucher, Mark Morelande, Keith Kastella, Alfred O. Hero III

1.	Introduction	59
2.	The Joint Multi-target Probability Density	62
3.	Particle Filter Implementation of JMPD	71
4.	Multi-target Tracking Experiments	85
5.	Conclusions	91

5
POMDP Approximation Using Simulation and Heuristics
Edwin K. P. Chong, Christopher M. Kreucher, Alfred O. Hero III

1.	Introduction	95
2.	Motivating Example	97
3.	Basic Principle: Q-value Approximation	98
4.	Control Architecture	101
5.	Q-value Approximation Methods	104
6.	Simulation Result	116

	7.	Summary and Discussion	118

6
Multi-armed Bandit Problems
Aditya Mahajan, Demosthenis Teneketzis

121

	1.	Introduction	121
	2.	The Classical Multi-armed Bandit	122
	3.	Variants of the Multi-armed Bandit Problem	134
	4.	Example	148
	5.	Chapter Summary	151

7
Application of Multi-armed Bandits to Sensor Management
Robert B. Washburn

153

	1.	Motivating Application and Overview	153
	2.	Application to Sensor Management	155
	3.	Example Application	162
	4.	Summary and Discussion	173

8
Active Learning and Sampling
Rui Castro, Robert Nowak

177

	1.	Introduction	177
	2.	A Simple One-dimensional Problem	179
	3.	Beyond 1d - Piecewise Constant Function Estimation	190
	4.	Final Remarks and Open Questions	199

9
Plan-in-Advance Learning
Xuejun Liao, Yan Zhang, Lawrence Carin

201

1.	Introduction	201
2.	Analytical Forms of the Classifier	203
3.	Pre-labeling Selection of Basis Functions ϕ	204
4.	Pre-labeling Selection of Data $\mathcal{X}_{\text{tr}}$	209
5.	Connection to Theory of Optimal Experiments	210
6.	Application to UXO Detection	212
7.	Chapter Summary	219

10
Sensor Scheduling in Radar
William Moran, Sofia Suvorova, Stephen Howard

221

1.	Introduction	221
2.	Basic Radar	222
3.	Measurement in Radar	233
4.	Basic Scheduling of Waveforms in Target Tracking	234
5.	Measures of Effectiveness for Waveforms	239
6.	Scheduling of Beam Steering and Waveforms	245
7.	Waveform Libraries	250
8.	Conclusion	255

11
Defense Applications
Stanton H. Musick

257

1.	Introduction	257
2.	Background	259
3.	The Contemporary Situation	260
4.	Dynamic Tactical Targeting (DTT)	262
5.	Conclusion	266

12	Appendices		269

Alfred O. Hero, Aditya Mahajan, Demosthenis Teneketzis, Edwin Chong

	1.	Information Theory	269
	2.	Markov Processes	273
	3.	Stopping Times	278

References	283
Index	305

Contributing Authors

Doron Blatt, DRW Holdings, Chicago, IL, USA

Lawrence Carin, Duke University, Durham, NC, USA

David A. Castañón, Boston University, Boston, MA, USA

Rui Castro, University of Wisconsin, Madison, WI, USA

Douglas Cochran, Arizona State University, Tempe, AZ, USA

Edwin K. P. Chong, Colorado State University, Fort Collins, CO, USA

Alfred O. Hero III, University of Michigan, Ann Arbor, MI, USA

Stephen Howard, Defence Science and Technology Organisation, Edinburgh, Australia

Keith Kastella, General Dynamics Michigan Research and Development Center, Ypsilanti, MI, USA

Christopher M. Kreucher, General Dynamics Michigan Research and Development Center, Ypsilanti, MI, USA

Xuejun Liao, Duke University, Durham, NC, USA

Aditya Mahajan, University of Michigan, Ann Arbor, MI, USA

William Moran, University of Melbourne, Melbourne, Australia

Mark Morelande, University of Melbourne, Melbourne, Australia

Stanton H. Musick, Sensors Directorate, Air Force Research Laboratory, Wright-Patterson Air Force Base, OH, USA

Robert Nowak, University of Wisconsin, Madison, WI, USA

Sofia Suvorova, University of Melbourne, Melbourne, Australia

Demosthenis Teneketzis, University of Michigan, Ann Arbor, MI, USA

Robert B. Washburn, Parietal Systems, Inc., North Andover, MA, USA

Yan Zhang, Innovation Center of Humana, Inc., Louisville, KY, USA

Symbol Index

General notation

$\mathbb{R}$	the real line
$\mathbb{R}^2$	the real plane
$\mathbf{x}$	bold lowercase letter denotes a vector as in $\mathbf{x} = [x_1, \ldots, x_2]$
s	complex variable, a vector $\Re(s) + j\Im(s)$ in the complex plane
s_*	complex conjugate $\Re(s) - j\Im(s)$ of complex variable s
$\mathbf{A}$	bold uppercase letter near beginning of alphabet denotes a matrix
$\mathbf{I}$	the identity matrix
T	vector or matrix transpose as in $\mathbf{x}^\mathsf{T}$
H	Hermitian transpose of a complex vector or matrix, $\mathbf{x}^H = (\mathbf{x}^*)^\mathsf{T}$
Π	permutation operator as in $\Pi([x_1, \ldots, x_M]) = [x_M, \ldots, x_1]$
$\langle \mathbf{x}, \mathbf{y} \rangle$	inner product between two vectors $\mathbf{x}$ and $\mathbf{y}$ in $\mathbb{R}^d$
$\|\mathbf{x}\|$	norm of a vector
$\|\mathbf{x}\|$	number of elements of vector x
$\|\mathcal{S}\|$	number of elements in the set $\mathcal{S}$
$I^A(x)$	indicator function of set A

Statistical notation

Y	uppercase letter near end of alphabet usually denotes a random variable
y	lowercase letter near end of alphabet usually denotes a realization
$\mathbf{Y}$	bold uppercase letter usually denotes a random vector
$\mathbb{E}$	statistical expectation as in $\mathbb{E}[\mathbf{Y}\mathbf{Y}^\mathsf{T}]$
P	probability measure as in $P(X \in [0,1])$
p_X	probability density or probability mass function of random variable X
$p(x)$	shorthand for $p_X(x)$
$\mathcal{D}$	information divergence as in α-divergence $\mathcal{D}_\alpha(p_X \| p_Y)$
J	Fisher information matrix

π_k	belief state (information state)
π	the belief state space, a space of density functions π
P_d, P_f	probability of detection, probability of false alarm
P_e	Bayes probability of error

Scheduling notation

a	a sensor action
$\mathcal{A}$	the space of sensor actions a
Y_k	measured sensor output at time k
$\mathbf{Y}^k$	vector of measurements, $\mathbf{Y}^k = [Y_1, \ldots, Y_k]^\mathsf{T}$
S_k	the state of nature to be determined from past sensor measurements
$\mathbf{X}_k$	state vector of a target in the plane, a special case of S_k
$R(S, a)$	reward due to taking action a in state s
γ_k	policy, a mapping from available measurements to sensor actions $\mathcal{A}$ at time k
V_γ	value function associated with policy γ
$\mathbb{E}_\gamma$	statistical expectation under a given policy γ
$<g, \pi>$	conditional expectation of g given the posterior density π
β	scheduling discount factor
Q_k	Q-function of reinforcement learning; an empirical approximation to value function

Chapter 1

OVERVIEW OF BOOK

Alfred O. Hero III
University of Michigan, Ann Arbor, MI, USA

David A. Castañón
Boston University, Boston, MA, USA

Douglas Cochran
Arizona State University, Tempe, AZ, USA

Keith Kastella
General Dynamics Michigan Research and Development Center, Ypsilanti, MI, USA

1. Introduction

Broadly interpreted, sensor management denotes the theory and application of dynamic resource allocation in a diverse system of sensors and sensing modalities. Over the past few years the problem of sensor management for active sensing has received an increasing amount of attention from researchers in areas such as signal processing, automatic control, statistics, and machine learning. Active sensing is recognized as an enabling technology for the next generation of agile, multi-modal, and multi-waveform sensor platforms to efficiently perform tasks such as target detection, tracking, and identification. In a managed active sensing system, the sequence of sensor actions, such as pointing angle, modality, or waveform, are selected adaptively based on information extracted from past measurements. When the adaptive selection rule is carefully designed such an on-line active approach to sensing can very significantly improve overall performance as compared to fixed off-line approaches. However, due to the classic curse of dimensionality, design and implementation of optimal active sensing strategies has been and remains very challenging.

Recently, several U.S. research funding agencies have supported efforts in areas related to sensor management. There has also been interest in operations research approaches to sensor management, e.g., multi-armed bandits. This has led to research activity in academia, government laboratories, and industry that has borne some fruit in specific technology areas. For example, several promising new methods to approximate optimal multistage sensor management strategies for target tracking have been developed and an understanding of design challenges and performance tradeoffs is beginning to emerge.

2. Scope of Book

This book provides an overview of the mathematical foundations of sensor management, presents several of its relevant applications, and lays out some of the principal challenges and open problems. Stochastic modeling and analysis play central roles in sensor management. This is because most sensing systems operate in a complicated noise and clutter degraded environment and thus the benefit of selecting one sensing modality instead of another cannot be determined with high certainty. Consequently, any useful theory and practice of sensor management must be based on a probabilistic formulation and statistical performance prediction. The nature of the chapters in this book reflect this fact[1].

The mathematics of sensor management grew out of the well-established fields of stochastic control theory, information theory, and sequential decision theory. These fields offer powerful design and analysis tools when reliable statistical models for the measurements are available, e.g. linear or non-linear Gauss-Markov process models. Three chapters in this book (Chapters 2, 3, 6) develop the theory of sensor management on these foundations.

More recently there has been interest in machine learning approaches to sensor management that are capable of learning from data to compensate for uncertainty in statistical models. Three chapters develop sensor management theory and algorithms in the context of active learning (Chapters 8 and 9) and reinforcement learning (Chapter 5), fields that have developed only recently.

---

[1] While some background in stochastics and decision theory is provided in the following chapters and appendices, the reader is assumed to have had exposure to probability theory at the level of an advanced undergraduate or early graduate course in stochastic processes.

A major axis for the practice of sensor management has been multiple target tracking using remote sensors such as radar, sonar, and electro-optical sensors. Four chapters develop sensor management for target detection, tracking, and classification using particle filtering (Chapter 4), multi-armed bandit theory (Chapter 7), embedded simulation (Chapter 5), and Kalman filtering and LFM waveform libraries (Chapter 10). Other applications covered in this book include landmine detection (Chapter 9), underwater mine classification (Chapter 2), active sampling and terrain mapping (Chapter 8), range-Doppler radar (Chapter 10), and terrain classification (Chapter 3).

3. Book Organization

In the large, the development of topics in this book follows a logical flow progressing from theory to application.

Chapter 2 introduces the mathematics of sensor management as a problem of sequential decision making that can be solved by stochastic control theory. The decisions are tied to control actions that the sensor can take, e.g., selecting a transmit waveform or steering a radar beam, based on previous sensor measurements. The chapter introduces Markov decision processes (MDP) and controlled Markov processes, partially observed Markov decision processes (POMDP), reward and value functions, and optimal control policies. Backwards induction is introduced and the Bellman equations are derived for recursively computing the optimal policy. As these equations are of high complexity methods of approximation to the optimal policy are discussed. Chapter 2 ends by applying this theory to sensor management for active acoustic underwater target classification.

In sensor management problems, the state of the controlled Markov process is almost always only partially observed, e.g., due to noise or clutter degradations. In this case the optimal sequential decisions are functions of the posterior density of the Markov state, called the belief state or the information state. One of the principal challenges is the choice of reward for making good decisions, i.e., the objective function that an optimal decision-making policy should maximize. Unfortunately, many real scenarios are multi-objective in nature. For example, over the course of a surveillance mission a commander may wish to manage his sensors in order to simultaneously detect new targets, track them and identify them. To handle this type of scenario, information theoretic measures have been proposed by many researchers as surrogate reward functions. Chapter 3 gives an overview of information theoretic approaches to sensor management and several examples are given that demonstrate information-based sensor management for single stage (myopic) decision making.

One of the earliest and most challenging applications of sensor management is multiple target tracking, a principal application area in this book. Chapter 4 presents background on Bayesian multiple target detection, tracking, and classification and develops information-driven particle filtering methods for approximating the posterior density of the target state, the key ingredient for the sensor manager. Then, building on the material presented in Chapters 2-4, Chapter 5 presents general algorithms for the sensor manager using value-to-go approximation and parallel rollout techniques. Then the focus of Chapter 5 turns to the deficiencies of myopic sensor management strategies when there are time-varying inter-visibility constraints between sensor and target, e.g., loss of line-of-sight to target in a highly accentuated terrain. Chapters 4 and 5 include studies of radar tracking performance using both simulated and real multi-track data.

As mentioned earlier, sensor management problems can be solved optimally using stochastic dynamic programming which usually results in computationally intensive solutions. For many applications of sensor management low computational complexity is more important than optimality. Such applications call for computationally efficient sub-optimal approaches. Multi-armed bandit formulations provide a framework of heuristically appealing low-complexity sub-optimal solutions to sensor management problems. The solution approach to multi-armed bandit problems is to determine an "index" for each alternative and choose the alternative with the highest index. Computing such an index is computationally easier than solving the Bellman equations for the entire problem. However such an *index policy* is not always optimal. Chapter 6 describes the formulation of the classical multi-armed bandit and some of its variants, and explains the intuition behind its solution approach. Chapter 7 explains how to apply the ideas of multi-armed bandits to sensor management for radar target tracking.

In Chapter 8 the emphasis turns to active learning theory in the context of adaptive sampling for spatio-temporal change detection. The authors of this chapter provide motivation by way of an airborne laser topographical mapping example. For this example, sensor management reduces to finding an optimal policy for sequential redirection of the laser beam in order to perform quickest detection of topographical step changes. Using statistical machine learning techniques the authors prove asymptotic optimality of a myopic probabilistic bisection method. As compared to standard fixed sampling strategies, e.g., Nyquist sampling, they establish that their adaptive sampling method requires exponentially fewer samples for equivalent change detection performance.

Chapter 9 continues along the adaptive sampling thread of Chapter 8. The authors propose an adaptive sampling method, called "plan-in-advance" sampling to improve performance of a sequential classification algorithm. Their method is closely related to D-optimal design in the theory of optimal design of experiments (DOE). Their proposed algorithm uses label-independent matching basis pursuit and a label-dependent weights selection procedure to sequentially select the most informative sample points and estimate the optimal classifier function. Chapter 9 concludes with an application to landmine detection and classification using electromagnetic induction (EMI) and magnetometric sensors.

Chapter 10 returns to the multiple target tracking theme of Chapters 4-5 but with different emphasis. The authors discuss sensor management strategies for the problem of waveform scheduling for multi-target tracking with Doppler radars. Waveform scheduling is a kind of sensor management "at the transmitter" and must account for radar spatial sensitivity variations in azimuth, elevation and range-Doppler. Consequently, radar waveform scheduling incorporates waveform design and thus requires more sophisticated models of Doppler-radar processing than were necessary in Chapters 4, 5 and 7. After reviewing the mathematics of radar Doppler processing the chapter turns to probabilistic data association (PDA) methods for multi-target tracking. It then addresses the problem of waveform design and selection using various optimization criteria including: the single noise covariance (SNM) matrix, the integrated clutter measure (ICM), and mutual information (MI).

The book concludes with a chapter on defense applications of sensor management. The focus of Chapter 11 is on intelligence, surveillance and reconnaissance (ISR) for vehicle detection and tracking. It provides a historical overview of ISR sensing systems and discusses a major US DoD research program (DARPA-DTT) in which sensor management led to significant improvements in system capabilities. The chapter ends with some perspectives on the remaining challenges in sensing and sensor management for national defense and security.

Most of the illustrative examples in this book are drawn from the domain of national defense and security. However, as pointed out in several of these chapters, the theory and methods described herein are applicable to a far wider range of sensing applications. We hope that this book will generate additional interest in the general area of sensor management and active sensing.

Chapter 2

STOCHASTIC CONTROL THEORY FOR SENSOR MANAGEMENT

David A. Castañón

Boston University, Boston, MA, USA

Lawrence Carin

Duke University, Durham, NC, USA

1. Introduction

Consider the following scenario: a physician examining a patient for breast cancer feels a hard area during an initial examination; she subsequently sends the patient to obtain a mammogram of the area, and sends the mammogram to a pathologist. The pathologist notes the presence of two local areas with potentially cancerous growths, but is unable to localize the areas accurately; he sends the patient to another imaging center, where a full 3-D Computed Tomography (CT) image is obtained and returned to the pathologist. The new CT image identifies accurately the location of the potential areas, but does not provide enough information to identify the nature of the growth. The pathologist performs two biopsies to extract sample cells from the two areas and examine them under a microscope to complete the diagnosis.

In a different setting, consider a player at a blackjack table in Las Vegas, playing two simultaneous games against a dealer. The player sees all the cards in his hands for both games, but only sees one of the dealer's two cards. The player asks for an extra card in his first game; after seeing the card, he asks for a second one. He sees this card and decides to stop and switch to the second game. After examining his cards, he chooses to stop asking for cards and let the dealer draw.

In a third setting, a phased-array radar is searching for new aircraft, while trying to maintain accurate track and classification information on known aircraft in its field of regard. The radar schedules a sequence of short pulses aimed at areas where new objects may appear, interleaved with short pulses aimed at positions where known objects are moving to update their position and velocity information using recursive estimation. Occasionally, the radar also introduces longer high range resolution (HRR) imaging waveforms into the mix and focuses these on known objects to collect HRR images of scatterers on the moving platforms, thereby providing information for estimating the object type. The interested reader is referred to Chapters 4, 5, 7 and 10 for radar applications and to Chapter 11 for some history and perspectives on defense applications of sensor management.

The above examples share a common theme: in each example, decisions are made sequentially over time. Each decision generates observations that provide additional information. In each example, the outcome of selecting a decision is uncertain; each subsequent decision is selected based on the previous observations, toward the purpose of achieving an objective that depends on the sequence of decisions. In each example, uncertainty is present, and the ultimate outcome of the decisions is unknown. In sum, these are sequential decision problems under uncertainty, where the choice of decisions can be adapted to the information collected.

Sequential decision problems under uncertainty constitute an active area of research in fields such as control theory [19–21, 23–25, 154, 251], statistics [243, 55, 34, 35], operations research [111, 70, 192, 157, 198], computer science [41, 227, 121, 16] and economics [17, 202], with broad applications to problems in military surveillance, mathematical finance, robotics, and manufacturing, among others. This chapter presents an overview of mathematical framework and techniques for representation and solution of sequential decision problems under uncertainty, with a focus on their application for problems of dynamic sensor management, where actions are explicitly selected to acquire information about an underlying unknown process.

The classical model for dynamic decisions under uncertainty is illustrated in the control loop in Figure 2.1(a). In such systems, information collected by sensors is used to design activities that change how the underlying system evolves in time. The problems of interest in this chapter differ from this model in a substantial manner, as illustrated in Figure 2.1(b). In sensor management, actions are not selected to change the evolution of a dynamical system; instead, they are selected to improve the available information concerning the system state. Thus, the focus is on controlling the evolution of information rather than state dynamics.

Stochastic Control Theory for Sensor Management

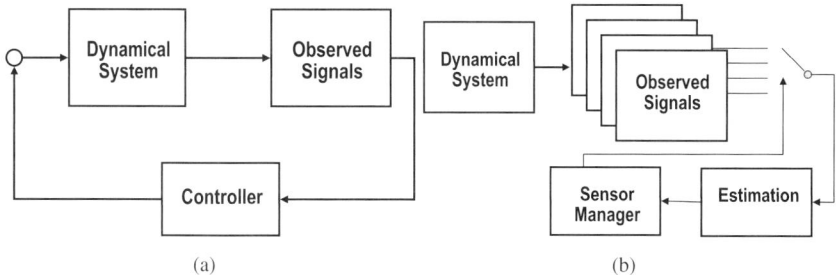

Figure 2.1. a) Feedback control loop b) Sensor management control

The foundation of sequential decision theory under uncertainty is a mathematical framework for representing the relationship between the underlying unknown quantities that may evolve in time, the relationship of observables to these unknowns, the objectives that measure the goals of the problem, and the effects that actions have on observables, the underlying unknown quantities, and the problem objectives. In this chapter, this representation is based on a formal probabilistic framework, with the following characteristics:

- Unknown quantities of interest are modeled as *states* of a dynamical system, modeled as a Markov process. A quick overview of Markov processes is included in Section 2 of the Appendix. At any time, the current state summarizes the statistical information needed to predict the evolution of future states.

- Observed quantities will also be modeled statistically, in terms of their relationship with the underlying state at the time observations are acquired.

- Actions may affect either the evolution of the state of the dynamical system, or the nature of the observation acquired. The latter will be more typical for sensor management. The choice of action may be constrained by the available information collected.

- Objectives will depend on the choice of actions and the specific dynamic trajectories of the state, with an additive structure over stages.

In the rest of this chapter, we differentiate between two classes of problems: *Markov Decision Problems* [193, 25, 198], where the observations provide enough information to determine exactly the current state of the system, and *Partially Observed Markov Decision Problems* [219, 218, 173, 248, 162], where the history of observations leaves residual uncertainty concerning the current state. We will overview the formulation for these classes of problems, selected key theoretical results, discuss algorithms for obtaining solutions. The chapter concludes with an example application to illustrate the methodology.

2. Markov Decision Problems

We restrict our discussion to decision problems that evolve using a discrete index, which we refer to as stages.

DEFINITION 2.1 *A Markov Decision Process (MDP) consists of*

- *A discrete stage index $k \in \{0, 1, \ldots, N\}$, $N \leq \infty$*
- *A set of possible states $\mathcal{S}_k$ for each stage index k*
- *An initial value for the state $s_0 \in \mathcal{S}_0$*
- *A set of possible actions $\mathcal{A}_k$*
- *A family of action constraints $\{\mathcal{A}_k(s) \subset \mathcal{A}_k\}$, for $s \in \mathcal{S}_k$.*
- *A state transition probability kernel $\mathcal{T}_k(ds'|s, a)$, where $\mathcal{T}_k(ds'|s, a) \equiv P(s_{k+1} \in ds'|s_k = s, a_k = a)$,*
- *A real-valued single stage reward function $R_k(s, a)$*

The spaces $\mathcal{S}_k$ and $\mathcal{A}_k$ are assumed to be metric spaces. The reward functions $R_k(s, a)$ and the transition kernels $\mathcal{T}_k(ds'|s, a)$ are assumed continuous functions of s, a. This is trivially satisfied when the spaces $\mathcal{S}_k, \mathcal{A}_k$ are discrete. Furthermore, we assume that, for each state, the admissible actions $\mathcal{A}_k(s)$ form a compact subset of $\mathcal{A}_k$. The resulting state evolves according to a Markov process given the actions $a_k, k = 0, \ldots, N - 1$, so that the effects of an action a_k taken in state s_k depend only on the current value of that state and not on the prior history of the state.[1]

Observations of past decisions, plus past and current values of the state, are available to select the choice of next decision. Let $\mathcal{I}_k$ denote the information available at stage k, defined as:

$$\mathcal{I}_k = \{s_0, a_0, \ldots, s_{k-1}, a_{k-1}, s_k\} \qquad (2.1)$$

---

[1]The reader is cautioned that this chapter does not adopt the upper/lower case notation to distinguish between random variables and their realizations, e.g., as in S_k and s_k, respectively. In this book, the upper/lower case convention is only used when its omission would risk confusing the reader.

A *policy* at stage k is a mapping $\gamma_k(\mathcal{I}_k)$, from the set of all possible information states $\{\mathcal{I}_k\}$ to $\mathcal{A}_k$. In this chapter, we define policies as deterministic mappings from available information into admissible actions. One can also define stochastic policies that map available information into probability kernels on $\mathcal{A}_k$, but such policies offer no better performance than deterministic policies for the MDP models discussed here. A policy is said to be *Markov* if the mapping γ_k depends only on the most recent value of the state, s_k. That is,

$$\gamma_k : \mathcal{S}_k \longrightarrow \mathcal{A}_k \qquad (2.2)$$

An *admissible* policy is a sequence $\underline{\gamma} \equiv \{\gamma_0, \ldots, \gamma_{N-1}\}$ with the property that $\gamma_k(\mathcal{I}_k) \in \mathcal{A}_k(s_k)$, so that the selected decisions given past information satisfy the constraints imposed by the current state value. An admissible policy generates a random state trajectory $s_k, k = 0, \ldots, s_N$ and an action trajectory $a_k, k = 0, \ldots, N - 1$. Associated with each trajectory is a total reward R which is assumed additive across stages:

$$R \equiv R_N(s_N) + \sum_{k=0}^{N-1} R_k(s_k, a_k) \qquad (2.3)$$

This additive structure can be exploited to develop efficient algorithms for solving MDPs, as discussed later. Under appropriate conditions on the sets $\mathcal{S}_k, \mathcal{A}_k$, the transition kernels $\mathcal{T}_k(ds'|s, a)$ and the reward functions $R_k(s, a)$, the policy $\underline{\gamma}$ will generate sequences of well-defined random variables corresponding to state, action and reward trajectories. These conditions involve measurability assumptions, and are satisfied in most applications of interest. As discussed in [25], these conditions will be satisfied whenever the sets $\mathcal{S}_k, \mathcal{A}_k$ are countable and the reward functions $R_k(s, a)$ are bounded. Conditions for more general spaces are beyond the scope of this chapter; see [27, 24] for additional details.

Given an admissible policy Γ, the total reward R becomes a well-defined random variable with expectation $\mathbb{E}_\Gamma[R]$. The objective of the problem is to select the admissible policy that maximizes the expected total reward

$$\max \mathbb{E}_{\underline{\gamma}} \left[R_N(s_N) + \sum_{k=0}^{N-1} R_k(s_k, a_k) \right] \qquad (2.4)$$

An important result in Markov Decision Processes is that, whenever an optimal admissible policy $\underline{\gamma}$ exists, there exist admissible Markov policies that achieve the same expected reward, and hence are also optimal. In the remainder of this section, we restrict our discussion to Markov policies $\gamma_k(s_k)$.

2.1 Dynamic Programming

The above formulation of a Markov decision problem has several important properties. First, the overall reward can be represented as an additive decomposition of individual rewards over stages. Second, the choice of admissible actions at each stage is not constrained by the states and actions that were generated at previous stages. Under these assumptions, Bellman's Principle of Optimality [19, 18, 25] applies:

DEFINITION 2.2 *Bellman's Principle of Optimality:*

Let γ^* be an optimal policy in a Markov Decision Problem. Assume that, when using γ^*, the state s_k is reached with positive probability, where $k < N$. Consider the subproblem starting from state s_k at stage K, with the expected reward

$$\max_{\gamma_k,\ldots,\gamma_{N-1}} \mathbb{E}\left[R_N(s_N) + \sum_{i=k}^{N-1} R_i(s_i, a_i) \big| s_k\right] \quad (2.5)$$

The policy $\{\gamma_k^*, \ldots, \gamma_{N-1}^*\}$ is an optimal policy for this subproblem.

Bellman's Principle of Optimality leads to the dynamic programming algorithm, defined as follows. Consider the subproblem starting from a particular state s_k at stage k, and consider an admissible policy $\underline{\gamma}$. Define the value of policy Γ starting at state s_k, stage k as

$$V_{\underline{\gamma}}(s_k, k) = \mathbb{E}_{\underline{\gamma}}\left[R_N(s_N) + \sum_{i=k}^{N-1} R_i(s_i, \gamma_i(s_i)) \big| s_k\right] \quad (2.6)$$

Define the optimal reward for the subproblem starting at stage K, state s_K as

$$V^*(s_k, k) = \max_{\underline{\gamma} \text{ admissible}} \mathbb{E}_{\underline{\gamma}}\left[R_N(s_N) + \sum_{i=k}^{N-1} R_i(s_i, a_i) \big| s_k\right] \quad (2.7)$$

The difficulty with (2.7) is that it represents a functional minimization over policies. The main result in dynamic programming is the Bellman equation, which provides a recursive solution for (2.7):

THEOREM 2.3 *For every initial state s_0, the optimal value $V^*(s_0, 0)$ is given by the backward recursion*

$$V^*(s, N) = R_N(s) \quad (2.8)$$

$$V^*(s,k) = \max_{a \in \mathcal{A}_k(s)} R_k(s,a) + \int_{s' \in \mathcal{S}_{k+1}} V^*(s', k+1) \mathcal{T}_k(ds'|a,s),$$
$$k = 0, \ldots, N-1 \quad (2.9)$$

If there exist policies $\gamma_k(s)$ such that

$$\gamma_k(s) \in \operatorname*{argmax}_{a \in \mathcal{A}_k(s)} R(s,a) + \int_{s' \in \mathcal{S}_{k+1}} V^*(s', k+1) \mathcal{T}_k(ds'|a,s), k = 0, \ldots, N-1 \quad (2.10)$$

then the policy $\underline{\gamma}^* = \{\gamma_0^*, \ldots, \gamma_{N-1}^*\}$ is an optimal policy.

Bellman's equation decomposes the functional optimization problem over sequences of admissible policies to a sequence of optimizations over admissible actions for each state at each stage.

2.2 Stationary Problems

In many problems of interest, the Markov Decision Problem (MDP) description is stage invariant: The sets $\mathcal{A}_k(s)$, $\mathcal{S}_k$, the reward functions $R_k(s,a)$ and the transition probability kernels $\mathcal{T}_k(ds'|s,a)$ are independent of the stage index. These problems are known as *stationary* MDPs; the number of stages N may be infinite, or may be a decision variable corresponding to choosing to stop the measurement process.

For stationary MDPs, define a *stationary* Markov policy $\Gamma = \{\gamma, \gamma, \ldots\}$, where the policy at any stage does not depend on the particular stage. We refer to stationary policy sequences in terms of the single stage policy γ, the policy that is used at every stage. Stationary MDP formulations often lead to optimal Markov policies which are also stationary, which allows for a simpler, time-invariant implementation of the optimal policy. There are three commonly used MDP formulations that lead to stationary policies with infinite horizon: discounted reward MDPs, total reward MDPs and average reward MDPs. The first two are commonly used models in sensor management; average reward models are seldom used because sensor management problems do not have statistics that are stage invariant over large numbers of stages. By choosing the discount factor or by rewarding stopping, one can approximately limit the horizon of the MDP problem to intervals where the statistics are stage invariant.

2.2.1 Infinite Horizon Discounted Problems.
Consider the case where the number of stages N is infinite. In order to keep the total reward finite, the reward R includes a nonnegative discount factor $\beta < 1$ for future

rewards, as

$$R = \sum_{k=0}^{\infty} \beta^k R(s_k, a_k) \qquad (2.11)$$

Assume that the rewards $R(s,a)$ are bounded, so that the total discounted cost R is finite. For these problems, Bellman's equation becomes

$$V^*(s) = \max_{a \in \mathcal{A}(s)} R(s,a) + \beta \int_{s' \in \mathcal{S}} V^*(s') \mathcal{T}(ds'|a,s) \qquad (2.12)$$

Note that (2.12) does not involve a recursion over stages, unlike (2.9). The connection is given by the following relationship. Define $V^0(s)$ to be a bounded, measurable function of $s \in \mathcal{S}$. Define the sequence of functions $V^n(s)$ as

$$V^n(s) = \max_{a \in \mathcal{A}(s)} R(s,a) + \beta \int_{s' \in \mathcal{S}} V^{n-1}(s') \mathcal{T}(ds'|a,s) \qquad (2.13)$$

Then, the sequence $V^n(s)$ converges to $V^*(s)$, the solution of Bellman's equation (2.12). Formally, let $\mathcal{S}$, $\mathcal{A}$ be complete metric spaces, and let $\mathcal{B}(\mathcal{S})$ denote the space of bounded, real-valued functions $f : \mathcal{S} \longrightarrow \mathbb{R}$ with the (essential) supremum norm $\|\cdot\|_\infty$. Define the dynamic programming operator $\mathbf{T} : \mathcal{B}(\mathcal{S}) \longrightarrow \mathcal{B}(\mathcal{S})$ as

$$\mathbf{T}f(s) = \max_{a \in \mathcal{A}(s)} R(s,a) + \beta \int_{s' \in \mathcal{S}} f(s') \mathcal{T}(ds'|a,s) \qquad (2.14)$$

and the fixed policy operator

$$\mathbf{T}_\gamma f(s) = R(s, \gamma(s)) + \beta \int_{s' \in \mathcal{S}} f(s') \mathcal{T}(ds'|\gamma(s), s) \qquad (2.15)$$

The following result characterizes the important property of the dynamic programming operator:

THEOREM 2.4 *Assume that $R(s,a)$ is bounded and $\beta < 1$. Then, the operator $\mathbf{T}$ is a contraction mapping with contraction coefficient β; i.e., for any functions $V, W \in \mathcal{B}(\mathcal{S})$,*

$$\|\mathbf{T}V - \mathbf{T}W\|_\infty \le \beta \|V - W\|_\infty \qquad (2.16)$$

The contraction mapping theorem [35] guarantees that the sequence $V^n(s)$ converges to a unique fixed point $V = \mathbf{T}(V)$ in $\mathcal{B}(\mathcal{S})$ as $n \to \infty$, where the existence is guaranteed by the completeness of the space $\mathcal{B}(\mathcal{S})$ from any initial estimate of V. This limit satisfies Bellman's equation (2.12). The main results in discounted dynamic programming are summarized below:

THEOREM 2.5 *Assume $R(s,a)$ is bounded and $\beta < 1$. Then,*

1. *For any bounded function $V \in \mathcal{B}(\mathcal{S})$,*

$$V^*(s) = \lim_{n \to \infty} (\mathbf{T}^n V)(s) \qquad (2.17)$$

2. *The optimal value function V^* satisfies Bellman's equation (2.12). Furthermore, the solution to the Bellman equation is unique in $\mathcal{B}(\mathcal{S})$.*

3. *For every stationary policy γ and for any $V \in \mathcal{B}(\mathcal{S})$, denote by $\mathbf{T}_\gamma$ the dynamic programming operator when policy γ is the only admissible policy. The expected value achieved by policy γ for each state s, denoted as $V_\gamma(s)$, is the unique solution in $\mathcal{B}(\mathcal{S})$ of the equation.*

$$V_\gamma(s) = (\mathbf{T}_\gamma V_\gamma)(s) = \lim_{n \to \infty} (\mathbf{T}_\gamma^n V)(s) \qquad (2.18)$$

4. *A stationary policy γ is optimal if and only if it achieves the maximum reward in the Bellman equation (2.12) for each $s \in \mathcal{S}$; i.e.,*

$$\mathbf{T}_\gamma V^* = \mathbf{T} V^* \qquad (2.19)$$

The last property in Proposition 2.5 provides a verification theorem for establishing the optimality of a strategy given the optimal value function.

Although most sensor management applications will not have an infinite horizon, a discounted infinite horizon model is often used as an approximation to generate stationary policies. The choice of discount factor in these problems sets an "effective" horizon that can be controlled to reflect the number of stages in the problem of interest.

2.2.2 Undiscounted Total Reward Problems.

Many sensor management applications are best formulated as total reward problems where the number of stages N can be infinite and no future discounting of value is used. For instance, consider a search problem where there is a finite number of areas to be searched, each of which may have an object present with a known probability in each area. When there is a cost associated with searching an area and a reward for finding objects in areas, an optimal strategy will search only the subset of areas where the expected reward of searching an area exceeds the expected cost of the search. In this problem, an admissible decision is to stop searching; however, the number of stages before the search is stopped depends on the stage evolution of the probabilities of areas containing objects, and is not specified *a priori*.

In undiscounted total reward problems, the accumulated reward may become unbounded. Hence, additional structure is needed in the rewards to guarantee that optimal strategies exist, and that the optimal value function remains finite. Following the exposition in [25], we specify two alternative assumptions for the undiscounted problems.

DEFINITION 2.6 *Assumption* **P**: *The rewards per stage satisfy:*

$$R(s,a) \leq 0, \text{ for all } s \in \mathcal{S}, a \in \mathcal{A}(s) \tag{2.20}$$

DEFINITION 2.7 *Assumption* **N**: *The rewards per stage satisfy:*

$$R(s,a) \geq 0, \text{ for all } s \in \mathcal{S}, a \in \mathcal{A}(s) \tag{2.21}$$

In discounted reward problems, the presence of a discount factor limits the effect of future rewards on the current choice of actions to a limited interval. In contrast, undiscounted problems must consider effects of long-term rewards. Thus, under Assumption **P**, the goal must be to bring the state quickly to a region where one can either terminate the problem or where the rewards approach 0. Under assumption **N**, the objective may be to avoid reaching a termination state for as long as possible.

As before, define the dynamic programming operator $\mathbf{T} : \mathcal{B}(\mathcal{S}) \longrightarrow \mathcal{B}(\mathcal{S})$ as

$$\mathbf{T}f(s) = \max_{a \in \mathcal{A}(s)} R(s,a) + \int_{s' \in \mathcal{S}} f(s')\mathcal{T}(ds'|a,s) \tag{2.22}$$

and the iteration operator for a stationary policy γ as $\mathbf{T}_\gamma$ as

$$\mathbf{T}_\gamma f(s) = R(s,\gamma(a)) + \int_{s' \in \mathcal{S}} f(s')\mathcal{T}(ds'|\gamma(a),s) \tag{2.23}$$

A key difference with the discounted rewards problem is that the operators $\mathbf{T}, \mathbf{T}_\gamma$ are no longer contractions. This raises issues as to the existence and uniqueness of optimal value functions and characterization of optimal strategies. The principal results for these problems are summarized below.

Under Assumption **P** or **N**, Bellman's equation holds, and becomes

$$V^*(s) = \max_{a \in \mathcal{A}(s)} R(s,a) + \int_{s' \in \mathcal{S}} V^*(s')\mathcal{T}(ds'|a,s) \tag{2.24}$$

Similarly, for any stationary policy γ, one has the property that

$$V_\gamma(s) = \mathbf{T}_\gamma V_\gamma(s) \tag{2.25}$$

Although Bellman's equation (2.24) holds, the solution may not be unique. However, the optimal value is either the largest solution (under Assumption **P**) or the smallest solution (under Assumption **N**). Specifically, under Assumption **P**, if V satisfies $V \leq 0$ and $V \leq \mathbf{T}V$, then $V \leq V^*$. Similarly, under Assumption **N**, if V satisfies $V \geq 0$ and $V \geq \mathbf{T}V$, then $V \geq V^*$.

Characterization of optimal strategies differs for the two cases **P** and **N**. Under Assumption **P**, a stationary policy γ is optimal if and only if it achieves the maximum reward in Bellman's equation (2.24):

$$\mathbf{T}_\gamma V^*(s) = \mathbf{T}V^*(s) \text{ for all } s \in \mathcal{S} \tag{2.26}$$

However, the sufficiency clause is not true under Assumption **N**. A different characterization of optimality is needed: under Assumption **N**, a stationary policy is optimal if and only if

$$\mathbf{T}_\gamma V_\gamma(s) = \mathbf{T}V_{\gamma(s)} \text{ for all } s \in \mathcal{S} \tag{2.27}$$

Another important consequence of losing the contraction property is that there may be no iterative algorithms for computing the optimal value function from arbitrary initial estimates. Fortunately, when the iteration is started with the right initial condition, one can still get convergence. Specifically, let $V^0(s) = 0$ for all $s \in \mathcal{S}$, and define the iteration

$$V^n = \mathbf{T}V^{n-1}, \; n = 1, 2, \ldots \tag{2.28}$$

Under Assumption **P**, the operator **T** generates a monotone sequence

$$V^0 \geq V^1 \geq \ldots \geq V^n \geq \cdots, \tag{2.29}$$

which has a limit (with values possibly $-\infty$) V^∞. Similarly, under Assumption **N**, the sequence generated by (2.28) yields

$$V^0 \leq V^1 \leq \ldots \leq V^n \leq \cdots \tag{2.30}$$

These limits become the optimal values under simple conditions, as indicated below:

THEOREM 2.8 *Under Assumption* **P**, *if* $V^\infty = \mathbf{T}V^\infty$ *for all* $s \in \mathcal{S}$, *and* $V^0 \geq V \geq V^*$, *then*

$$\lim_{n \to \infty} \mathbf{T}^n V^0 = V^* \tag{2.31}$$

Under Assumption **N**, *if* $V^0 \leq V \leq V^*$,

$$\lim_{n \to \infty} \mathbf{T}^n V = V^* \tag{2.32}$$

Conditions that guarantee that $V^\infty = TV^\infty$ under Assumption **N** are compactness of the action sets $\mathcal{A}(s)$ for each s [25].

Important classes of sensor management problems that satisfy the assumptions in this subsection are optimal stopping problems, which are described in the Appendix, Section 3. In these problems, one must choose between a finite number of sensing actions, each of which has a cost that depends on the action and state $c(a, s)$, or stopping the sensing problem and receive a cost $s(a)$. For instance, the dynamic classification problems using multimodal sensing considered in [47, 48] fit this structure: there is a cost for using sensing time for each mode, and when the decision to stop sensing is made, the system undergoes a cost related to the probability of misclassification of each target. A similar formulation was used in [119] for underwater target classification using multiple views. This example is discussed in Section 5.

One can formulate these optimal stopping problems as maximization problems, redefining rewards $R(a, s) = -c(a, s)$, plus adding an additional termination state t to the state space $\mathcal{S}$, corresponding to the action of termination. Once the problem reaches a termination state, the only possible action is to continue in this state, incurring no further rewards. The resulting problem fits the model of Assumption **P** above.

2.3 Algorithms for MDPs

Bellman's equation (2.9) provides a recursive algorithm for computation of the optimal value function in finite horizon MDPs. For infinite horizon problems, Bellman's equation represents a functional equation for the optimal value function $V^*(s)$. The most common algorithm for computing $V^*(s)$ is known as *value iteration*. It consists of starting from a guess at the value function $V^0(s)$ and generating a sequence $V^n(s)$ using the iteration (2.19). For discounted reward problems, Theorem 2.5 establishes that this sequence converges to $V^*(s)$ from any initial condition. For total reward problems, the value iteration algorithm converges to $V^*(s)$ provided the initial condition $V^0(s)$ is chosen appropriately, as in Theorem 2.8.

In discounted reward problems, an optimal policy γ^* can be generated using (2.19) as

$$\gamma^*(s) \in \arg\max_{a \in \mathcal{A}} R(s, a) + \beta \int_{s' \in \mathcal{S}} V^*(s') \mathcal{T}(ds'|a, s). \tag{2.33}$$

For total reward problems under Assumption **P**, a similar characterization follows from (2.26). Assumption **P** is the most appropriate model for sensor management with stopping criteria, as illustrated in the example later in this chapter.

Another approach for the solution of discounted reward problems is *policy iteration*. In policy iteration, the algorithm starts with a stationary policy γ^0. Given a policy γ^n, Bellman's equation for a single policy is then solved to obtain the optimal value function $V_{\gamma^n}(s)$, as

$$V_{\gamma^n}(s) = R(s, \gamma^n(s)) + \beta \int_{s' \in \mathcal{S}} V_{\gamma^n}(s') \mathcal{T}(ds'|\gamma^n(s), s) \quad (2.34)$$

A new policy γ^{n+1} is then generated as

$$\gamma^{n+1}(s) \in \arg\max_{a \in \mathcal{A}} R(s, a) + \beta \int_{s' \in \mathcal{S}} V_{\gamma^n}(s') \mathcal{T}(ds'|a, s) \quad (2.35)$$

The policy iteration algorithm requires many fewer iterations than value iteration to converge. However, each iteration is more complex, as it requires the solution of a functional equation to obtain the value of a single policy. An approximate form of policy iteration is often used, where (2.34) is solved approximately using a few value iterations.

3. Partially Observed Markov Decision Problems

The primary assumption underlying the MDP theory discussed in the previous section is that the state of the system, s_k, is observed perfectly at each stage k. In many sensor management applications, the full state is not observed at each stage; instead, some statistics related to the underlying state are observed, which yield uncertain information about the state. These problems are known as Partially Observed Markov Decision Problems (POMDPs) as the observations yield only partial knowledge of the true state. POMDPs are also known as partially observable Markov decision processes.

DEFINITION 2.9 *A Partially Observed Markov Decision Process (POMDP) consists of*

- *A discrete stage index $k \in \{0, 1, \ldots, N\}, N \leq \infty$*
- *A finite set of possible states $\mathcal{S}_k$ for each stage index k, with cardinality $|\mathcal{S}_k|$*
- *An initial probability distribution $\pi_0(s)$ over the finite set $\mathcal{S}_0$, where $\pi_0(s) \equiv \mathrm{P}(s_0 = s)$.*

- A finite set of possible actions $\mathcal{A}_k$ for each stage index k
- State transition probability matrices $\mathcal{T}_k(s'|s,a)$, where $\mathcal{T}_k(s'|s,a) \equiv P(s_{k+1} = s'|s_k = s, a_k = a)$,
- A real-valued single stage reward function $R_k(s,a)$ and an overall objective J which is additive across stages
- A finite set of possible observations $\mathcal{Y}_k$ for each stage k
- An observation probability likelihood $\mathcal{Q}_k(y|s,a)$, where

$$\mathcal{Q}_k(y|s,a) \equiv P(y_k = y|s_k = s, a_k = a)$$

Unlike the MDP model, the sets $\mathcal{S}_k$, $\mathcal{A}_k$ and the observation sets $\mathcal{Y}_k$ are assume to be finite. The initial distribution $\pi_0(s)$ and the transition probability distributions $\mathcal{T}(s'|s,a)$ define a controlled Markov chain given a sequence of decisions $a_0, a_1, \ldots$. The observations y_k are assumed to be conditionally independent for different k, with distribution depending only on the current state s_k and the current action a_k. Note that the choice of action a_k affects the generation of observations; this model represents the sensor management problem, where choice of sensing actions determine what information is collected.

We assume that past decisions, plus past values of the observations, are available to select the choice of next decision. Let $\mathcal{I}_k$ denote the information available at stage k for selecting action a_k, defined as:

$$\mathcal{I}_k = \{a_0, y_0, \ldots, a_{k-1}, y_{k-1}\} \tag{2.36}$$

Note that this information does not include any observations of past states. Due to the finite assumption on the action and observation spaces, the number of possible information sets $\mathcal{I}_k$ is also finite. As in MDPs, a *policy* at stage k is a deterministic mapping $\gamma_k(\mathcal{I}_k)$, from the set of all possible information states $\{\mathcal{I}_k\}$ to $\mathcal{A}_k$. Such policies are *causal*, in that they select current actions based on past information only. Similarly, an *admissible policy* for POMDPs is a sequence $\Gamma \equiv \{\gamma_0, \ldots, \gamma_{N-1}\}$ with the property that $\gamma_k(\mathcal{I}_k) \in \mathcal{A}_k$.

An admissible policy generates a random state trajectory $s_k, k = 0, \ldots, s_N$, an observation trajectory $y_k, k = 0, \ldots, N-1$ and an action trajectory $a_k, k = 0, \ldots, N-1$. The causal sequence corresponds to the following: Given information $\mathcal{I}_k$, the policy generates an action $a_k = \gamma_k(I_k)$. Given this action a_k, a new observation y_k is collected, and the state transitions from s_k to s_{k+1}. The action a_k and the observation y_k are added to the information set $\mathcal{I}_k$ to generate $\mathcal{I}_{k+1}$. This causal chain is illustrated in figure 2.2. The policy γ will generate sequences of well-defined random variables corresponding to the state, action and reward trajectories.

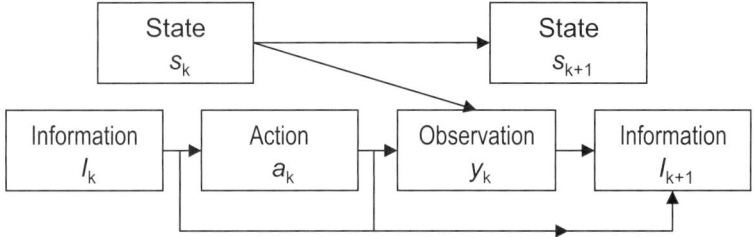

Figure 2.2. Illustration of sequential actions and measurements

Similar to the MDP formulation, there is a total reward R associated with these trajectories that is additive across stages:

$$R \equiv R_N(s_N) + \sum_{k=0}^{N-1} R_k(s_k, a_k) \qquad (2.37)$$

Given an admissible policy Γ, the total reward R becomes a well-defined random variable with expectation $\mathbb{E}_\Gamma[R]$. The objective of POMDP problem is to select an admissible policy that maximizes the expected total reward

$$\max \mathbb{E}_{\underline{\gamma}} \left[R_N(s_N) + \sum_{k=0}^{N-1} R_k(s_k, a_k) \right] \qquad (2.38)$$

3.1 MDP Representation of POMDPs

The POMDP problem can be converted to a standard MDP problem, where the underlying state corresponds to the information sets $\mathcal{I}_k$. Note that these sets take values in discrete sets, and evolve according to the simple evolution

$$\mathcal{I}_{k+1} = \mathcal{I}_k \cup \{a_k, y_k\} \qquad (2.39)$$

To show the equivalent MDP problem, consider the objective of maximizing the expected reward. For any policy Γ, the smoothing property of conditional expectations yields

$$\mathbb{E}_{\underline{\gamma}} \left[R_N(s_N) + \sum_{k=0}^{N-1} R_k(s_k, a_k) \right] \qquad (2.40)$$

$$= \mathbb{E}_{\underline{\gamma}} \left[\mathbb{E}_{\underline{\gamma}}[R_N(s_N)|\mathcal{I}_N] + \sum_{k=0}^{N-1} \mathbb{E}_{\underline{\gamma}}[R_k(s_k, a_k)|\mathcal{I}_k, a_k] \right]$$

Define the equivalent reward function $\hat{R}(\mathcal{I}_k, a_k)$ as

$$\hat{R}(\mathcal{I}_k, a_k) = \mathbb{E}_\gamma [R_k(s_k, a_k) | \mathcal{I}_k, a_k] \qquad (2.41)$$

Note that the conditional expectation in (2.41) does not depend on the specific strategies γ, since all the past action values and past observation values are part of $\mathcal{I}_k$. To show equivalence to an MDP, we have to establish that (2.39) generates a Markov transition probability kernel that describes the random evolution of $\mathcal{I}_{k+1}$ given $\mathcal{I}_k$. This requires that

$$P(\mathcal{I}_{k+1} | \mathcal{I}_k, a_k, \mathcal{I}_{k-1}, \ldots, \mathcal{I}_0) = P(\mathcal{I}_{k+1} | \mathcal{I}_k, a_k) \qquad (2.42)$$

To show this is true, note that the only conditionally random component of $\mathcal{I}_{k+1}$ given $\mathcal{I}_k, a_k$ is the observation y_k. Thus, one has to show that

$$P(y_k | a_k, \mathcal{I}_k, \mathcal{I}_{k-1}, \ldots, \mathcal{I}_0) = P(y_k | a_k, \mathcal{I}_k), \qquad (2.43)$$

which follows because $\mathcal{I}_j \subset \mathcal{I}_k, j < k$.

The above argument establishes that a POMDP is equivalent to an MDP with states corresponding to the information sets $\mathcal{I}_k$, and Markov dynamics corresponding to (2.39), and Markov transition probabilities given by (2.42). Hence, all of the results discussed in the previous section can be applied to this problem. In particular, for N finite, there exists an optimal value function $V^*(\mathcal{I}_k, k)$ that satisfies Bellman's equation

$$V^*(\mathcal{I}_k, k) = \max_{a \in \mathcal{A}_k} \hat{R}(\mathcal{I}_k, a) + \mathbb{E}_{y_k} [V^*(\mathcal{I}_k \cup \{y_k, a\}, k+1)] \qquad (2.44)$$

and the optimal strategies satisfy

$$\gamma_k^*(\mathcal{I}_k) \in \underset{a \in \mathcal{A}_k}{\operatorname{argmax}}\ \hat{R}(\mathcal{I}_k, a) + \mathbb{E}_{y_k} [V^*(\mathcal{I}_k \cup \{y_k, a\}, k+1)] \qquad (2.45)$$

However, it is difficult to extend this characterization to infinite horizon problems because the set of possible information sets $\mathcal{I}_k$ grows larger as k increases. An alternative parameterization is possible using the concept of sufficient statistics for control [225, 24, 25]. A *sufficient statistic* is a function $h_k(\mathcal{I}_k)$ such that the maximization in (2.45) is achieved by a function of $h_k(\mathcal{I}_k)$ instead of all of $\mathcal{I}_k$. To show that a statistic is sufficient for control, it is enough to show that the right hand side of (2.40) depends on $\mathcal{I}_k$ only through $h_k(\mathcal{I}_k)$. A sufficient statistic for stochastic control problems such as POMDPs is the posterior distribution $\pi_k(s) = P(s_k = s | \mathcal{I}_k)$, the conditional probability that the state s_k takes the value s given past information $\mathcal{I}_k$.

Given a bounded function $g : \mathcal{S}_k \to \mathbb{R}$, we use the notation

$$<g, \pi_k> = \mathbb{E}\left[g(s_k)|\mathcal{I}_k\right] = \sum_{s \in \mathcal{S}_k} g(s)\pi_k(s). \qquad (2.46)$$

Since $\mathcal{S}_k$ is a finite set, these bounded functions are real-valued $|\mathcal{S}_k|$ dimensional vectors, and $< \cdot >$ is an inner product. Define the function $r_k(a_k)(s_k) \equiv R(s_k, a_k)$. Then, (2.40) becomes

$$\mathbb{E}_{\underline{\gamma}}\left[R_N(s_N) + \sum_{k=0}^{N-1} R_k(s_k, a_k)\right] = \mathbb{E}_{\underline{\gamma}}\left[<R_N, \pi_N> + \sum_{k=0}^{N-1} <r(a_k), \pi_k>\right],$$

which establishes that the sequence $\pi_k, k = 0, \ldots, N$ is a sufficient statistic for POMDPs.

Given the independence assumptions of the POMDP model, one can compute a controlled Markov evolution for the sufficient statistic π_k as follows:

$$\pi_k(s_k) = \mathrm{P}(s_k|\mathcal{I}_k) = \mathrm{P}(s_k|\mathcal{I}_{k-1}, a_{k-1}, y_{k-1})$$

$$= \sum_{s_{k-1}} \mathrm{P}(s_k, s_{k-1}|\mathcal{I}_{k-1}, a_{k-1}, y_{k-1})$$

$$= \sum_{s_{k-1}} \mathrm{P}(s_k|s_{k-1}, \mathcal{I}_{k-1}, a_{k-1}, y_{k-1}) \mathrm{P}(s_{k-1}|\mathcal{I}_{k-1}, a_{k-1}, y_{k-1})$$

$$= \sum_{s_{k-1}} \mathcal{T}(s_k|s_{k-1}, a_{k-1}) \mathrm{P}(s_{k-1}|\mathcal{I}_{k-1}, a_{k-1}, y_{k-1}), \qquad (2.47)$$

where the last equality follows from the Markov evolution of s_k given the actions a_k. We can further simplify the right hand side using Bayes' rule as

$$\mathrm{P}(s_{k-1}|\mathcal{I}_{k-1}, a_{k-1}, y_{k-1}) = \frac{\mathrm{P}(s_{k-1}, y_{k-1}|\mathcal{I}_{k-1}, a_{k-1})}{\mathrm{P}(y_{k-1}|\mathcal{I}_{k-1}, a_{k-1})}$$

$$= \frac{\mathrm{P}(y_{k-1}|s_{k-1}, a_{k-1})\mathrm{P}(s_{k-1}|\mathcal{I}_{k-1})}{\mathrm{P}(y_{k-1}|\mathcal{I}_{k-1}, a_{k-1})}$$

$$= \frac{\mathcal{Q}_{k-1}(y_{k-1}|s_{k-1}, a_{k-1})\pi_{k-1}(s_{k-1})}{\sum_{\sigma \in \mathcal{S}_{k-1}} \mathcal{Q}_{k-1}(y_{k-1}|\sigma, a_{k-1})\pi_{k-1}(\sigma)} \qquad (2.48)$$

The resulting evolution is given by

$$\pi_k(s) = \sum_{s_{k-1} \in \mathcal{S}_{k-1}} \mathcal{T}_{k-1}(s|s_{k-1}, a_{k-1}) \frac{\mathcal{Q}_{k-1}(y_{k-1}|s_{k-1}, a_{k-1})\pi_{k-1}(s_{k-1})}{\sum_{\sigma \in \mathcal{S}_{k-1}} \mathcal{Q}_{k-1}(y_{k-1}|\sigma, a_{k-1})\pi_{k-1}(\sigma)}$$

$$\equiv \hat{\mathcal{T}}_{k-1}(\pi_{k-1}, y_{k-1}, a_{k-1}) \qquad (2.49)$$

This evolution is a function of the causal dependencies depicted in Figure 2.2. A different order can be used (e.g. [173, 248, 162]) where the information set $\mathcal{I}_k$ includes the observation y_k, which is generated as depending on the states s_k and action a_{k-1}. The reason for the difference is that, in sensor management problems, decisions are chosen primarily to control the measurements obtained, and not the underlying state s_k. In contrast, standard POMDP formulations focus on using decisions to control the underlying state evolution.

Using sufficient statistics allows us to define an equivalent MDP with state π_k at stage k, and objectives (2.44). In the POMDP literature, these states are referred to as *information states* or *belief states* (See Appendix, Section 2). In terms of these information states, Bellman's equation (2.44) becomes

$$V^*(\pi, k) = \max_{a \in \mathcal{A}_k} <r_k(a), \pi> + \sum_{y \in \mathcal{Y}} V^*(\hat{\mathcal{T}}_k(\pi, y, a), k+1) \, \mathrm{P}(y|\mathcal{I}_k, a), \quad (2.50)$$

where

$$\mathrm{P}(y|\mathcal{I}_k, a) \equiv P_k(y|\pi_k, a) = \sum_{s' \in \mathcal{S}_k} \mathcal{Q}_k(y|s', a)\pi_k(s') \quad (2.51)$$

3.2 Dynamic Programming for POMDPs

The use of sufficient statistics allows for a constant-dimension representation of the underlying MDP state as the horizon increases. The information state now takes values in $\pi_{|\mathcal{S}_k|}$, the simplex of probability distributions on $\mathcal{S}_k$. When $\mathcal{S}_k \equiv \mathcal{S}$ and $\mathcal{A}_k \equiv \mathcal{A}$ are constant in k, and the transition probabilities $\mathcal{T}_k$, measurement probabilities $\mathcal{Q}_k$ and rewards $R(s, a)$ do not depend on k, one can define stationary problems as in MDPs with infinite horizons, with or without discounting, and apply the MDP theory to the POMDP problem with the information state representation. The resulting discounted cost Bellman equation with discount factor β is

$$V^*(\pi) = \max_{a \in \mathcal{A}} <r(a), \pi> + \beta \sum_{y \in \mathcal{Y}} V^*(\hat{\mathcal{T}}_k(\pi, y, a)) \, \mathrm{P}(y|\pi, a) \quad (2.52)$$

The equivalent MDP using information states has a special structure where the immediate reward associated with an action is a linear function of the state. Sondik [219, 218, 220] exploited this property to obtain a unique characterization for the solution of Bellman's equation (2.50). Sondik observed that

$$V^*(\pi, N) = <r_N, \pi> \quad (2.53)$$

is a linear function of π. Analyzing the recursion, this leads to the conjecture that, for $k < N$, there exists a set of real-valued vectors $\mathcal{H}_k$, of dimension $|\mathcal{S}_k|$,

such that
$$V^*(\pi, k) = \max_{h \in \mathcal{H}_k} <h, \pi>, \quad (2.54)$$

which implies that $V^*(\pi, k)$ is a piecewise linear, convex function of π. This can be established by induction, as it is true for $k = N$. Assuming the inductive hypothesis that such a representation is valid at $k+1$, and $\mathcal{H}_{k+1}$ is known, Bellman's equation yields

$$V^*(\pi, k) = \max_{a \in \mathcal{A}_k} <r_k(a), \pi> + \sum_{y \in \mathcal{Y}_k} V^*(\hat{\mathcal{T}}_k(\pi, y, a), k+1) P_k(y|\pi, a)$$

$$= \max_{a \in \mathcal{A}_k} <r_k(a), \pi> + \sum_{y \in \mathcal{Y}_k} \max_{h \in \mathcal{H}_{k+1}} <h, \hat{\mathcal{T}}_k(\pi, y, a)> P_k(y|\pi, a)$$

$$= \max_{a \in \mathcal{A}_k} <r_k(a), \pi> + \sum_{y \in \mathcal{Y}_k} \max_{h \in \mathcal{H}_{k+1}} <h, \sum_{s' \in \mathcal{S}_k} \mathcal{T}_k(\cdot|s, a) \mathcal{Q}_k(y|s, a) \pi(s)>$$

where the denominator in $\hat{\mathcal{T}}$ cancels the multiplication by $P_k(y|\pi, a)$. This can further be simplified as

$$V^*(\pi, k) = \max_{a \in \mathcal{A}_k} \sum_{y \in \mathcal{Y}_k} \max_{h \in \mathcal{H}_{k+1}} < \frac{r_k(a)}{|\mathcal{Y}_k|} + \sum_{\sigma \in \mathcal{S}_k} h(\sigma) \mathcal{T}_k(\sigma|\cdot, a) \mathcal{Q}_k(y|\cdot, a), \pi>, \quad (2.55)$$

where $|\mathcal{Y}_k|$ is the number of possible observation values. Note that the sum of a finite number of piecewise linear, convex functions of π is also a piecewise linear convex function, and the maximum of a finite number of piecewise linear convex functions is again a piecewise linear, convex function, which establishes the induction. Furthermore, a new set $\mathcal{H}_k$ containing the needed linear support functions can be computed as

$$\mathcal{H}_k = \{h \in \mathcal{B}(\mathcal{S}_k) : h(s) = R_k(a, s) + \sum_{y \in \mathcal{Y}_k} \sum_{\sigma \in \mathcal{S}_{k+1}} h^y_{k+1}(\sigma) \cdot$$
$$\mathcal{T}_k(\sigma|s, a) \mathcal{Q}_k(y|s, a) \text{ for some } h^y_{k+1} \in \mathcal{H}_{k+1}, a \in \mathcal{A}_k\} \quad (2.56)$$

The set $\mathcal{H}_k$ defined above contains far more linear support functions than are necessary to define $V^*(\pi, k)$ in (2.54); specifically, there are many functions $h \in \mathcal{H}_k$ for which there is no information state π such that $V^*(\pi, k) = <h, \pi>$. Thus, efficient algorithms for solution of POMDP problems focus on finding small subsets of $\mathcal{H}_k$ which are sufficient for defining $V^*(\pi, k)$. The details of these algorithms are beyond the scope of this overview chapter, and can be found in review articles such as [162, 45, 46, 159, 189, 173]. The key in all of these algorithms is that, given a specific information state π at stage k

and the set $\mathcal{H}_{k+1}$ of support functions, one can use (2.55) to construct the linear support function h for which $V^*(\pi, k) = <h, \pi>$. By judiciously choosing the information states π for which this is done, one can generate a minimal set of support functions for defining $V^*(\pi, k)$ recursively. In general, the number of linear support pieces still grows exponentially with the number of stages, limiting the application of numerical techniques to small horizon problems or problems with special structure.

4. Approximate Dynamic Programming

The dynamic programming algorithms described above often require extensive computation to obtain an optimal policy. For MDPs, one has to compute a value function indexed by the number of states, which could be uncountable. For POMDPs with finite state spaces, the value functions depend on the information states, which are probability vectors of dimension equal to the number of states. This has led to a number of *Approximate Dynamic Programming* (ADP) techniques [16, 28, 26] that are used to reduce the required computations. Such approximations form the basis for the results in Chapters 7 and 5. We discuss the nature of these approximations briefly for the case of discounted infinite horizon MDPs.

The foundation for most ADP techniques is the characterization of optimal strategies in (2.33)

$$\gamma^*(s) \in \arg\max_{a \in \mathcal{A}} R(s,a) + \beta \int_{s' \in \mathcal{A}} V^*(s') \mathcal{T}(ds'|a,s) \qquad (2.57)$$

If the optimal value function $V^*(s)$ were available, one can compute the optimal action at the current state s by performing the above maximization. ADP techniques compute an approximation $\tilde{V}^*(s)$ to the optimal value function $V^*(s)$ and use (2.57) to generate decisions for each state.

There are three classes of ADP techniques commonly used in the literature. *Offline learning* techniques [16, 230] use simulation and exploratory strategies such as temporal difference learning to learn functional approximations $\tilde{V}^*(s)$ to the optimal value function. These are typically used for problems where the dynamical model is well-understood; a powerful application was demonstrated by Tesauro [230] in the context of backgammon. *Rollout* techniques [26, 207] use real-time simulation of suboptimal policies to evaluate an approximation to the future expected reward in (2.57); these are typically used when the problem instance is not known *a priori*, so that simulations are not readily implemented. *Problem Approximation* techniques [49, 48, 245, 205] use the exact value function computed for an approximate problem with special structure as surrogates for the optimal value function. These techniques exploit special classes of stochastic control problems that have simple solutions, such as the multi-armed bandit problems of Chapter 6 or one-step lookahead problems.

The effectiveness of ADP depends on the choice of technique and the problem structure. Specific ADP techniques tailored to sensor management problems are discussed in greater detail in Chapters 5 and 7.

5.   Example

We conclude this chapter with an example consisting of selecting measurements to classify underwater elastic targets at unknown orientations using active acoustic sensors, described in greater detail in [119]. The scattered fields from each target depend on target type and the target-sensor orientation [200]. Typically, there are contiguous ranges of orientations for which the scattering physics is relatively stationary for each target type. Assuming that the targets of interest are rotationally symmetric, and the scattered fields are observed in a plane bisecting the axis of symmetry, the scattered fields at a fixed radial distance are characterized by a single orientation angle ϕ and the target type.

We model this problem as a POMDP. The underlying discrete states S_k consist of five target types and five discrete orientation bins from 0 to 90°. Actions correspond to taking a measurement of the object from a given angular position in a fixed coordinate system. Assuming that the objects are moving, there is a Markov model for transitions from one relative orientation bin to another given measurements, since moving the sensor to a different angular location will change the relative orientation. Furthermore, there will be some random relative angular changes created by target motion. This is captured by a finite state Markov chain model that depends on the chosen action (measurement position).

A sensing action a at stage k corresponds to selecting a change in relative measurement angle $\Delta\Phi$ for movement of the sensor position from its previous position. As discussed above, there is a state transition probability $\mathcal{T}_k(s'|s,a)$ associated with this action. Furthermore, this action generates an observation y_k which is related to the underlying state: the true object type and the quantized relative observation angle. To continue the development of the POMDP formulation, one must describe the finite set of possible values $\mathcal{Y}$, and the observation probability likelihoods $\mathcal{Q}_k(y|s,a)$. For this application, we collected measured scattered fields for each target as a function of relative angle, with data sampled in 1° increments. Figure 2.3 shows a plot of the magnitude of the discrete Fourier transform of the measured scattered fields, for two of the five targets, as a function of relative sensing angle ϕ. The time-domain scattered fields from each target were processed using matching pursuits [200, 201, 169] to extract a set of feature vectors. The feature vectors were collected across all target-sensor orientations and target types, and vector quantization (VQ) was performed [96], leading to a finite number of possible observation values $\mathcal{Y}$. The error statistics of the vector quantization algorithm were used to generate the observation likelihoods $\mathcal{Q}(y|s,a)$, which were assumed to be stationary. For these experiments, the number of VQ codes (possible observations) was 25. The number of possible observation directions was discretized to 11 directions, at increments of $5°$, with a maximum displacement of $50°$.

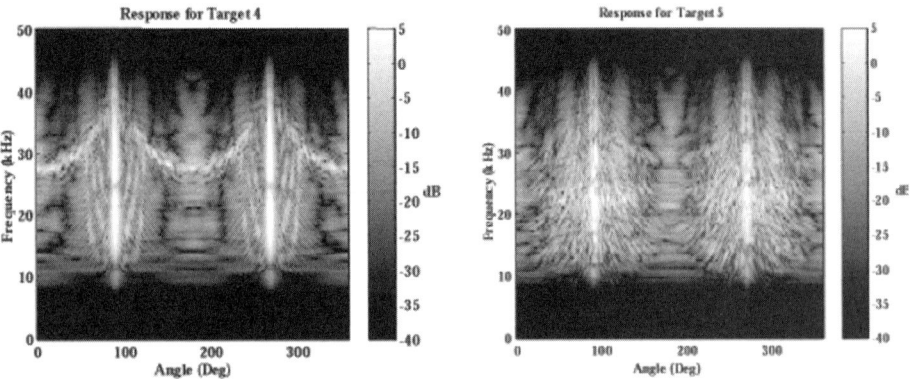

Figure 2.3. Scattered fields (magnitude) as a function of sensing angle.

After performing k observation actions, starting at stage 0, the information state π_k can be computed as in (2.49). To complete the POMDP description, one needs a description of the objective. Let C_{uv} denote the cost of declaring the object under interrogation to be target u, when in reality it is target v, where u and v are members of the five targets of interest. Given the collected information, one can choose to make a classification decision based on the current information, or to continue collecting information. If one chooses to make a classification decision, the selected label will be the one that minimizes the Bayes cost given the available information:

$$\text{Target class} = \arg\min_u \sum_{v=1}^{5} C_{uv} \sum_{s \in \mathcal{S}_v} \pi_k(s), \qquad (2.58)$$

where $\mathcal{S}_v$ is the set of discrete states associated with target v. In terms of the maximization formulation discussed previously, a classification decision incurs a negative reward, corresponding to the expected Bayes cost. Note that this increases the set of potential control actions at any stage $\mathcal{A}_k$, as one can choose to either select a new measurement, or make one of 5 classification decisions. Making a classification decision places the state into an absorbing state, from which there are no further transitions or rewards associated with the state.

In addition to the classification costs, there are costs associated with any sensing action, which may depend on the cost of moving the relative angle displacement $\Delta\phi$. This cost will be independent of the underlying state s_k of the system, and is set to the value 1 in the results below, independent of the angle displacement. Using negative costs as rewards, this implies that, for sensing actions a, $R_k(s,a) \equiv -1$ for all $s \in \mathcal{S}$. For classification actions $a = u$, $R_k(s,a) \equiv -C_{u,v}$ if $s \in \mathcal{S}_v$. The classification costs C_{uv} have a uniform penalty for errors as $C_{uv} = C_c$ for all $u \neq v$, and $C_{uu} = -10$ (a reward of 10 is obtained upon correct classification); the error penalty will be varied in the experiments. In terms of overall reward R, the goal is to maximize the discounted infinite horizon reward

$$R = \mathbb{E}\left[\sum_{k=0}^{\infty} \beta^k R(s_k, a_k)\right], \qquad (2.59)$$

where the discount factor β is chosen to be 0.99.

Denote by s_0 the special state in $\mathcal{S}$ which results from making a final classification decision. Note that $\pi_k(s_0) = 0$ until a classification decision is made, and $\pi_k(s_0) = 1$ for all k after a classification decision is made. Let $\mathcal{A}_m \subset \mathcal{A}$ be the admissible decisions to collect further measurements, and $\mathcal{A}_c \subset \mathcal{A}$ be the admissible decisions to make a classification.

Bellman's equation for this POMDP is given by (2.50), for all π such that $\pi(s_0) = 0$, as

$$V^*(\pi) = \max[\max_{a \in \mathcal{A}_m} <R(s,a), \pi>, -1 + \beta \max_{a \in \mathcal{A}_c} \sum_{y \in \mathcal{Y}} V^*(\hat{T}(\pi, y, a))\, \mathrm{P}(y|\pi, a)]$$

This can be solved using the value iteration algorithm, starting from the piecewise linear approximation $V(\pi) = \max_{a \in \mathcal{A}_m} <R(s,a), \pi>$. After a finite number of iterations, the value function will be a piecewise linear, convex function (cf. (2.55)).

In the results below, the optimal value function was computed approximately using the point-based value iteration (PBVI) algorithm [189], which limits the number of linear support functions to those which support the value function at a given finite set of information states $\pi \in \pi_{|\mathcal{S}|}$. This yields a lower bound on the optimal value function, which converges to the optimal value function as the discrete set of information states increases to fill the space of information states [162].

The POMDP formulation discussed above yields a non-myopic multi-stage sensor management policy, mapping belief states into actions that trade off immediate rewards for acquiring information to make better decisions. It is an adaptive stopping policy, in that the decision to make a final classification depends on the information state. However, its computational complexity can be significant unless the number of discrete linear support functions is restricted.

Alternative sensor management policies with reduced computation requirements can be developed using different POMDP formulations. For instance, one can consider an alternative adaptive policy, generated by a one-step lookahead POMDP: at each stage k, given an information state π_k that has not yet reached the classification state, select action a_k as

$$a_k(\pi_k) = \arg\max[\max_{a \in \mathcal{A}_m} < R(s,a), \pi_k >, -1+$$
$$\beta \max_{a \in \mathcal{A}_c} \sum_{y \in \mathcal{Y}} \max_{a \in \mathcal{A}_m} < R(s,a), \hat{\mathcal{T}}(\pi_k, y, a)) > \mathrm{P}(y|\pi, a)]$$

That is, decide the current choice of action by evaluating the benefit of making a classification decision at k, or taking one additional measurement and making a classification decision afterward. At every new stage k, this problem is solved to determine the current action a_k; this approach is known in control theory as *receding horizon control* or *model-predictive control*, and yields a low-complexity approximation to the longer horizon control problems.

Another approach to generating a sensor management policy is to use a finite horizon POMDP formulation that takes a fixed number of observations T before making a classification decision. The resulting policy does not stop adaptively, but instead performs a classification decision at stage T.

Figure 2.4(a) shows the classification accuracy achieved versus expected number of measurements taken for three algorithms: The POMDP algorithm with adaptive stopping, the one-step lookahead adaptive stopping algorithm, and the non-adaptive stopping algorithm with fixed number of views. Each point in the graph for the adaptive algorithms was generated by varying the cost of an erroneous classification C_c from 15 to 150. For the non-adaptive stopping algorithm, the number of measurement actions was varied. The results represent the average performance of the strategies using Monte Carlo simulations, averaging over possible initial conditions of the target type and orientation and measurement values.

Figure 2.4(a) highlights the advantages of adaptive stopping policies over the fixed stopping policy. Adaptive stopping exploits the availability of good measurements in determining whether additional measurements would be valuable. The results also highlight a small increase in performance when using the

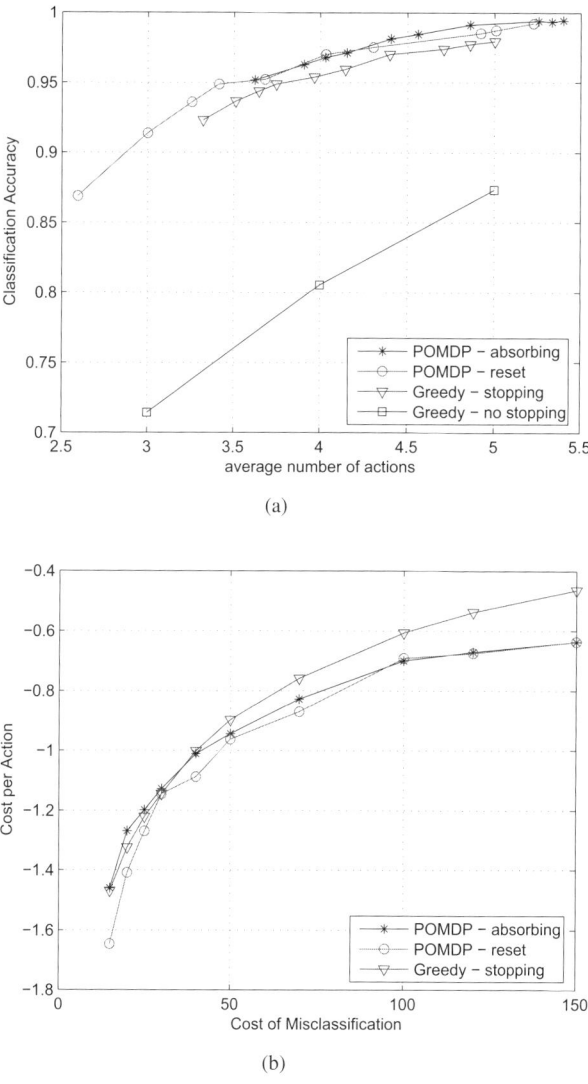

Figure 2.4. a) Classification performance of the different sensor scheduling algorithms as a function of the average number of actions. b) Classification performance versus classification error cost for adaptive stopping policies.

full POMDP adaptive policy versus the one-step lookahead (single stage) policy, as expected.

Figure 2.4(b) presents classification performance for the two adaptive stopping algorithms versus the misclassification cost C_c. Note that the one-step lookahead algorithm underperforms the full horizon algorithm, as it underestimates the amount of information that can be acquired in the future (because it is one-step lookahead). As a consequence, the one-step lookahead algorithm tends to make classification decisions earlier, resulting in a loss of classification performance.

6. Conclusion

This chapter presented an overview of the models and algorithms used for sequential decision making under uncertainty, with a focus on sensor management models. Accumulated information is modeled as a Markov state that evolves in response to selection of sensor actions. Using a reward structure that is additive over time, we discussed the application of stochastic dynamic programming to characterize both the optimal rewards and optimal strategies in these classes of problems. We also presented alternative formulations of rewards, from finite horizon rewards to infinite horizon discounted and undiscounted rewards.

The models presented in this chapter form the foundation for sensor management applications that plan over temporal sequences of sensor actions and adapt to the information observed. Such models are exploited in Chapters 5, 6, 7, 9 and 10 for addressing specific sensor management applications. The next chapter develops information theoretic reward functions that are combined with the joint particle filtering methods of Chapter 4 to implement approximate POMDP sensor management algorithms discussed in Chapter 5.

Chapter 3

INFORMATION THEORETIC APPROACHES TO SENSOR MANAGEMENT

Alfred O. Hero III

University of Michigan, Ann Arbor, MI, USA

Christopher M. Kreucher

General Dynamics Michigan Research and Development Center, Ypsilanti, MI, USA

Doron Blatt

DRW Holdings, Chicago, IL, USA

1. Introduction

A good sensor management algorithm should only schedule those sensors that extract the highest quality information from the measurements. In recent years, several developers of new sensor management algorithms have used this compelling "folk theorem" as a fundamental design principle. This principle relies on tools of information theory to quantify and optimize the information collection capability of a sensor management algorithm. Any design methodology that uses such a guiding principle can be called an *information theoretic* approach to sensor management. This chapter reviews several of these approaches and explains the relevant information theory behind them.

A principal motivation behind information theoretic sensor management systems is that the system should be able to accommodate changing priorities in the mission of the sensing system, e.g., target detection, classification, or identification, as situational awareness evolves. Simply stated, the principal advantage of the information theoretic approach is that it simplifies system design by separating it into two independent tasks: information collection and risk/reward optimization. The sensor manager can therefore focus on the first task, optimizing information extraction, leaving the more complicated mission-specific details of risk-optimal tracking or classification to a downstream algorithm. In other words, information theoretic sensor management approaches substitute a mission-independent surrogate reward function, the extracted information, for the mission-specific reward function of the standard POMDP approach described in Chapter 2.

Information theoretic approaches to selecting between different sources (sensors) of measurement have a long history that can be traced back to R. A. Fisher's theory of experimental design [82, 81], a field which remains very active to this day. In this setting one assumes a parametric model for the measurement generated by each sensor. Selecting a sensor is equivalent to selecting the likelihood function for estimating the value of one of the parameters. The mean (negative) curvature of the log likelihood function, called the Fisher information, is adopted as the reward. The optimal design is given by the choice of sensor that maximizes the Fisher information. When combined with a maximum likelihood parameter estimation procedure, the mission-specific part of the problem, this optimal design can be shown to asymptotically minimize the mean squared error as the number of measurements increases. Such an approach can be easily generalized to multiple parameters and to scheduling a sequence of sensor actions. However, it only has guaranteed optimality properties when a large amount of data is available to accurately estimate these parameters.

If one assumes a prior distribution on the parameters one obtains a Bayesian extension of Fisher's approach to optimal design. In the Bayesian setting each sensor's likelihood function induces a posterior density on the true value of the parameter. Posterior densities that are more concentrated about their mean values are naturally preferred since their lower spread translates to reduced parameter uncertainty. A monotone decreasing function of an uncertainty measure can be used as a reward function. The standard measure of uncertainty is the variance of the posterior, which is equal to the mean squared error (MSE) of the parameter estimator that minimizes MSE. A closely related reward is the Bayesian Fisher information, which is an upper bound on 1/MSE and has been proposed by [105]. These measures are only justifiable when minimum MSE makes sense, i.e., when the parameters are continuous and when the posterior is smooth (twice differentiable). An alternative measure of spread is the entropy of the posterior. Various measures of entropy can be defined for either discrete or continuous parameters and these will be discussed below.

In active sensor management where sensors are adaptively selected as measurements are made a more sensible strategy might be to maximize the rate of decrease of parameter uncertainty over time. This decrease in uncertainty is more commonly called the *information gain* and several relevant measures have been proposed including: the change in Shannon entropy of successive posteriors [109, 108]; the Kullback divergence [204, 125, 165]; and the Rényi divergence [144] between successive posteriors. Shannon and Rényi divergence have strong theoretical justification as they are closely related to the rate at which the probability of decision error of the optimal algorithm decreases to zero.

The chapter is organized as follows. We first give relevant information theory background in Section 2, describing various information theoretic measures that have been proposed for quantifying information gains in sensing and processing systems. Then in Section 3 we describe methods of single stage optimal policy search driven by information theoretic measures. In Section 5 we show that in a precise theoretical sense that information gain can be interpreted as a proxy for any measure of performance. Finally, we turn to illustrative examples including sensor selection for multi-target tracking applications in Section 6 and waveform selection for hyperspectral imaging applications in Section 7.

2. Background

The field of information theory was founded by Claude Shannon in the mid-twentieth century [208] and has had an enormous impact on science in general, and in particular on communications, signal processing, and control. Shannon's information theory has been the basis for breakthroughs in digital communications, data compression, and cryptography. A central result of information theory is the data processing theorem: physical processing of a signal cannot increase the amount of information it carries (see Section 1 of the Appendix). Thus one of the main applications of information theory is the design of signal processing and communication systems that preserve the maximum amount of information about the signal. Among many other design tools, information theory has led to optimal and sub-optimal techniques for source coding, channel coding, waveform design, and adaptive sampling. The theory has also given general tools for assessing the fundamental limitations of different measurement systems, e.g., sensors, in achieving particular objectives such as detection, classification, or tracking. These fundamental limits can all be related to the amount of information gain associated with a specific measurement method and a specific class of signals. This leads to information gain methods of sensor management when applied to systems for which the user has the ability to choose among different types of sensors to detect an unknown signal. This topic will be discussed in the next section of this chapter.

Information theory provides a way of quantifying the amount of signal-related information that can be extracted from a measurement by its entropy, conditional entropy, or relative entropy. These information measures play central roles in Shannon's theory of compression, encryption, and communication and they naturally arise as the primary components of information theoretic sensor management.

2.1 α-Entropy, α-Conditional Entropy, and α-Divergence

The reader is referred to Section 1 of the Appendix for background on Shannon's entropy, conditional entropy, and divergence. Here we discuss the more general α-class of entropies and divergence used in the subsequent sections of this chapter.

Let Y be a measurement and S be a quantity of interest, e.g., the position of a target or the target identity (i.d.). We assume that Y and S are random variables with joint density $p_{Y,S}(y,s)$ and marginal densities p_Y and p_S, respectively. By convention, a random variable is denoted by a capital letter, e.g., Y, and its lowercase counterpart, e.g., y, denotes a realization. To simplify notation we sometimes will drop the subscript on densities, e.g., $p_S(s)$ replaced by $p(s)$, when there is little danger of ambiguity.

As discussed in Section 1 of the Appendix, the Shannon entropy of S, denoted $\mathcal{H}(S)$, quantifies uncertainty in the value of S before any measurement is made, called the prior uncertainty in S. High values of $\mathcal{H}(S)$ imply high uncertainty about the value of S. A more general definition than Shannon entropy is the alpha-entropy, introduced by I. Csiszár and A. Rényi [195]:

$$\mathcal{H}_\alpha(S) = \frac{1}{1-\alpha} \log \mathbb{E}[p_S^{\alpha-1}(S)] = \frac{1}{1-\alpha} \log \int p_S^\alpha(s) ds, \qquad (3.1)$$

where we constrain α to $0 < \alpha < 1$. The alpha-entropy (3.1) reduces to the Shannon entropy (12.1) in the limit as α goes to one:

$$\mathcal{H}_1(S) \stackrel{\text{def}}{=} \lim_{\alpha \to 1} \mathcal{H}_\alpha(S) = -\int p_S(s) \log p_S(s) ds.$$

The conditional α-entropy of S given Y is the average α-entropy of the conditional density $p_{S|Y}$ and is related to the uncertainty of S after the measurement Y is made, called the posterior uncertainty. Similarly to (3.1) the conditional alpha-entropy is defined as

$$\mathcal{H}_\alpha(S|Y) = \frac{1}{1-\alpha} \log \mathbb{E}[p_{S|Y}^{\alpha-1}(S|Y)] \tag{3.2}$$
$$= \frac{1}{1-\alpha} \log \int \int p_{S|Y}^\alpha(s|y) p_Y(y) ds dy,$$

where again $0 < \alpha < 1$. A high quality measurement will yield a posterior density with low $\mathcal{H}_\alpha(S|Y)$ and given the choice among many possible sensors, one would prefer the sensor that yields a measurement that induces the lowest possible conditional entropy. This is the basis for entropy-minimizing sensor management strategies.

We will also need the point conditioned α-entropy of S given Y defined as

$$\mathcal{H}_\alpha(S|Y=y) = \frac{1}{1-\alpha} \log \mathbb{E}[p_{S|Y}^{\alpha-1}(S|Y)|Y=y] = \frac{1}{1-\alpha} \log \int p_{S|Y}^\alpha(s|y) ds.$$

As contrasted with the point conditioned α-entropy $\mathcal{H}_\alpha(S|Y=y)$, which is a function of the conditional density function $p_{S|Y}$ and y, $\mathcal{H}_\alpha(S|Y)$ is a function of the joint density function $p_{S,Y}$ but not a function of Y. These two entropies are related by $\mathcal{H}_\alpha(S|Y) = \int \mathcal{H}_\alpha(S|Y=y) p_Y(y) dy$.

A few comments about the role of the parameter $\alpha \in [0,1]$ are necessary. As compared to $p_{S|Y}$, $p_{S|Y}^\alpha$ is a function with reduced dynamic range. Reducing α tends to make $\mathcal{H}_\alpha(S|Y)$ more sensitive to the shape of the density $p_{S|Y}$ in regions of s where $p_{S|Y}(s|y) \ll 1$. As will be seen in Section 6, this behavior of $\mathcal{H}_\alpha(S|Y)$ can be used to justify different choices for α in measuring the reduction in posterior uncertainty due to taking different sensor actions.

As discussed in the Appendix, Section 1, given two densities p,q of a random variable S the Kullback-Leibler divergence $\mathrm{KL}(p\|q)$ is a measure of similarity between them. Rényi's generalization, called the *Rényi alpha-divergence*, is

$$\mathcal{D}_\alpha(p\|q) = \frac{1}{\alpha - 1} \log \int \left(\frac{p(s)}{q(s)}\right)^\alpha q(s) ds, \tag{3.3}$$

$0 < \alpha < 1$. There are other definitions of alpha-divergence that can be advantageous from a computational perspective, see e.g., [102], [6, Section 3.2] and [229]. These alpha-divergences are special cases of the information divergence, or f-divergence [66], and all have the same limiting form as $\alpha \to 1$. Taking this limit we obtain the Kullback-Leibler divergence

$$\mathrm{KL}(p\|q) = \mathcal{D}_1(p\|q) \stackrel{\mathrm{def}}{=} \lim_{\alpha \to 1} \mathcal{D}_\alpha(p\|q).$$

In the sequel the conditional α-divergence will be extensively used in defining the information gain. Let $p_{X|Y}(x|y)$ and $p_{X|Z}(x|z)$ be two conditional densities of a random variable X given random variables Y and Z, respectively. The point conditioned α-divergence given Y and Z is defined as

$$\mathcal{D}_\alpha(p_{X|Y}(\cdot|y)\|p_{X|Z}(\cdot|z)) = \frac{1}{\alpha-1}\log\int\left(\frac{p_{X|Y}(x|y)}{p_{X|Z}(x|z)}\right)^\alpha p_{X|Z}(x|z)dx,$$

and a shorthand notation will sometimes be used for this quantity: $\mathcal{D}_\alpha(Y,Z)$.

The simple multivariate Gaussian model arises frequently in applications. When p_0 and p_1 are multivariate Gaussian densities over the same domain with mean vectors μ_0, μ_1 and covariance matrices Λ_0, Λ_1, respectively, [106]:

$$\mathcal{D}_\alpha(p_1\|p_0) \qquad (3.4)$$
$$= -\frac{1}{2(1-\alpha)}\log\frac{|\Lambda_0|^\alpha|\Lambda_1|^{1-\alpha}}{|\alpha\Lambda_0+(1-\alpha)\Lambda_1|} + \frac{\alpha}{2}\Delta\mu^\mathsf{T}(\alpha\Lambda_0+(1-\alpha)\Lambda_1)^{-1}\Delta\mu$$

where $\Delta\mu = \mu_1 - \mu_0$ and $|A|$ denotes the determinant of square matrix A.

2.2 Relations Between Information Divergence and Risk

While defined independently of any specific mission objective, e.g., making the right decision concerning target presence, it is natural to expect a good sensing system to exploit all the information available about the signal. Or, more simply put, one cannot expect to make accurate decisions without good quality information. Given a probability model for the sensed measurements and a specific task, this intuitive notion can be made mathematically precise.

2.2.1 Relation to Detection Probability of Error: the Chernoff Information.

Let S be the indicator function of some event, i.e., a realization $S = s$ is either 0 or 1, and $\mathrm{P}(S=1) = 1 - \mathrm{P}(S=0) = p$, for known parameter $p \in [0,1]$. Relevant events could be that a target is present in a particular cell of a scanning radar or that the clutter is of a given type. After observing the sensor output Y it is of interest to decide whether this event occurred or not and this can be formulated as testing between the hypotheses

$$H_0 : S = 0 \qquad (3.5)$$
$$H_1 : S = 1 \;.$$

A test of H_0 vs. H_1 is a decision rule ϕ that maps Y onto $\{0,1\}$ where if $\phi(Y) = 1$ the system decides H_1; otherwise it decides H_0. The 0-1 loss associated with ϕ is the indicator function $L_{0-1}(S, \phi(Y)) = \phi(Y)(1 - S) + (1-\phi(Y))S$. The optimal decision rule that minimizes the average probability of error $P_e = \mathbb{E}[L_{0-1}(S, \phi)]$ is the maximum *a posteriori* (MAP) detector denoted ϕ^*

$$\phi^*(Y) = \begin{cases} 1, & \text{if } \frac{p(S=1|Y)}{p(S=0|Y)} > 1 \\ 0, & \text{otherwise} \end{cases}.$$

The average probability of error P_e^* of the MAP detector satisfies the Chernoff bound [64, Section 12.9]

$$P_e^* \geq \exp\left(\log \int p_{Y|S}^{\alpha^*}(y|1) p_{Y|S}^{1-\alpha^*}(y|0) \, dy\right), \quad (3.6)$$

where

$$\alpha^* = \operatorname*{argmin}_{0 \leq \alpha \leq 1} \int p_{Y|S}^{\alpha}(y|1) p_{Y|S}^{1-\alpha}(y|0) dy.$$

The exponent in the Chernoff bound is identified as a scaled version of the α-divergence $\mathcal{D}_{\alpha^*}\left(p_{Y|S}(\cdot|1) \| p_{Y|S}(\cdot|0)\right)$, called the *Chernoff exponent* or the *Chernoff information*, and it bounds the minimum log probability of error. For the case of n conditionally i.i.d. measurements $Y = [Y_1, \ldots, Y_n]^\mathsf{T}$ given S, the Chernoff bound becomes tight as $n \to \infty$ in the sense that

$$-\lim_{n \to \infty} \frac{1}{n} \log P_e = (1 - \alpha^*) \mathcal{D}_{\alpha^*}\left(p_{Y|S}(\cdot|1) \| p_{Y|S}(\cdot|0)\right).$$

Thus, in this i.i.d. case, it can be concluded that the minimum probability of error converges to zero exponentially fast with rate exponent equal to the Chernoff information.

2.2.2 Relation to Estimator MSE: the Fisher Information.

Now assume that S is a scalar signal and consider the squared loss $(S - \hat{S})^2$ associated with an estimator $\hat{S} = \hat{S}(Y)$ based on measurements Y. The corresponding risk $\mathbb{E}[(S - \hat{S})^2]$ is the estimator mean squared error (MSE), which is minimized by the conditional mean estimator $\hat{S} = \mathbb{E}[S|Y]$

$$\min_{\hat{S}} \mathbb{E}[(S - \hat{S})^2] = \mathbb{E}[(S - \mathbb{E}[S|Y])^2].$$

The minimum MSE obeys the so-called Bayesian version of the Cramér-Rao Bound (CRB) [236]:

$$\mathbb{E}[(S - \mathbb{E}[S|Y])^2] \geq \frac{1}{\mathbb{E}[J(S)]},$$

or, more generally, the Bayesian CRB gives a lower bound on the MSE of any estimator of S

$$\mathbb{E}[(S-\hat{S})^2] \geq \frac{1}{\mathbb{E}[J(S)]}, \quad (3.7)$$

where $J(s)$ is the conditional *Fisher information*

$$J(s) = \mathbb{E}\left[-\frac{\partial^2 p_{S|Y}(S|Y)}{\partial S^2}\bigg|S=s\right]. \quad (3.8)$$

When the prior distribution of S is uniform over some open interval $J(s)$ reduces to the standard Fisher information for non-random signals

$$J(s) = \mathbb{E}\left[-\frac{\partial^2 p_{Y|S}(Y|S)}{\partial S^2}\bigg|S=s\right]. \quad (3.9)$$

2.3 Fisher Information and Information Divergence

The Fisher information $J(s)$ can be viewed as a local approximation to the information divergence between the conditional densities $p_s = p_{Y|S}(y|s)$ and $p_{s+\Delta} = p_{Y|S}(y|s+\Delta)$ in the neighborhood of $\Delta = 0$. Specifically, let s be a scalar parameter. A straightforward Taylor development of the α-divergence (3.3) gives

$$\mathcal{D}_\alpha(p_s \| p_{s+\Delta}) = \frac{\alpha}{2} J(s)\Delta^2 + o(\Delta^2).$$

The quadratic term in Δ generalizes to $\frac{\alpha}{2}\underline{\Delta}^\mathsf{T}\mathbf{J}(s)\underline{\Delta}$ in the case of a vector perturbation $\underline{\Delta}$ of a vector signal $\underline{s}$, where $\mathbf{J}(s)$ is the Fisher information matrix [6]. The Fisher information thus represents the curvature of the divergence in the neighborhood of a particular signal value s. This gives a useful interpretation for optimal sensor selection. Specializing to the weak signal detection problem, $H_0 : S = 0$ vs. $H_1 : S = \Delta$, we see that the sensor that minimizes the signal detector's error rate also minimizes the signal estimator's error rate. This equivalence breaks down when the signal is not weak, in which case there may exist no single sensor that is optimal for both detection and estimation tasks.

3. Information-Optimal Policy Search

At time $t = k$, consider a sensor management system that direct the sensors to take one of M actions $a \in \mathcal{A}$, e.g., selecting a specific sensor modality, sensor pointing angle, or transmitted waveform. The decision to take action a is made only on the basis of past measurements $\mathbf{Y}^{k-1} = [Y_0, \ldots, Y_{k-1}]^\mathsf{T}$ and affects the distribution of the next measurement Y_k. This decision rule is a mapping of $\mathbf{Y}^{k-1}$ to the action space $\mathcal{A}$ and is called a policy $\gamma = \gamma(\mathbf{Y}^{k-1})$.

As explained in Chapter 2, the selection of an optimal policy involves the specification of the reward (or risk) associated with different actions. Recall that in the POMDP setting a policy generates a "(state, measurement, action)" sequence $\{(S_0, Y_0, a_0), (S_1, Y_1, a_1), \ldots\}$ and the quality of the policy is measured by the quality of the sequence of rewards $\{R(S_1, a_0), R(S_2, a_1), \ldots, \}$. In particular, under broad assumptions the optimal policy that maximizes the discounted rewards is determined by applying Bellman's dynamic programming algorithm to the sequence of expected rewards

$$\{\mathbb{E}[R(S_1, a_0)|Y_0], \mathbb{E}[R(S_2, a_1)|Y_0, Y_1], , \ldots, \}.$$

For simplicity, in this section we will restrict our attention to the case of $k = 1$, i.e., an optimal single stage myopic policy seeking to maximize the predicted reward $\mathbb{E}[R(S_1, a_0)|Y_0]$. The basis for information gain approaches to sensor management is the observation that the expected reward depends on the action a_0 only through the *information state*, also called the *belief state* (See Appendix, Section 2)

$$\mathbb{E}[R(S_1, a_0)|Y_0] = \int R(s,a) p_{S_1|Y_0, a_0}(s|Y_0, a_0) ds = \int R(s,a) \pi_1(s) ds.$$

The spread of the information state over state space indicates the uncertainty associated with predicting the future state S_1 given the past measurement Y_0 and the action a_0 dictated by the policy. Information gain strategies try to choose the policy that achieves maximum reduction in uncertainty of the future state. There are several information measures that can capture uncertainty reduction in the information state.

Perhaps the simplest measure of uncertainty reduction is the expected reduction in the variance of the optimal state estimator after an action a_0 is taken

$$\Delta U(a_0) = \mathbb{E}[(S_1 - \mathbb{E}[S_1|Y_0])^2|Y_0] - \mathbb{E}[(S_1 - \mathbb{E}[S_1|Y_1, Y_0, a_0])^2|Y_0, a_0]$$

This measure is directly related to the expected reduction in the spread of the future information state $p_{S_1|Y_1, a_o}$ relative to that of the observed information state $p_{S_1|Y_0, a_o}$ due to action a_0. ΔU measures reduction in spread using the mean-squared error norm, denoted $\|\epsilon\| = \sqrt{\mathbb{E}[|\epsilon|^2|Y_0, a_0]}$, where ϵ is a prediction error. A wide variety of other types of norms can also be used, e.g., the absolute error norm $\|\epsilon\| = \mathbb{E}\left[|\epsilon| \mid Y_0, a_0\right]$, to enhance robustness or otherwise emphasize/de-emphasize the tails of the posterior density. The optimum policy γ^* will minimize the expected prediction error norm. For example, in the case of the mean-squared error, the optimal myopic policy is

$$\gamma^*(Y_0) = \underset{a_0}{\operatorname{argmin}} \; \mathbb{E}\left[\|S_1 - \mathbb{E}[S_1|Y_1, Y_0, a_0]\|^2 \,\big|\, Y_0, a_0\right]. \tag{3.10}$$

Another natural measure of uncertainty reduction is the expected change in entropy due to scheduling sensor action $a = a_0$

$$\Delta U(a_0) = \mathcal{H}_\alpha(S_1|Y_0 = y_0) - \int \mathcal{H}_\alpha(S_1|Y_1 = y_1, Y_0 = y_0) p_{Y_1|Y_0}(y_1|y_0, a_0) dy_1$$

Note that only the second term on the right hand side depends on the sensor action a_0, which determines the conditional density function $p_{Y_1|Y_0,a_0}$.

The entropy-optimal policy γ^* is obtained by replacing the error norm squared $\|S_1 - \mathbb{E}[S_1|Y_1, Y_0, a_0]\|^2$ in (3.10) with the function $p_{S_1|Y_1,Y_0,a_0}^\alpha$ for $\alpha \in (0, 1)$ (Rényi entropy) or with $-\log p_{S_1|Y_1,Y_0,a_0}$ (Shannon entropy) for $\alpha = 1$. The Shannon entropy version of this policy search method was used by Hintz [109] in solving sensor management problems for target tracking applications.

The expected information divergence, called expected information gain, is another measure of uncertainty reduction:

$$\mathrm{IG}_\alpha(a_0) = \mathbb{E}\left[\mathcal{D}_\alpha\left(p_{S_1|Y_1,Y_0,a_0} \| p_{S_1|Y_0}\right) \big| Y_0, a_0\right]. \tag{3.11}$$

This can be expressed as the conditional expectation $\mathbb{E}[\mathrm{IG}_\alpha(Y_1, a_0)|Y_0, a_0]$ of the Y_1-dependent (unobservable) information gain:

$$\mathrm{IG}_\alpha(Y_1, a_0) = \tag{3.12}$$
$$\frac{1}{1-\alpha} \log \int \left(\frac{p_{S_1|Y_1,Y_0,a_0}(s_1|Y_1,Y_0,a_0)}{p_{S_1|Y_0}(s_1|Y_0)}\right)^\alpha p_{S_1|Y_0}(s_1|Y_0) ds_1.$$

When α approaches one the gain measure (3.11) reduces to the Kullback-Leibler divergence studied by Schmaedeke and Kastella [204], Kastella [125], Mahler [165], Zhao [261] and others in the context of sensor management.

When $p_{S_1|Y_0}$ is replaced by the marginal density p_{S_1}, as occurs when S_1, Y_0 are independent, the α-divergence (3.12) reduces to the α-mutual information (α-MI). As α converges to one the α-MI converges to the standard Shannon MI (see Section 1.4 of the Appendix). The Shannon MI has been applied to problems in pattern matching, registration, fusion and adaptive waveform design. The reader is referred to Chapter 10, Section 7 for more details on the latter application of MI.

Motivated by the sandwich bounds in Section 5, a definition of expected information gain that is more closely related to average risk can be defined

$$\Delta \overline{\mathrm{IG}}_\alpha(a_0) \qquad (3.13)$$
$$= \frac{1}{\alpha - 1} \log \mathbb{E}\left[\left(\frac{p_{S_1|Y_1,Y_0,a_0}(S_1|Y_1,Y_0,a_0)}{p_{S_1|Y_0}(S_1|Y_0)} \right)^\alpha \Big| Y_0, a_0 \right],$$

which can be expressed as an expectation involving the Y_1-dependent information gain (3.12):

$$\Delta \overline{\mathrm{IG}}_\alpha(a_0) = \frac{1}{\alpha - 1} \log \mathbb{E}\left[e^{-(1-\alpha) \mathrm{IG}_\alpha(Y_1, a_0)} \Big| Y_0, a_0 \right].$$

The choice of an appropriate value of α can be crucial to obtaining robust IG-optimal sensor scheduling policies. This issue will be addressed in Section 6.

4. Information Gain Via Classification Reduction

A direct relation between optimal policies for sensor management and associated information divergence measures can be established using a recent result of Blatt and Hero [38] for reducing the search for an optimal policy to an equivalent search for an optimal classifier. This strategy is called classification reduction of optimal policy search (CROPS) and leads to significantly more flexibility in finding approximations to optimal sensor management policies (see [39] and [38] for examples). The focus of this section is to show how CROPS leads us to a direct link between optimal policy search and information divergence measures.

The process of obtaining this relation is simple. An average reward maximizing policy is also a risk minimizing policy. A risk minimizing policy is equivalent to a classifier that minimizes a certain weighted probability of error (the risk) for a related label classification problem. After a measure transformation the weighted probability of error is equivalent to an unweighted probability of error, which is related to information divergence via the Chernoff bound. For simplicity of presentation, here we concentrate on binary action space and single stage policies. The general case is treated in [39].

Let the binary action space $\mathcal{A}$ consist of the two actions a_0, a_1 and define the associated (random) rewards $R_0 = R(S_1, a_0)$ and $R_1 = R(S_1, a_1)$, respectively, when state S_1 is observed after taking the specified action. Straightforward algebraic manipulations yield the following expression for the reward associated with policy γ [38]:

$$R(S_1, \gamma(Y_0)) = b - |R_0 - R_1| I(\gamma(Y_0) \neq C)$$

where $I(A)$ denotes the indicator of the event A, C is the random label $C = \operatorname{argmax}_{a=a_0,a_1} R(S_1, a)$, and b is a constant independent of γ. With this expression the optimal single stage policy satisfies

$$\operatorname{argmax}_\gamma \mathbb{E}[R(S_1, \gamma(Y_0))|Y_0] = \operatorname{argmin}_\gamma \tilde{\mathbb{E}}[I(\gamma(Y_0) \neq C)|Y_0], \quad (3.14)$$

where, defining $g(R_0, R_1, Y_0) = I(\gamma(Y_0) \neq C)$, $\tilde{\mathbb{E}}[g(R_0, R_1, Y_0)|Y_0]$ denotes conditional expectation

$$\tilde{\mathbb{E}}[g(R_0, R_1, Y_0)|Y_0] = \int\int g(r_0, r_1, Y_0) \tilde{p}_{R_0,R_1|Y_0}(r_0, r_1|Y_0) dr_0 dr_1,$$

and $\tilde{p}_{R_0,R_1|Y_0}$ is the "tilted" version of the density $p_{R_0,R_1|Y_0}$

$$\tilde{p}_{R_0,R_1|Y_0}(r_0, r_1|y_0) = w(r_0, r_1) p_{R_0,R_1|Y_0}(r_0, r_1|y_0),$$

with weight factor

$$w(r_0, r_1) = \frac{|r_0 - r_1|}{\mathbb{E}\left[|R_0 - R_1| \,|\, Y_0 = y_0\right]}.$$

The relation (3.14) links the optimal reward maximizing policy γ to an optimal error-probability minimizing classifier of the random label C with posterior label probabilities: $P(C = i|Y_0) = \tilde{\mathbb{E}}[I(C = i)|Y_0]$, $i = 0, 1$. Furthermore, by Chernoff's bound (3.6), the average probability of error of this optimal classifier has error exponent:

$$(1 - \alpha^*) \mathcal{D}_{\alpha^*}(p_0 \| p_1), \quad (3.15)$$

where $p_0 = p(Y_0|C = 0)$ and $p_1 = p(Y_0|C = 1)$ are conditional densities of the measurement Y_0 obtained by applying Bayes' rule to $P(C = 0|Y_0)$ and $P(C = 1|Y_0)$, respectively. This provides a direct link between optimal sensor management and information divergence: the optimal policy is a Bayes optimal classifier whose probability of error decreases to zero at rate proportional to the information divergence (3.15).

5. A Near Universal Proxy

Consider a situation where a target is to be detected, tracked and identified using observations acquired sequentially according to a given sensor selection policy. In this situation one might look for a policy that is "universal" in the sense that the generated sensor sequence is optimal for all three tasks. A truly universal policy is not likely to exist since no single policy can be expected to simultaneously minimize target tracking MSE and target misclassification probability, for example. Remarkably, policies that optimize information gain are near universal: they perform nearly as well as task-specific optimal policies for a wide range of tasks. In this sense the information gain can be considered as a proxy for performance for any of these tasks.

The fundamental role of information gain as a near universal proxy has been demonstrated both by simulation and by analysis in [142] and we summarize these results here. First we give a mathematical relation between marginalized alpha divergence and any task based performance measure. The key result is the following simple bound linking the expectation of a non-negative random variable to weighted divergence. Let U be an arbitrary random variable, let p and q be densities of U, and for any bounded non-negative (risk) function g define $\mathbb{E}_p[g(U)] = \int g(u)p(u)du$. Assume that q dominates p, i.e. $q(u) = 0$ implies $p(u) = 0$. Then, defining $w = \text{ess inf } g(u)$ and $W = \text{ess sup } g(u)$, Jensen's inequality immediately yields

$$w\mathbb{E}_q\left[\left(\frac{p(U)}{q(U)}\right)^{\alpha_1}\right] \leq \mathbb{E}_p[g(U)] \leq W\mathbb{E}_q\left[\left(\frac{p(U)}{q(U)}\right)^{\alpha_2}\right], \quad (3.16)$$

where $\alpha_1 \in [0,1)$ and $\alpha_2 > 1$. Equality holds when $p = q$. This simple bound sandwiches any bounded risk function between two weighted alpha divergences.

Using the notation in Section 3 of this chapter, (3.16) immediately yields an inequality that sandwiches the predicted risk after taking an action a_0 by the expected information gain of form (3.13) with two different values of the Rényi divergence α parameter

$$we^{-(1-\alpha_1)\Delta\overline{\text{IG}}_{\alpha_1}(a_0)} \leq \mathbb{E}[g(S_1)|Y_1, a_0] \leq We^{-(1-\alpha_2)\Delta\overline{\text{IG}}_{\alpha_2}(a_0)}, \quad (3.17)$$

where $w = \inf_{y_0} \mathbb{E}[g(S_1)|Y_0 = y_0]$, $W = \sup_{y_0} \mathbb{E}[g(S_1)|Y_0 = y_0]$. This inequality is tight when α_1 and α_2 are close to one, and the conditional risk $\mathbb{E}[g(S_1)|Y_0]$ is only weakly dependent on the current measurement Y_0.

When the state variable S is a vector $\mathbf{S}$ one may only be concerned with estimation of certain of its elements. For example, for target tracking, the target state may be described by position velocity and acceleration but only the position of the target is of interest. In such cases, the state can be partitioned into variables of interest $\mathbf{U}$ and nuisance parameters $\mathbf{V}$. For simplicity assume that the state consists of two scalar variables $\mathbf{S} = [U, V]$. If only U is of interest the risk function will be specified as constant with respect to V, i.e., $g(\mathbf{S}) =$

$g(U)$. According to (3.16), the appropriate sandwich inequality is modified from (3.17) by replacing **S** by U. Specifically, the expected information gain (3.13) in the resultant bounds on the right and left of the inequality in (3.17) are replaced by expected IG expressions of the form

$$\Delta \overline{\mathrm{IG}}_\alpha(a_0) = \qquad (3.18)$$
$$\frac{1}{\alpha-1} \log \mathbb{E}\left[\left(\frac{p_{U_1|Y_1,Y_0,a_0}(U_1|Y_1,Y_0,a_0)}{p_{U_1|Y_0}(U_1|Y_0)}\right)^\alpha \bigg| Y_0, a_0\right],$$

which, as contrasted to the information gain (3.11), is expected IG between marginalized versions of the posterior densities, e.g.,

$$p_{U_1|Y_0}(U_1|Y_0) = \int p_{S_1|Y_0}(U_1, v_1|Y_0) dv_1,$$

where U_1, U_2 are state components: $\mathbf{S}_1 = [U_1, V_1]$. We call the divergence (3.18) the marginalized information gain (MIG).

The sandwich inequality (3.17) is a theoretical result that suggests that the expected information gain (3.13) is a near universal proxy for arbitrary risk functions. Figure 3.1 quantitatively confirms this theoretical result for a simple single target tracking and identification example. In this simulation the target moves through a 100×100 cell grid according to a two dimensional Gauss-Markov diffusion process (see Section 6.1 for details). The moving target is one of 10 possible target types. At each time instant a sensor selects one of two modes, identification mode or tracking mode, and one of the 10,000 cells to query. In identification mode the sensor has higher sensitivity to the target type, e.g., a high spatial resolution imaging sensor, while in tracking mode the sensor has higher sensitivity to target motion, e.g., a moving target indicator (MTI) sensor. The output of these sensors was simply an integer-valued decision function taking values from 0 to 10. Output "0" denotes the "no target present" decision, output "Not 0" the "target present" decision, and output "k", $k \in \{1, \ldots, 10\}$ the "target is present and of class k" decision. The parameters (false alarm and miss probability, confusion matrix) of the sensor were selected to correspond to a realistic multi-function airborne surveillance system operating at 10dB SNR and to exhibit the tradeoff between tracking and identification performance.

The optimal target tracker and classifier are non-linear and intractable, as the measurement is non-Gaussian while state dynamics are Gaussian, and they were implemented using a particle filter as described in Chapter 4 of this book. Several policies for making sequential decisions on sensor mode and pointing direction were investigated: (1) a pure information gain (IG) policy that maximizes divergence between predicted posterior distributions of the four dimensional target state (position and velocity); (2) the marginalized IG (MIG)

Information Theoretic Approaches to Sensor Management

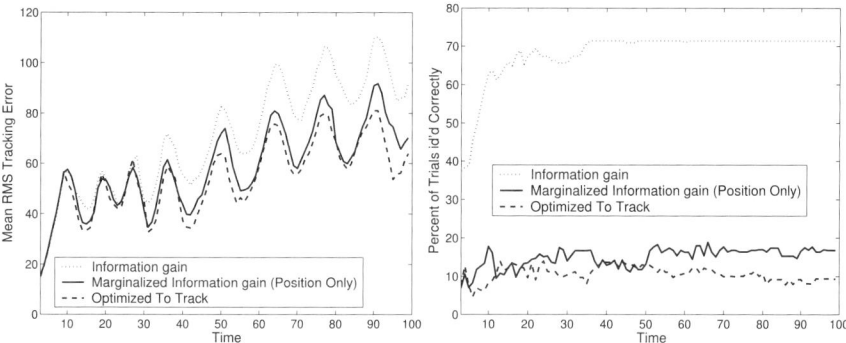

Figure 3.1. Left: comparisons between target position tracking RMS error of a tracker that acquires data using sensor selection policies optimized under the information gain (IG), marginalized information gain (MIG), and tracking (RMS) error reward functions. Right: same as left for target classification performance. (Adapted from Figure 1 in [142] - ©2005 IEEE used with permission).

policy that maximizes the predicted divergence between posteriors of the two dimensional sub-state corresponding to position coordinate only; (3) a policy (RMS) that minimizes predicted tracking mean squared error.

Studying the topmost curve in the left panel of Figure 3.1, it is evident that the IG optimized policy is not optimal for target tracking since it deprives the optimal non-linear tracker of the best sequence of measurements. On the other hand, even though it is based only on information gain, the MIG optimized policy is nearly optimal for tracking as measured by the performance of the optimal tracker that uses MIG generated data. Conversely, from the right panel of Figure 3.1, we see that the IG policy does a much better job at classifying the target type. Thus, as predicted by the theoretical near-universality results in this section, the IG policy achieves a good compromise between tracking and classification tasks.

6. Information Theoretic Sensor Management for Multi-target Tracking

In this section, we illustrate the efficacy of a specific information theoretic approach to sensor management which is based on the alpha divergence. We specialize to a multiple target tracking application consisting of estimating positions of a collection of moving ground targets using a sensor capable of interrogating portions of the surveillance region. Here the sensor management problem is one of selecting, on-line, the portion of the surveillance region to be interrogated. This is done by computing the expected gain in information, as measured by the Rényi divergence, for each candidate action and scheduling the one of highest value.

6.1 The Model Multi-target Tracking Problem

The model problem is constructed to simulate a radar or EO platform, e.g., JSTARS, whose task is to track a collection of moving ground targets. The targets are assumed to be moving in a planar surveillance region. Specifically, there are ten ground targets moving in a $5km \times 5km$ region. Each target is described by its own 4 dimensional state vector $\mathbf{X}(t)$ corresponding to target position and velocity and assumed to follow the 2D diffusion model: $\dot{\mathbf{X}}_i(t) = \rho \mathbf{X}_i(t) + \mathbf{B}W_i(t)$, where ρ is the diffusion coefficient, $\mathbf{B} = [0, 0, \sigma_i, \sigma_i]^\mathsf{T}$, and $W(t)$ is a zero-mean, unit variance white Gaussian noise process. The probability distribution of the target states is estimated on-line from sensor measurements via the joint multi-target probability density (JMPD) model (see Chapter 4. This captures the uncertainty present in the estimate of the states of the targets. The target trajectories were extracted from GPS data on vehicle positions collected as part of a battle training exercise at the U.S. Army's National Training Center.

The sensor simulates a moving target indicator (MTI) system in that at any time t_k, $k = 1, 2, \ldots$, it lays a beam down on the ground that is one resolution cell (1 meter) wide and 10 resolution cells deep. The sensor is at a fixed location above the targets and there are no obscurations that would prevent a sensor from viewing a region in the surveillance area. The objective of the sensor manager is to select the specific 10 meter^2 MTI strip of ground to acquire. When measuring a cell, the imager returns either a 0 (no detection) or a 1 (a detection) which is governed by a probability of detection (P_d) and a per-cell false alarm rate (P_f). The signal to noise ratio (SNR) links these values together. In this illustrative example, we assume $P_d = 0.5$ and $P_f = P_d^{(1+\text{SNR})}$, which is a model for a Doppler radar using envelope detection (thresholded Rayleigh distributed returns - see Chapter 4, Section 2.5). When there are T targets in the same cell the detection probability increases according to $P_d(T) = P_d^{\frac{1+\text{SNR}}{1+T\ \text{SNR}}}$; however the detector is not otherwise able to discriminate or spatially resolve the targets. Each time a beam is formed, a vector of measurements (a vector of zeros and ones corresponding to non-detections and detections) is returned, one measurement for each of the ten resolution cells.

6.2 Rényi Divergence for Sensor Scheduling

As explained above, the goal of the sensor scheduler is to choose which portion of the surveillance region to measure at each time step. This is accomplished by computing the value for each possible sensing action as measured by the Rényi alpha-divergence (3.3). In this multi-target tracking application we use the JMPD (see Chapter 4) to capture uncertainty about the current multi-target state $\mathbf{X}^k = [\mathbf{x}_1^k, \ldots, \mathbf{x}_{T}^k]^\intercal$ and number T^k of targets conditioned on all the previous measurements $\mathbf{Y}^{k-1} = \{Y_1, \ldots, Y_{k-1}\}$. In the expression for the JMPD the integer m ($m = 1, \ldots, M$) will refer to the index of a possible sensing action $a_m \in \{a_1, \ldots, a_M\}$ under consideration, including but not limited to sensor mode selection and sensor beam positioning.

First, given all the measurements, the conditional Rényi divergence between the JMPD $p(\mathbf{X}^k, T^k | \mathbf{Y}^{k-1})$ and the updated JMPD $p(\mathbf{X}^k, T^k | \mathbf{Y}^k)$ must be computed. Therefore, we need the point conditioned alpha-divergence

$$\mathcal{D}_\alpha(\mathbf{Y}^k) = \mathcal{D}_\alpha \left(p(\cdot|\mathbf{Y}^k) \| p(\cdot|\mathbf{Y}^{k-1}) \right) \qquad (3.19)$$

$$= \frac{1}{\alpha - 1} \log \int p^\alpha(\mathbf{x}^k, T^k | \mathbf{Y}^k) p^{1-\alpha}(\mathbf{x}^k, T^k | \mathbf{Y}^{k-1}) d\mathbf{x}^k \ .$$

Using Bayes' rule, this takes the form

$$(3.20)$$

$$\frac{1}{\alpha - 1} \log \frac{1}{p^\alpha(Y_k | \mathbf{Y}^{k-1}, a_m)} \int p^\alpha(Y_k | \mathbf{x}^k, T^k, a_m) p(\mathbf{x}^k, T^k | \mathbf{Y}^{k-1}) d\mathbf{x}^k \ .$$

Our aim is to choose the sensing action to take *before actually receiving* the measurement Y_k. Specifically, we would like to choose the action that makes the divergence between the current density and the density after a new measurement as large as possible. This indicates that the sensing action has maximally increased the information content of the measurement updated density, $p(\mathbf{X}^k, T^k | \mathbf{Y}^k)$, with respect to the density before a measurement was made, $p(\mathbf{X}^k, T^k | \mathbf{Y}^{k-1})$. However, we cannot choose the action that maximizes the divergence as we do not know the outcome of the action before taking it. An alternative, as explained in Section 5, is to calculate the expected value of (3.20) for each of the M possible sensing actions and choose to take the action that maximizes the expectation. Given past measurements $\mathbf{Y}^{k-1}$, the expected value of (3.20) may be written as an integral over all possible measurement outcomes $Y_k = y$ when performing sensing action a_m as

$$\mathbb{E}[\mathcal{D}_\alpha(\mathbf{Y}^k) | \mathbf{Y}^{k-1}, a_m] = \int p(y | \mathbf{Y}^{k-1}, a_m) \mathcal{D}_\alpha \left(y, \mathbf{Y}^{k-1}) \right) dy \ . \qquad (3.21)$$

In analogy to (3.13) we refer to this quantity as the expected information gain associated with sensor action a_m.

6.3 Multi-target Tracking Experiment

The empirical study shown in Figure 3.2 shows the benefit of the information theoretic sensor management method. In this figure, we compare the performance of the information theoretic method where sensing locations are chosen based on expected information gain, a periodic method where the sensor is sequentially scanned through the region, and two heuristic methods based on interrogating regions where the targets are most likely to be given the kinematic model and the estimated positions and velocities at the previous time step (see [146] for a more detailed explanation). We compare the performance by looking at root-mean-square (RMS) error versus number of sensor resources available ("looks"). All tests use the true SNR (= 2) and are initialized with the true number of targets and the true target positions.

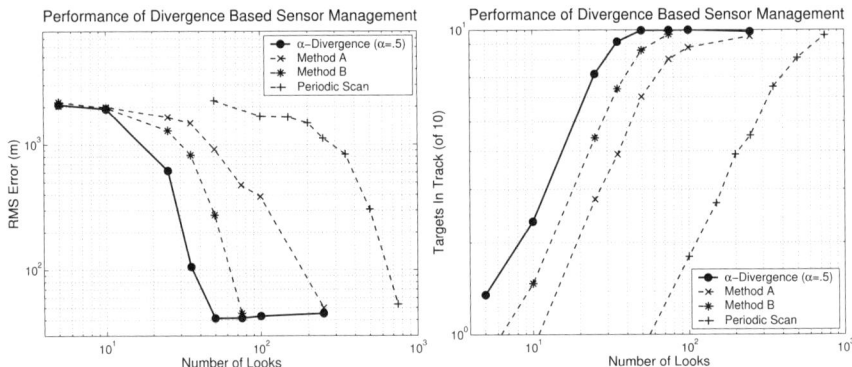

Figure 3.2. A comparison of the information-based method to periodic scan and two other methods. The performance is measured in terms of the (median) RMS error versus number of looks and the (average) number of targets in track. The α-divergence strategy out-performs the other strategies, and at 35 looks performs similarly to non-managed with 750 looks. (Left panel is Figure 6 of [146] - ©2005 IEEE used with permission).

6.4 On the Choice of α

The Rényi divergence has been used in many applications, including content-based image retrieval, image georegistration, and target detection [107, 106]. These studies provide guidance as to the optimal choice of α.

Information Theoretic Approaches to Sensor Management   51

In the georegistration problem [106] it was empirically determined that the value of α leading to highest resolution clusters around either $\alpha = 1$ or $\alpha = 0.5$ corresponding to the KL divergence and the Hellinger affinity respectively. The determining factor appears to be the degree of to which the two densities under consideration are different. If the densities are very similar then the indexing performance of the Hellinger affinity distance ($\alpha = 0.5$) was observed to be better than that of the KL divergence ($\alpha = 1$). Furthermore, the asymptotic analysis of [106] shows that $\alpha = 0.5$ provides maximum discrimination between two similar densities. This value of α provides a weighting which stresses the tails, or the minor differences, between two distributions. In target tracking applications with low measurement SNR and slow target dynamics with respect to the sampling rate, the future posterior density can be expected to be only a small perturbation on the current posterior density, justifying the choice of $\alpha = 0.5$.

Figure 3.3 gives an empirical comparison of the performance under different values of α. All tests assume the detector has receiver operating characteristic (ROC) $P_d = 0.5$ and $P_f = P_d^{(1+\text{SNR})}$ with SNR $= 2$. See Equation (4.24) of Chapter 4 for justification of this choice of ROC for radar multiple target tracking detectors. We find that $\alpha = 0.5$ performs best here as it does not lose track on any of the 10 targets during any of the 50 simulation runs. Both cases of $\alpha \approx 1$ and $\alpha = 0.1$ cause frequent loss of track of targets.

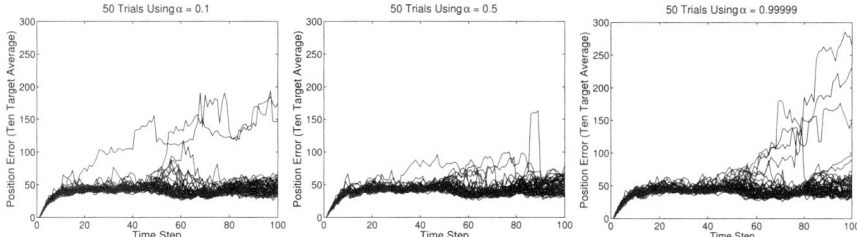

Figure 3.3. A comparison of sensor management performance under different values of α. On simulations involving ten real targets, $alpha = 0.5$ leads to the best tracking performance. (Figure 5 of [146] - ©2005 IEEE used with permission).

6.5 Sensitivity to Model Mismatch

Here we present empirical results regarding the performance of the algorithm under model mismatch. As discussed in Chapter 4, computation of the JMPD and information gain requires accurate models of target kinematics and the sensor. In practice, these models may not be well known. Figure 3.4 shows the effect of mismatch between the assumed target kinematic model and the true model. Specifically, Figure 3.4 shows the sensitivity to mismatch in the assumed diffusion coefficient ρ and the noise variance $\sigma_i = \sigma$, equivalently, the sensor SNR, relative to their true values used to generate the data. The vertical axis of the graph shows the true coefficient of diffusion of the target under surveillance. The horizontal axis shows the mismatch between the filter estimate of kinematics and the true kinematics (matched = 1). The color scheme shows the relative degradation in performance present under mismatch (< 1 implies poorer performance than the matched filer). The graph in Figure 3.4 shows (a) how a poor estimate of the kinematic model affects performance of the algorithms, and (b) how a poor estimate of the sensor SNR affects the algorithm. In both cases, we find that the information gain method is remarkably robust to model mismatch, with a graceful degradation in performance as the mismatch increases.

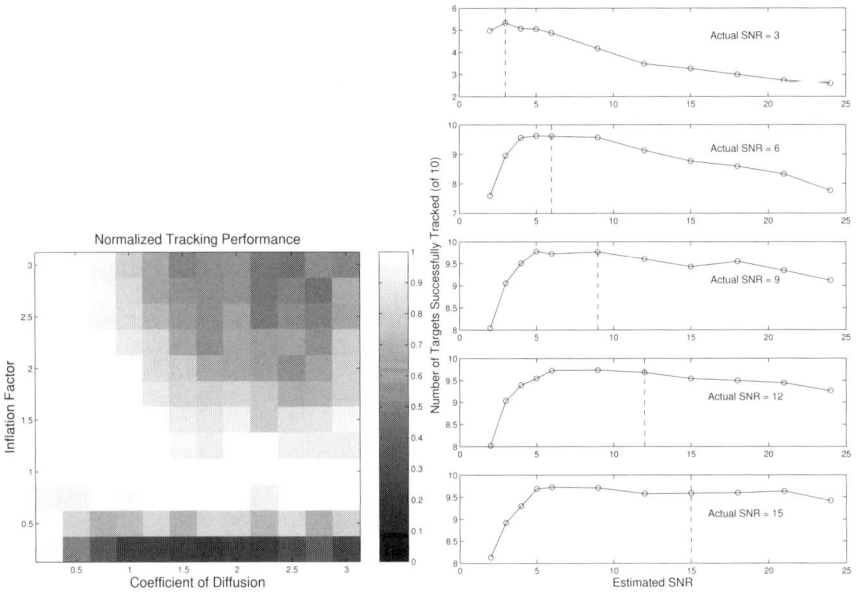

Figure 3.4. Left: Performance degradation when the kinematic model is mismatched. Performance degrades gradually, particularly for high SNR targets.

6.6 Information Gain vs Entropy Reduction

Another information theoretic method that is very closely related to maximizing the Rényi divergence is maximization of the expected change in Rényi entropy. This method proceeds in a manner nearly identical to that outlined above, with the exception that the metric to be maximized is the expected change in entropy rather than the expected divergence. It might be expected that choosing sensor actions to maximize the decrease in entropy may be better than trying to maximize the divergence since entropy is a more direct measure of concentration of the posterior density. On the other hand, unlike the divergence, the Rényi entropy difference is not sensitive to the dissimilarity of the old and new posterior densities and does not have the strong theoretical justification provided by the error exponent (3.15) of Section 3. Generally we have found that if there is a limited amount of model mismatch, maximizing Rényi either divergence or entropy difference gives equivalent sensor scheduling results. For example, for this multi-target tracking application, an empirical study (Figure 3.5) shows that the two methods yield very similar tracking performance.

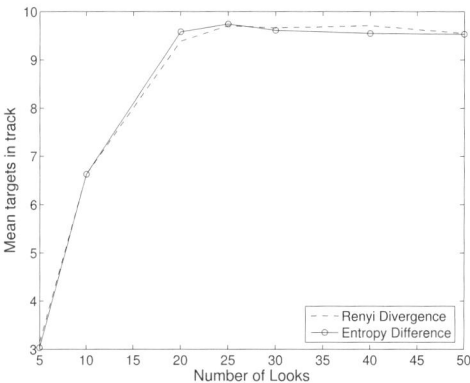

Figure 3.5. Performance, in terms of the number of targets successfully tracked, when using a sensor management strategy predicated on either the Rényi divergence or the change in entropy.

7. Terrain Classification in Hyperspectral Satellite Imagery

A common problem in passive or active radar sensing is the waveform selection problem. For more detailed discussion of active radar waveform design and selection the reader is referred to Chapter 10. Different waveforms have different capabilities depending on the target type, the clutter type and the propagation characteristics of the medium. For example, accurate discrimination between some target and clutter scenarios may require transmission of the full set of available waveforms while in other scenarios one may get by with only a few waveforms. In many situations the transmitted energy or the processed energy are a limited resource. Thus, if there is negligible loss in performance, reduction of the average number of waveforms transmitted is desirable. the problem of selection of an optimal subset of the available waveforms is relevant. This is an instance of an optimal resource allocation problem for active radar.

This section describes a passive hyperspectral radar satellite-to-ground imaging application for which the available waveforms are identified with different spectral bands. To illustrate the methods discussed in this chapter we use the Landsat satellite image dataset [222]. This dataset consists of a number of examples of ground imagery and is divided into training data (4435 examples) and test data (2000 examples) segments. The ground consists of six different classes of earth: "red soil," "cotton crop," "grey soil," "damp grey soil," "soil with vegetation subtle," and "very damp grey soil." For each patch of the ground, the database contains a measurement in each of four spectral bands, ranging from visible to infra-red. Each band measures emissivity at a particular wavelength in each of the 6435 pixelated 3×3 spatial regions. Thus the full four bands give a 36 dimensional real valued feature vector. Furthermore, the database comes with ground truth labels which associates each example with the type of earth corresponding to the central pixel of each 3×3 ground patch.

We consider the following model problem. Assume that the sensor is only allowed to measure a number $b < 4$ of the four spectral bands. When we come upon an unidentified type of earth the objective is to choose the collection of bands that will give the most information about the class. Thus here the state variable x is the unknown class of the terrain type. We assume that at initialization only prior information extracted from the data in the training set is available. Specifically, denote by $p(x)$ the prior probability mass function for class x, $x \in \{1, 2, 3, 4, 5, 7\}$. The set of relative frequencies of class labels in the training database implies that $p(x = 1) = 0.24$, $p(x = 2) = 0.11$, $p(x = 3) = 0.22$, $p(x = 4) = 0.09$, $p(x = 5) = 0.11$, and $p(x = 7) = 0.23$. Denote by $p(Y|x = c)$ the multivariate probability density function of the (9b)-dimensional measurement $Y = [Y_{i_1}, \ldots, Y_{i_b}]^\mathsf{T}$ of the 3×3 terrain patch when selecting the combination $B = \{i_1, \ldots, i_b\}$ of spectral bands and when the class label of the terrain is $x = c$.

7.1 Optimal Waveform Selection

Here we explore optimal off-line waveform selection based on maximizing information gain as compared to waveform selection based on minimizing misclassification error P_e. The objective in off-line waveform selection is to use the training data to specify a single best subset of b waveform bands that entails minimal loss in performance relative to using all 4 waveform bands. Off-line waveform selection is to be contrasted to on-line approaches that account for the effect on future measurements of waveform selection based on current measurements Y_0. We do not explore the on-line version of this problem here. Online waveform design for this hyperspectral imaging example is reported in [39] where the classification reduction of optimal policy search (CROPS) methodology of Section 4 is implemented to optimally schedule measurements to maximize terrain classification performance.

7.1.1 Terrain Misclassification Error. The misclassification probability of error P_e of a good classifier is a task specific measure of performance that we use as a benchmark for studying the information gain measure. For the Landsat dataset the k-nearest neighbor (kNN) classifier with $k = 5$ has been shown to perform significantly better than other more complex classifiers [101] when all four of the spectral bands are available. The kNN classifier assigns a class label to a test vector $\mathbf{z}$ by taking a majority vote among the labels of the k closest points to $\mathbf{z}$ in the training set. The kNN classifier is non-parametric, i.e., it does not require a model for the likelihood function $\{p(\mathbf{z}_B | x = c)\}_{c=1}^{6}$. However, unlike model-based classifiers that only require estimated parameter values obtained from the training set, the full training set is required to implement the kNN classifier. The kNN classifier with $k = 5$ was implemented for all possible combinations of 4 bands to produce the results below (Section 7.1.3).

7.1.2 Terrain Information Gain. To compute the information gain we assume a multivariate Gaussian model for the likelihood $p(Y_k | X_k)$ and infer its parameters, i.e., the mean and covariance, from the training data. Since x is discrete valued the Rényi divergence using the combination of bands B is simply expressed:

$$\mathbb{E}[\mathcal{D}_\alpha(\mathbf{z}_B)] = \int_{\mathbf{z}_B} p(\mathbf{z}_B) \frac{1}{\alpha - 1} \log \sum_{x=1}^{6} p^\alpha(x) p^{1-\alpha}(x|\mathbf{z}_B) d\mathbf{z}_B \ , \quad (3.22)$$

where

$$p(x|\mathbf{z}_B) = \frac{p(x|\emptyset) p(\mathbf{z}_B|x)}{p(\mathbf{z}_B)} \ . \quad (3.23)$$

All of the terms required to compute these integrals are estimated by empirical moments extracted from the training data. The integral must be evaluated numerically as, to the best of our knowledge, there is no closed form.

7.1.3 Experimental Results. For the supplied set of Landsat training and test data we find the expected gain in information and misclassification error P_e as indicated in the Tables below.

Table 3.1. Expected gain in information and Pe of kNN classifier when only a single band can be used. The worst band, band 3, provides the minimum expected gain in information and also yields the largest P_e. Interestingly, the single bands producing maximum information gain (band 4) and minimum P_e (band 2) are different.

Single Band	Mean Info Gain	P_e(kNN)
1	0.67	0.379
2	0.67	0.347
3	0.45	0.483
4	0.75	0.376

Table 3.2. Expected gain in information and P_e of kNN classifier when a pair of bands can be used. The band pair (1,4) provides the maximum expected gain in information followed closely by the band pair (2,4), which is the minimum P_e band pair.

Band Pair	Mean Info Gain	P_e(kNN)
1,2	0.98	0.131
1,3	0.93	0.134
1,4	1.10	0.130
2,3	0.90	0.142
2,4	1.08	0.127
3,4	0.95	0.237

Table 3.3. Expected gain in information and P_e of kNN classifier when only three bands can be used. Omitting band 3 results in the highest expected information gain and lowest P_e.

Band Triple	Mean Info Gain	P_e(kNN)
2,3,4	1.17	0.127
1,3,4	1.20	0.112
1,2,4	1.25	0.097
1,2,3	1.12	0.103

These numbers are all to be compared to the "benchmark" values of information gain, 1.30, and the misclassification probability, 0.96, when all 4 spectral bands are available. Some comments on these results will be useful. First, if one had to throw out a single band, use of bands 1,2,4 entails only a very minor degradation in P_e from the benchmark and the best band to eliminate (band 3) is correctly predicted by the information gain. Second, the small discrepancies between the ranking of bands by P_e and information criteria can be explained by several factors: 1) the kNN is non-parametric while the information gain imposes a Gaussian assumption on the measurements; 2) the information gain is only related to P_e indirectly, through the Chernoff bound;

3) the kNN classifier is not optimal for this dataset - indeed recent results show that a 10% to 20% decrease in P_e is achievable using dimensionality reduction techniques [194, 101]. Finally, as these results measure average performance they are dependent on the prior class probabilities, which have been determined from the relative frequencies of class labels in the training set. For a different set of priors the results could change significantly.

8. Conclusion and Perspectives

The use of information theoretic measures for sensor management and waveform selection has several advantages over task-specific criteria. Foremost among these is that, as they are defined independently of any estimation of classification algorithm, information theoretic measures decouple the problem of sensor selection from algorithm design. This allows the designer to hedge on the end-task and go after designing the sensor management system to optimize a more generic criterion such as information gain or Fisher information. In this sense, information measures are similar to generic performance measures such as front end signal-to-noise ratio, instrument sensitivity, or resolution. Furthermore, as we have shown here, information gain can be interpreted as a near universal proxy for any performance measure. On the other hand, information measures are more difficult to compute in general situations since, unlike SNR, they may involve evaluating difficult non-analytical integrals of functions of the measurement density. There remain many open problems in the area of information theoretic sensor management, the foremost being that no general real-time information theory yet exists for systems integrating sensing, communication, and control.

A prerequisite to implementation of information theoretic objective functions in sensor management is the availability of accurate estimates of the posterior density (belief state) of the state given the measurements. In the next chapter a general joint particle filtering approximation is introduced for constructing good approximations for the difficult problem of multiple target tracking with target birth/death and possibly non-linear target state and measurement equations. In Chapter 5 this approximation will be combined with the information gain developed in this chapter to perform sensor management for multiple target trackers. Information theoretic measures are also applied in Chapter 10 for adaptive radar waveform design.

Chapter 4

JOINT MULTI-TARGET PARTICLE FILTERING

Christopher M. Kreucher

General Dynamics Michigan Research and Development Center, Ypsilanti, MI, USA

Mark Morelande

University of Melbourne, Melbourne, Australia

Keith Kastella

General Dynamics Michigan Research and Development Center, Ypsilanti, MI, USA

Alfred O. Hero III

University of Michigan, Ann Arbor, MI, USA

1. Introduction

In this chapter we review the Joint Multi-target Probability Density (JMPD) for target tracking and sensor management applications and show how it can be evaluated using particle filtering methods. The JMPD is a belief state, i.e., the posterior density given past measurements, for multiple targets. Specifically, it specifies the probability that: 1) there are T targets present, $T = 0, \ldots, \infty$; and 2) given T, the targets are in states $\mathbf{x}_1, \ldots \mathbf{x}_T$. Because it captures the probability distribution across target number, target state and target type, the JMPD serves as the essential starting point for any multiple target tracking and sensor management task. Indeed, as discussed in Chapter 2 and Chapter 3, respectively, the JMPD specifies the crucial information required for implementation of POMDP or information theoretic sensor management algorithms.

Consider an Intelligence, Surveillance and Reconnaissance (ISR) system tasked with detecting, tracking and identifying all ground vehicles within a 100 square kilometer region. For perspectives on ISR in defense systems the reader is referred to Chapter 11. The system consists of one or more Ground Moving Target Indicator (GMTI) radars deployed on a set of airborne, near-space or space platforms. The system has agile radar beams that can be selectively pointed and operated in a number of different resolution modes. The sensor manager must automatically determine the best beam pointing directions and resolution modes. This is a sensor scheduling problem.

When the system is first deployed there is great uncertainty about how many targets are present as well as their locations and identities, so it operates in a low-resolution wide-area search mode. This wide-area search mode produces a number of tentative detections which are revisited using a higher resolution mode that has improved signal-to-noise+clutter ratio (SNCR). This reduces the uncertainty regarding the number of targets but the system still has great uncertainty regarding their locations and identities. Once targets are detected, they will be tracked and localized. Then the sensors will be tasked to produce high range resolution profiles or inverse synthetic aperture radar (ISAR) images that are used for automatic target identification. The JMPD captures the uncertainties of target positions and number of targets given past measurements.

At each stage in this process the sensor manager deploys the sensors so as to minimize the uncertainty regarding the targets in the scene. It may appear that this problem can be neatly solved by moving through a sequence of operations: detect, localize, track, identify. However, in reality one has a complex task mixture. Some tracking operations may also provide detection information on new targets; identification tasks include localization information and so on. Additionally, the system must constantly search for new targets that may have entered the scene. This requires a single entity that captures the information quality for all aspects of the problem: this is the role of the joint multi-target probability density. If we can develop numerically tractable methods to evaluate this JMPD, then its information content can be evaluated using methods developed in Chapter 3 and we can predict how the JMPD information content will change on average for a wide set of alternative sensing actions (alternative modes, beam pointing angles and so on). The optimal sensor manager is then the one that produces the largest expected increase in information for each action.

There are two features of the JMPD belief state approximation developed in this Chapter that should be emphasized: 1) the permutation symmetry of the JMPD has significant impact on algorithm design; and 2) the ability of the JMPD to perform multi-target tracking without the need for data association. To see why we require that the JMPD be symmetric under permutation of target indices, consider the scenario that a jeep is near hill A and a tank is near hill B. When we construct a probability density to describe this situation, the ordering of the arguments in the density cannot effect its value. If $\mathbf{x}_1 = \{\text{jeep}, x_1\}$ and $\mathbf{x}_2 = \{\text{tank}, x_2\}$, where the x_i are position and velocity states of the two objects, then we require that the probability density satisfy $p(\mathbf{x}_1, \mathbf{x}_2) = p(\mathbf{x}_2, \mathbf{x}_1)$. Proper treatment of this permutation symmetry has a significant impact on the development of efficient particle sampling methods for the JMPD.

Most target tracking solutions developed from about 1960 to the mid-1990's relied heavily on the use of the Kalman filter, linear Gaussian state dynamics, linear Gaussian measurements, and data association methods for assigning different measurements to different tracks. Recent progress on this approach is described in Chapter 10. Unlike methods depending on Kalman filtering and data association, developed in Chapter 10, the particle filtering approximation to the JMPD belief state described in this chapter is a fully Bayesian approach that can handle non-linear states, non-Gaussian measurements, and multiple targets-per-measurement.

Until recently, the literature in multi-target tracking was focused on Kalman filtering-based techniques such as multiple hypothesis tracking (MHT) and joint probabilistic data association (JPDA) [33, 13, 14], again discussed in Chapter 10. The fully Bayesian perspective on multiple target tracking adopted in this Chapter is more recent. Stone [223] developed a mathematical theory of multiple target tracking from a Bayesian point of view and Srivistava, Miller [172], and Kastella [124] did early work in this area. The issue with fully Bayesian approaches is computation of the JMPD - which suffers from the curse of dimensionality as the number of targets increases. The particle filtering approximation to the JMPD developed in this chapter is one way to reduce the complexity of the fully Bayesian approach.

Particle filters have been previously applied by others to extend MHT multi-target tracking approaches to non-linear and non-Gaussian target dynamics. In [113], Hue introduces the probabilistic multiple hypothesis tracker (PMHT), which is a blend between the traditional MHT and particle filtering. Others have blended JPDA and particle filtering ideas [122, 40].

The use of particle filters in fully Bayesian multi-target tracking problems also has a recent history. By directly approximating the JMPD, the BraMBLe [114] system, the independent partition particle filter (IPPF) of Orton and Fitzgerald [187] and the work of Maskell [168] eliminate the need for data association while retaining the flexibility of non-linear and non-Gaussian modeling.

The approach described in this chapter builds on these fully Bayesian particle filtering approaches and adds a significant innovation with respect to numerics: the adaptive particle proposal method. By utilizing an adaptive sampling scheme that exploits independence when present, the particle filtering JMPD method described here provides significant computational advantages over brute-force methods.

In our work (as in [187]) each particle encapsulates multiple targets simultaneously. Put another way, instead of using one particle per target we use *one particle per scenario*. That is, a particle encodes a hypothesis about the entire multi-target state – which includes the number of targets and the state (position, velocity, etc.) of each target.

Finally, the fully Bayesian approach is distinguished from traditional approaches of MHT and JPDA as well as the approaches of Hue [113, 112] and others [122, 206, 74], which require thresholded measurements (detections) and a measurement-to-track association procedure. Further, by estimating the joint multi-target density rather than a many single target densities, our method explicitly models target correlations. These two features together, combined with the tractable numerical implementation discussed here, make the JMPD method a quite broadly applicable approach.

2. The Joint Multi-target Probability Density

The JMPD is the probability density on a jump-diffusion system of the type introduced for target-tracking applications in [223, 172]. The basic building block is the single target state space s. For ground target applications the target inertial state is $\mathbf{x} = [x, \dot{x}, y, \dot{y}]^\mathsf{T} \in s = \mathbb{R}^4$, where $\mathbf{z}^\mathsf{T}$ denotes transpose of vector $\mathbf{z}$. At any instant the inertial state of the surveillance volume is determined by $\mathbf{X} \in \mathcal{S} \equiv \emptyset \cup \bigcup_{T=1}^{\infty} s^T$ where $\emptyset$ is a system with no targets and s^T is a T-target system characterized by the concatenated target state $\mathbf{X} = [\mathbf{x}_1^\mathsf{T}, \ldots, \mathbf{x}_T^\mathsf{T}]^\mathsf{T}$. The non-negative integer T is the number of targets in the scene. For fixed T the target motion element of the dynamics undergoes a diffusive evolution obeying an Itô equation of the type used in many non-linear filtering applications. Since T can vary in time, it makes discrete jumps as targets enter or depart the scene. This is the "jump" part of the jump-diffusion dynamics.

For tracking and sensor management applications we construct the posterior density conditioned on a set of observations $\mathbf{y}^i$ occurring at times τ^i. Then

$\mathbf{Y}^k$ is the collection of measurements up to and including time τ^k, $\mathbf{Y}^k = \{\mathbf{y}^1, \mathbf{y}^2, ... \mathbf{y}^k\}$. Each observation or "scan" $\mathbf{y}^i$ may be a single measurement or a vector of measurements made at time i. The posterior $p(\mathbf{X}^k, T^k | \mathbf{Y}^k)$ is what we mean by the "Joint Multi-target Probability Density" (JMPD). The JMPD is the belief state as defined in Sec. 2.5 of the Appendix.

For simplicity, we will typically suppress time indices when they are not important to the discussion. Furthermore, when it is clear by context, we may write the JMPD $p(\mathbf{X}, T|\mathbf{Y}) = p(\mathbf{x}_1, \ldots, \mathbf{x}_T | \mathbf{Y})$ as simply $p(\mathbf{X}|\mathbf{Y})$ and similarly for transition densities and likelihood functions. For example,

- $p(\emptyset|\mathbf{Y})$, is the posterior probability density for no targets in the surveillance volume
- $p(\mathbf{x}_1|\mathbf{Y})$, is the posterior probability density for one target with state $\mathbf{x}_1$
- $p(\mathbf{x}_1, \mathbf{x}_2, \mathbf{x}_3|\mathbf{Y})$, is the posterior probability density for three targets with respective states $\mathbf{x}_1$, $\mathbf{x}_2$ and $\mathbf{x}_3$

The subset of the state space $\mathcal{S}$ corresponding to a fixed number T of targets is called the T-target sector of $\mathcal{S}$. The parts $\mathbf{x}_t$ of the state vector corresponding to single targets induces a partition of $\mathbf{X}$. The JMPD permutation symmetry discussed in the introduction can now be made precise: the JMPD is symmetric on each sector under partition permutation,

$$p(\mathbf{x}_1, ..., \mathbf{x}_T | \mathbf{Y}^k) = p(\mathbf{x}_{\Pi(1)}, ..., \mathbf{x}_{\Pi(T)} | \mathbf{Y}^k), \quad (4.1)$$

where Π is any of the $T!$ permutations of the T labels, $\Pi : i \to \Pi(i)$.

We can gain insight into the role of this permutation symmetry by examining the entropy of the JMPD compared with an unsymmetrical density. As defined in Section 1 of the Appendix, the entropy is

$$\mathcal{H}(p) \equiv -\sum_{T=0}^{\infty} \int p(\mathbf{X}|\mathbf{Y}) \log(p(\mathbf{X}|\mathbf{Y})) d\mathbf{x}_1 \cdots d\mathbf{x}_T. \quad (4.2)$$

For illustration consider the case of two well-localized targets restricted to the real line with separation $2d$ given by the Gaussian sum

$$\begin{aligned} p(x_1, x_2) =\ & \delta_{2,T}(1/4\pi)\big[\exp\left(-[(x_1-d)^2+(x_2+d)^2]/2\right) \\ & + \exp\left(-[(x_1+d)^2+(x_2-d)^2]/2\right)\big], \end{aligned} \quad (4.3)$$

where $\delta_{i,j}$ is the Kronecker δ-function. When $d \gg 1$ (4.3) gives two well-separated peaks near $(x_1, x_2) = \pm(d, -d)$ and the entropy can be approximately evaluated as $H_{d\gg 1} \approx 1 + \log(4\pi)$. On the other hand as $d \to 0$ the peaks coalesce to form a single mono-modal distribution, reducing the entropy to $H_{d=0} = 1 + \log(2\pi)$.

If we model this 2-target problem using the permutation non-symmetric expression $p_{\text{NS}}(x_1, \ldots, x_T) = \delta_{2,T} 1/2\pi \exp\left(-[(x_1 - d)^2 + (x_2 + d)^2]/2\right)$ direct calculation shows that the entropy is $H_{\text{NS}} = 1 + \log(2\pi)$ independent of the target separation. For the properly symmetrized density there is a relative reduction in entropy when targets are close together. This corresponds to a reduction in the relevant phase space for the density in order to develop efficient particle sampling schemes. To treat this aspect of the problem, we have developed sampling techniques that explicitly account for the permutation symmetry. This is one of the key features of the JMPD approach that differentiates it from other popular trackers such as the Multiple Hypothesis Tracker (MHT) or Joint Probability Density Association (JPDA) tracker.

2.1 General Bayesian Filtering

Construction of a filter to update the posterior density as measurements come in proceeds according to the usual rules of Bayesian filtering. Define the aggregate target state at time k as the vector $[\mathbf{X}^k, T^k]$. Under a Markovian model, the conditional density of the current target state given the past states is given by $p(\mathbf{X}^k, T^k | \mathbf{X}^{k-1}, T^{k-1})$ and will be referred to as the kinematic prior (KP). The kinematic prior describes probabilistically how the state of the multi-target system evolves over time. It includes models of target motion, target birth and death, and any additional prior information that may exist such as terrain- and road-constraints. The time-updated density is computed via the *time update* equation:

$$p(\mathbf{X}^k, T^K | \mathbf{Y}^{k-1}) = \sum_{T^{(k-1)}=0}^{\infty} \int d\mathbf{X}^{k-1} p(\mathbf{X}^k, T^k | \mathbf{X}^{k-1}, T^{k-1}) p(\mathbf{X}^{k-1}, T^{k-1} | \mathbf{Y}^{k-1}). \quad (4.4)$$

Observe that the number of targets in the scene can change with time update when $T^k \neq T^{(k-1)}$ to capture the effect of target birth and death. The *measurement update* equation uses Bayes' rule to update the posterior density with a new measurement $\mathbf{y}^k$:

$$p(\mathbf{X}^k, T^k | \mathbf{Y}^k) = \frac{p(\mathbf{y}^k | \mathbf{X}^k, T^k) p(\mathbf{X}^k, T^k | \mathbf{Y}^{k-1})}{p(\mathbf{y}^k | \mathbf{Y}^{k-1})}. \quad (4.5)$$

2.2 Non-Linear Bayesian Filtering for a Single Target

Non-linear Bayesian filtering generalizes the linear filtering equations that govern the Kalman filter for linear Gaussian dynamics and measurements. A convenient starting point is the continuous-discrete non-linear filtering theory developed in Jazwinsky [118, Ch. 5]. To apply this theory, let $\mathbf{x}_\tau$ denote the vector of inertial coordinates (position, velocity) of one of the targets indexed over the continuous time variable τ. In this notation the discrete time state vector $\mathbf{x}^k = \mathbf{x}_{\tau_k}$ is the time sample of $\mathbf{x}_\tau$ at discrete time $\tau = \tau_k$, $k = 1, 2, \ldots$. We assume that the continuous time state vector $\mathbf{x}_\tau$ evolves according to the Itô stochastic differential equation

$$d\mathbf{x}_\tau = \mathbf{f}(\mathbf{x}_\tau, \tau)d\tau + \mathbf{g}(\mathbf{x}_\tau, \tau)d\beta_\tau, \quad \tau \geq \tau_0, \tag{4.6}$$

where $\mathbf{f}$ is a known vector valued non-linear diffusion function, $\mathbf{g}$ is a matrix valued noise scaling function, and $\beta_\tau, \tau \geq \tau_0$ is a vector valued Brownian motion process with covariance $\mathbb{E}[d\beta_\tau d\beta_\tau^\top] = \mathbf{Q}(\tau)d\tau$. Note that both $\mathbf{f}$ and $\mathbf{g}$ can be non-linear functions of $\mathbf{x}_\tau$ in which case the target state evolves as a non-Gaussian process.

Using the Itô state equation model (4.6) and measurement update equation (4.5) the continuous time posterior density $p(\mathbf{x}_\tau | \mathbf{Y}^k)$ for $\tau \in [\tau_k, \tau_{k+1}]$ of the state can be obtained by solving the Fokker-Planck Equation (FPE)

$$\frac{\partial}{\partial \tau} p(\mathbf{x}_\tau | \mathbf{Y}^k) = L\left(p(\mathbf{x}_\tau | \mathbf{Y}^k)\right), \quad \tau_k \leq \tau < \tau_{k+1}, \tag{4.7}$$

with initial condition

$$p(\mathbf{x}_\tau | \mathbf{Y}^k)\Big|_{\tau=\tau_k} = p(\mathbf{x}^k | \mathbf{Y}^k),$$

obtained from (4.5). In (4.7) the linear differential operator L is specified by the functions $\mathbf{f}$ and $\mathbf{g}$.

$$L(p) \equiv -\sum_{i=1}^{n} \frac{\partial(\mathbf{f}_i p)}{\partial \mathbf{x}_i} + \frac{1}{2}\sum_{i,j=1}^{n} \frac{\partial((\mathbf{g}\mathbf{Q}\mathbf{g}^\top)_{ij} p)}{\partial \mathbf{x}_i \partial \mathbf{x}_j}, \tag{4.8}$$

with i, j indexing components of $\mathbf{x}_\tau$.

In the case that the measurements $\mathbf{Y}^k$ are Gaussian and linear in the state $\mathbf{x}^k$, and the Itô diffusion functions $\mathbf{f}$ and $\mathbf{g}$ are linear functions, the posterior density is Gaussian and the measurement and time update equations of the Bayesian filter can be solved by the linear Kalman filter for estimating the state [118].

2.3 Accounting for Target Birth and Death

The FPE operator (4.8) determines how the single target probability density evolves in time due to the stochastic target kinematics. The time evolution of the full multi-target JMPD is simply a linear superposition of FPE operators acting on the partitions of each T-target segment of the concatenated inertial state vector $\mathbf{X}$,

$$\frac{\partial}{\partial t} p(\mathbf{X})\Big|_{\text{FPE}} = \sum_{t=1}^{T} L_t \left(p(\mathbf{X}) \right). \tag{4.9}$$

The other aspect of temporal evolution that must be addressed is the effect of changes in the number of targets, sometimes referred to as target birth and death. Here we adopt a Markovian birth-death process model. Births of targets occur with birth rate $\Lambda^+(\mathbf{x}, t)$, where the arguments $\mathbf{x}$ and t denote the locations and times at which a target emerges. For simplicity assume that the birth rate is constant over time so that $\Lambda^+(\mathbf{x}, t) = \Lambda^+(\mathbf{x})$.

Target birth must maintain the JMPD permutation symmetry property. This requires explicit symmetrization through summation over all T-target permutations

$$\begin{aligned}
\frac{\partial}{\partial t} p(\mathbf{X})\Big|_{T-1 \to T} &= \frac{1}{T!} \sum_{\Pi} p\left(\mathbf{x}_{\Pi_1}, \ldots, \mathbf{x}_{\Pi_{T-1}}\right) \Lambda^+(\mathbf{x}_{\Pi_T}) \quad (4.10)\\
&= \frac{1}{T} \sum_{\rho} p\left(\mathbf{x}_{\rho_1}, \ldots, \mathbf{x}_{\rho_{T-1}}\right) \Lambda^+(\mathbf{x}_{\rho_T}),
\end{aligned}$$

where ρ is the subset of permutations $\rho_i : (1, \ldots, T) \to (1, \ldots, i-1, T, i+1, \ldots, T-1)$, $i = 1, \ldots, T$. Since $p(\mathbf{x}_1, \ldots, \mathbf{x}_{T-1})$ is permutation symmetric, only T terms are required in this summation. Target birth also contributes a *loss term* in the T-target density through transitions from T-target states to $(T+1)$-target states

$$\frac{\partial}{\partial t} p(\mathbf{X})\Big|_{T \to T+1} = -\sum_{t=1}^{T} \int d\mathbf{x}_t \Lambda^+(\mathbf{x}_t) p(\mathbf{x}_1, \ldots, \mathbf{x}_T). \tag{4.11}$$

Target death is treated similarly defining the state-space dependent death rate $\Lambda^-(\mathbf{x})$. The T-target JMPD sector decreases when the T targets are present and one dies while it increases when there are $T+1$ targets and one dies. These contribute time dependencies

$$\left.\frac{\partial}{\partial t}p(\mathbf{X})\right|_{T \to T-1} = -\sum_{t=1}^{T} \Lambda^-(\mathbf{x}_t) p(\mathbf{x}_1, \ldots, \mathbf{x}_T) \qquad (4.12)$$
$$= -T\Lambda^-(\mathbf{x}_T) p(\mathbf{x}_1, \ldots, \mathbf{x}_T),$$

and

$$\left.\frac{\partial}{\partial t}p(\mathbf{X})\right|_{T+1 \to T} = +\sum_{t=1}^{T+1} \int d\mathbf{x}_t \Lambda^-(\mathbf{x}_t) p(\mathbf{x}_1, \ldots, \mathbf{x}_{T+1}) \qquad (4.13)$$
$$= (T+1) \int d\mathbf{x}_{T+1} \Lambda^-(\mathbf{x}_{T+1}) p(\mathbf{x}_1, \ldots, \mathbf{x}_{T+1}).$$

Combining these terms leads to the full multi-target Fokker-Planck equation

$$\frac{\partial}{\partial t}p(\mathbf{X}) = \sum_{t=1}^{T} L_t(p(\mathbf{X})) \qquad (4.14)$$

$$+\frac{1}{T}\sum_{\rho} p(\mathbf{x}_{\rho_1}, \ldots, \mathbf{x}_{\rho_{T-1}}) \Lambda^+(\mathbf{x}_{\rho_T}) - \sum_{t=1}^{T} \int d\mathbf{x}_t \Lambda^+(\mathbf{x}_t) p(\mathbf{x}_1, \ldots, \mathbf{x}_T)$$

$$-T\Lambda^-(\mathbf{x}_T) p(\mathbf{x}_1, \ldots, \mathbf{x}_T) + (T+1) \int d\mathbf{x}_{T+1} \Lambda^-(\mathbf{x}_{T+1}) p(\mathbf{x}_1, \ldots, \mathbf{x}_{T+1}).$$

2.4 Computing Renyi Divergence

Chapter 3 discusses several alternative information theoretic metrics that can be used in sensor management applications. The expected Renyi divergence can be evaluated for JMPD and used to predict the information gain expected for a set of alternative sensing actions. For sensor management the relevant quantity is the point conditioned divergence between the predicted density $p(\mathbf{X}^k, T^k | \mathbf{Y}^{k-1})$ and the updated density after a measurement is made, $p(\mathbf{X}^k, T^k | \mathbf{Y}^k)$, (see Sec. 2.1 of Chapter 3) given by

$$\mathcal{D}_\alpha(\mathbf{Y}^k) = \frac{1}{\alpha - 1} \log \sum_{T^k=0}^{\infty} \int d\mathbf{X}^k p(\mathbf{X}^k, T^k | \mathbf{Y}^k)^\alpha p(\mathbf{X}^k, T^k | \mathbf{Y}^{k-1})^{1-\alpha}. \quad (4.15)$$

We follow the principle that measurements should be selected that maximize the information gain, i.e., the divergence between the post-updated density, $p(\mathbf{X}^k, T^k | \mathbf{Y}^k)$, and the pre-updated density, $p(\mathbf{X}^k, T^k | \mathbf{Y}^{k-1})$. To do this, let

a ($1 \leq a \leq A$) index feasible sensing actions such as sensor mode selection and sensor beam positioning. The expected value of equation (4.15) can be written as an integral over all possible outcomes $\mathbf{y}_k$ when performing sensing action a:

$$\mathbb{E}[\mathcal{D}_\alpha(\mathbf{Y}^k)|\mathbf{Y}^{k-1}, a] = \int p(\mathbf{y}|\mathbf{Y}^{k-1}, a)\mathcal{D}_\alpha\left(p(\cdot|\mathbf{Y}^k)\|p(\cdot|\mathbf{Y}^{k-1})\right) d\mathbf{y}. \tag{4.16}$$

2.5 Sensor Modeling

Implementing the Bayes update (4.5) requires evaluating the measurement likelihood function $p(\mathbf{y}|\mathbf{X})$. We use an association-free model instead of the more common associated measurement model. In the associated measurement model, e.g. JPDA, an observation vector consists of M measurements, denoted $\mathbf{y} = [y_1, \ldots, y_M]^\mathsf{T}$ where $\mathbf{y}$ is composed of threshold exceedances, i.e., valid detections and false alarms. The model usually assumes that each measurement is generated by a single target (the valid measurements) or by clutter and noise (false alarms). The valid measurements are related (possibly non-linearly) to the target state in a known way. If measurement m is generated by target t, then it is a realization of the random process $y_m \sim H_t(\mathbf{x}_t, w_t)$. False alarms have a known distribution independent of the targets and the targets have known detection probability P_d (often modeled as constant for all targets). The origin of each measurement is unknown so a significant portion of any algorithm based on the associated measurement model goes to determining how the measurements correspond to possible targets either through some sort of likelihood weighting (MHT and PDA) or a maximum likelihood assignment process (e.g., multidimensional assignment).

The associated measurement model is widely used and a number of successful tracking systems are based on it. Its practical advantage is that it breaks the tracking problem into two disjoint sub-problems: data association and filtering. While the data association problem is challenging, filtering can be performed using a linearized approach, such as the extended Kalman filter, which is quite efficient. However, there are two disadvantages to the associated measurement model. First, it is based on an artificial idealization of how sensors work in that it assumes each valid detection comes from a single target. This makes it challenging to treat measurement interactions amongst close targets. Second, it requires solution of the data association problem, which usually consumes a large amount of computing resources.

While most data association systems ultimately rely on some variant of the Kalman filter, the use of non-linear filtering methods such as the particle filter frees us to explore new approaches such as association-free methods for computing the full Bayesian posterior density. This type of model has been used in track-before-detect algorithms, such as the "Unified Data Fusion" work of Stone et al. [223] and in [124]. There are several advantages to the association-free method. First, it requires less idealization of the sensor physics and can readily accommodate issues such as merged measurements, side-lobe interference amongst targets and velocity aliasing. Second, it eliminates the combinatorial bottleneck of the associated-measurement approach. Finally, it simplifies the processing of unthresholded measurements to enable improved tracking at lower target SNR.

The starting point for developing an association-free model is the recognition that nearly all modern sensor systems produce multidimensional arrays of pixelized data in some form. The sensor measures return energy from a scene, digitizes it and performs signal processing to produce measurements consisting of an array real or complex amplitudes. This can be 1-dimensional (a bearing-only passive acoustic or electronic sensing measures system), 2-dimensional (an electro-optical imager), 3-dimensional (the range, bearing, range-rate measurements of a GMTI system), or higher dimensional data.

The measurement likelihood $p(\mathbf{y}|\mathbf{X})$ describes how amplitudes in the pixel array depend on the state of all of the targets and background in the surveillance region. To be precise, a sensor scan consists of M pixels, and a measurement $\mathbf{y}$ consists of the pixel output vector $\mathbf{y} = [y_1, \ldots, y_M]^\mathsf{T}$, where y_i is the output of pixel i. The measurement y_i can be an integer, real, or complex valued scalar, a vector or even a matrix, depending on the sensor. If the data are thresholded, then each y_i will be either a 0 or 1. Note that for thresholded data, $\mathbf{y}$ consists of both threshold exceedances and non-exceedances. The failure to detect a target at a given location can have as great an impact on the posterior distribution as a detection.

In the simulation studies described below we model pixel measurements as conditionally independent, yielding

$$p(\mathbf{y}|\mathbf{X}) = \prod_i p(y_i|\mathbf{X}). \qquad (4.17)$$

Let $\chi_i(\mathbf{x}_t)$ denote the indicator function for pixel i, defined as $\chi_i(\mathbf{x}_t) = 1$ when a target in state $\mathbf{x}_t$ projects into sensor pixel i (i.e., couples to pixel i) and $\chi_i(\mathbf{x}_t) = 0$ when it does not. A pixel can couple to multiple targets and single target can contribute to the output of multiple pixels, say, by coupling through side-lobe responses. The indicator function for the joint multi-target state is the logical disjunction

$$\chi_i(\mathbf{X}) = \bigvee_{t=1}^{T} \chi_i(\mathbf{x}_t) \qquad (4.18)$$

The set of pixels that couple to $\mathbf{X}$ is

$$i_\mathbf{X} = \{i | \chi_i(\mathbf{X}) = 1\} \qquad (4.19)$$

For the pixels that do not couple to any targets the measurements are characterized by the background distribution, denoted $p_0(y_i)$ (this can generally depend on where the pixel is within the scene but here we assume a constant background). With this, (4.17) becomes

$$p(\mathbf{y}|\mathbf{X}) = \prod_{i \in i_\mathbf{X}} p(y_i|\mathbf{X}) \prod_{i \notin i_\mathbf{X}} p_0(y_i) \propto \prod_{i \in i_\mathbf{X}} \frac{p(y_i|\mathbf{X})}{p_0(y_i)} \qquad (4.20)$$

In the last step of (4.20) we have dropped the $\mathbf{X}$-independent factor $\prod_i p_0(y_i)$ since it makes no contribution to the JMPD measurement update.

To completely specify the measurement likelihood we must determine how targets couple to the individual pixels. In our simulations the sensor response within pixel i is uniform for targets in i and vanishes for targets outside pixel i. This is equivalent to modeling the point-spread function as a boxcar. It is convenient to define the occupation number $n_i(\mathbf{X})$ for pixel i as the number of targets in $\mathbf{X}$ that lie in i. The single target signal-noise-ratio (SNR), assumed constant across all targets, is denoted λ. We assume that when multiple targets lie within the same pixel their amplitudes add non-coherently. Then the effective SNR when there are n targets in a pixel is $\lambda_n = n\lambda$ and we use $p_n(y_i)$ to denote the pixel measurement distribution. In this model the measurement distribution in pixel i depends only on its occupation number and (4.20) becomes

$$p(\mathbf{y}|\mathbf{X}) \propto \prod_{i \in i_\mathbf{X}} \frac{p_{n_i(\mathbf{X})}(y_i)}{p_0(y_i)}. \qquad (4.21)$$

The effect of the sensor on the measurement likelihood can be determined by detailed modeling, e.g. studying the radar ambiguity function and radar processing noise statistics for different waveforms as in Chapter 10. We adopt a simpler approximation here which reduces the effect of the sensor to two scalar parameters: the range-azimuth-elevation cell resolution and the SNR. In particular, we assume a cell-averaged scalar Rayleigh-distributed measurement corresponding to envelope detected signals under a Gaussian noise model. Such a model has been used to model interfering targets in a monopulse radar system [36, 233] and to model clutter and target returns in turbulent environments

[95]. Rayleigh models are also often used for diffuse fading channels. In the unthresholded case, the likelihood function is

$$p_n(y) = \frac{y}{1+n\lambda} \exp\left(-\frac{y^2}{2(1+n\lambda)}\right). \tag{4.22}$$

When the tracker only has access to thresholded measurements, we use a constant false-alarm rate (CFAR) model for the sensor. The background false alarm rate is set to a level $P_f \in [0,1]$ by selecting a threshold η such that:

$$\mathrm{P}(y > \eta | \text{clutter alone}) = \int_\eta^\infty p_n(y) dy = P_f. \tag{4.23}$$

Under the Rayleigh model the detection probability is

$$P_{d,n} = P_f^{\frac{1}{1+n\lambda}}, \tag{4.24}$$

where n is the number of targets in the cell. This extends the usual relation $P_d = P_f^{\frac{1}{1+\lambda}}$ for thresholded Rayleigh random variables at SNR λ [33].

Note that this simple thresholded Rayleigh model can be easily extended to account for other sensor characteristics, e.g. its non-Gaussian noise, amplitude saturation, or other non-linearities. Once the likelihood function has been specified, as in (4.20), the posterior density (JMPD) can be updated with new measurements via (4.5). However, as the likelihood function (4.20) is not Gaussian this requires some form of function approximation. In the next section the particle filtering approximation to the JMPD is described.

3. Particle Filter Implementation of JMPD

We begin with a brief review of the Sampling Importance Resampling (SIR) particle filter for single targets. This can be generalized directly to produce a SIR JMPD particle filter for multiple targets. The salient feature of the SIR filter is that it uses the kinematic prior as the so-called importance density used to propose new particle states in the time update step.

The SIR filter is a relatively inefficient particle filter and the JMPD SIR filter requires a very large number of particles to track even a modest number of targets. This is largely due to the high dimensionality of the multi-target phase space. Also, when targets are close together we have seen that the entropy of the JMPD is reduced relative to that of a permutation non-symmetric multi-target density. The smaller entropy means that the relevant phase space region is further reduced, placing a great premium on efficient sampling methods.

While the kinematic prior automatically preserves permutation symmetry, care is required when using more sophisticated schemes and has led us to develop Independent Partition (IP), Coupled Partition (CP), Adaptive Partition (AP) and Joint Sampling (JS) methods detailed below.

3.1 The Single Target Particle Filter

To implement a single target particle filter, the single target posterior density, $p(\mathbf{x}|\mathbf{Y})$ is approximated by a set of N_{part} weighted samples (particles) [10, 93]:

$$p(\mathbf{x}|\mathbf{Y}) \approx \sum_{p=1}^{N_{\text{part}}} w_p \delta_D(\mathbf{x} - \mathbf{x}_p), \qquad (4.25)$$

where δ_D is the Dirac delta function and the weight w_p is a function of $\mathbf{Y}$. Here, as elsewhere, for simplicity we have dropped the time index k and denote $\mathbf{x} = \mathbf{x}^k$ as the state and $\mathbf{Y} = \mathbf{Y}^k = [\mathbf{y}^1, \ldots, \mathbf{y}^k]$ as all available observations at time k.

The time update (4.4) and measurement update (4.5) are simulated by the following three step recursion of Table 4.1. First, the particle locations at time k are generated using the particle locations $\mathbf{x}_p$ at time $k-1$ and the current measurements $\mathbf{y}^k$ by sampling from an importance density, denoted $q(\mathbf{x}^k|\mathbf{x}^{k-1}, \mathbf{y}^k)$. The design of the importance density is a well studied area [73] since it plays a key role in the efficiency of the particle filter algorithms. It is known that the optimal importance density (OID) is $p(\mathbf{x}^k|\mathbf{x}^{k-1}, \mathbf{y}^k)$, but this is usually prohibitively difficult to sample from. The kinematic prior $p(\mathbf{x}^k|\mathbf{x}^{k-1})$ is a simple but suboptimal choice for the importance density.

The second step in the implementation of the particle filter is to update particle weights according to [10].

$$w_p^k = w_p^{k-1} \frac{p(\mathbf{y}^k|\mathbf{x}_p^k) p(\mathbf{x}_p^k|\mathbf{x}_p^{k-1})}{q(\mathbf{x}_p^k|\mathbf{x}_p^{k-1}, \mathbf{y}^k)}. \qquad (4.26)$$

When using the kinematic prior as the importance density, the weight equation reduces to $w_p^k = w_p^{k-1} p(\mathbf{y}^k|\mathbf{x}_p^k)$.

Finally, particle resampling is performed to prevent particle degeneracy. Without resampling, the variance of the particle weights increases with time, yielding a single particle with all the weight after a small number of iterations [72]. Resampling may be done on a fixed schedule or adaptively, based on variance of the weights. The particle filter algorithm that uses the kinematic prior as the importance density $q(\mathbf{x}_p^k|\mathbf{x}_p^{k-1}, \mathbf{y}^k)$ and resamples at each time step is called the sampling importance resampling (SIR) algorithm.

Table 4.1. SIR Single Target Particle Filter (Table I from [145] which is ©2005 IEEE - used with permission).

1. For each particle p, p = 1, $\cdots$, N_{part},
   (a) **Particle proposal**: Sample $\mathbf{x}_p^k \sim q(\mathbf{x}^k|\mathbf{x}^{k-1}, \mathbf{y}^k) = p(\mathbf{x}|\mathbf{x}_p^{k-1})$
   (b) **Particle weighting**: Compute $w_p^k = w_p^{k-1} p(\mathbf{y}|\mathbf{x}_p)$ for each p.
2. **Weight normalization**: Normalize w_p^k to sum to one over p
3. **Particle resampling**: Resample N_{part} particles with replacement from $\mathbf{x}_p$ based on the distribution defined by w_p

3.2 The Multi-target Particle Filter

To implement the particle filter for approximating the multiple target JMPD we must sample from the surveillance volume belief state defined by (4.5). We approximate the joint multi-target probability density $p(\mathbf{X}|\mathbf{Y})$ by a set of N_{part} weighted samples. For p = 1, ..., N_{part}, particle p has T_p targets and is given by

$$\mathbf{X}_p = [\mathbf{x}_{p,1}^\mathsf{T}, \ldots, \mathbf{x}_{p,T_p}^\mathsf{T}]^\mathsf{T}. \quad (4.27)$$

Defining

$$\delta(\mathbf{X} - \mathbf{X}_p) = \begin{cases} 0 & T \neq T_p \\ \delta_D(\mathbf{X} - \mathbf{X}_p) & \text{otherwise} \end{cases}, \quad (4.28)$$

the particle filter approximation to the JMPD is given by a set of particles X_p and corresponding weights w_p as

$$p(\mathbf{X}|\mathbf{Y}) \approx \sum_{p=1}^{N_{\text{part}}} w_p \delta(\mathbf{X} - \mathbf{X}_p), \quad (4.29)$$

where $\sum w_p = 1$. This is analogous to a multiple hypothesis tracker in that different particles in the sample may correspond to different hypotheses for the number T_p of targets in the surveillance region.

With these definitions the SIR particle filter extends directly to JMPD filtering, as shown in Table 4.2. This simply proposes new particles at time k using the particles at time $k-1$ and the target kinematics model (4.6) while (4.21) is used in the weight update. Target birth and death given in (4.14) corresponds to probabilistic addition and removal of partitions within particles.

Table 4.2. SIR Multi-target Particle Filter (Table II from [145] which is ©2005 IEEE - used with permission).

1. For each particle p, p = 1, ..., N_{part},
   (a) **Particle proposal**: Sample $\mathbf{X}_p^k \sim q(\mathbf{X}, T | \mathbf{X}_p^{k-1}, T_p^{k-1}, \mathbf{y}^k) = p(\mathbf{X}, T | \mathbf{X}_p^{k-1}, T_p^{k-1})$
   (b) **Particle weighting**: Compute $w_p^k = w_p^{k-1} p(\mathbf{y} | \mathbf{X}_p^k)$ for each p.
2. **Weight normalization**: Normalize w_p^k to sum to one over p
3. **Particle resampling**: Resample N_{part} particles with replacement from $\mathbf{X}_p^k$ based on w_p^k

3.3 Permutation Symmetry and Improved Importance Densities for JMPD

The probability of generating a high-likelihood particle proposal using the kinematic prior of the SIR filter decreases as the number of partitions in a particle grows. This is due to the fact that the likelihood of multi-target proposal is roughly the product of the likelihoods for each partition on its own (this is made precise below). As a result, if a few partitions fall in low-likelihood regions the entire joint proposal likelihood is low. This suggests that improved JMPD importance densities can be developed by first generating high-likelihood proposals for each partition and then combining high-likelihood partition proposals to generate multi-target proposals. The key challenge is that, due to the JMPD permutation symmetry, there is no unique ordering of the partitions in the JMPD particles: the single target state vector corresponding to a particular target can appear in different partitions in different particles. Even if we initialize a filter with ground truth so that all of the particles correspond to the same target ordering, the order generally changes in some particles as targets approach each other and there is measurement-to-target association uncertainty (we refer to this as "partition swapping"). The key to addressing this issue is to impose a particular ordering on the partitions in all of the targets. Then the correspondence between partitions in different particles is well defined, allowing us to develop partition-based proposal schemes that significantly improve the efficiency of the JMPD-PF. This partition ordering does not violate the permutation symmetry requirement of the JMPD if we approximate it by an appropriately symmetrized function of the sample particles.

To gain more insight into this issue, consider the JMPD for two targets in one dimension introduced in (4.3). For widely separated, well-localized targets the JMPD has two distinct peaks in the (x_1, x_2) plane. Samples from $p(x_1, x_2)$ can fall in either the upper or lower diagonal half-plane. For convenience we can choose to approximate p using samples with a particular permutation symmetry, say only those in the upper half plane $x_2 \geq x_1$, explicitly symmetrizing as needed. Now as the targets approach each other, the peaks of the JMPD approach the diagonal $x_1 = x_2$. Even if we initially approximate the density with particles that only lie in the upper half-plane, when the targets approach each other two peaks coalesce along the diagonal. This effect leads to the entropy reduction noted in Section 2. If we generate random samples from the JMPD where the targets are close together, some samples will lie in the upper half-plane while others are in the lower half-plane. This suggests the following strategy: When the targets are widely separated we can impose a fixed partition ordering on the particles and use that ordering to construct efficient proposals. When the targets are close together, we must account for the fact that different samples may have different partitions as we construct proposals. Proposals that are more efficient than kinematic proposals can be constructed, but we incur additional overhead when the targets are close together.

3.4 Multi-target Particle Proposal Via Individual Target Proposals

While the kinematic prior is simple to implement, it requires a large number of particles (see Figure 4.5 below). The kinematic prior does not account for the fact that a particle represents many targets. Targets that are far apart in measurement space behave independently and should be treated as such. Another drawback to kinematic prior is that current measurements are not used when proposing new particles. These considerations taken together result in a very inefficient use of particles and therefore require large numbers of particles to successfully track.

To overcome the deficiencies mentioned above, we have employed alternative particle proposal techniques which bias the proposal process toward the measurements and allow for factorization of the target state when permissible. These strategies propose each partition (target) in a particle separately, and form new particles as the combination of the proposed partitions. We describe several methods here, beginning with the independent partitions (IP) method of [187] and the coupled partitions (CP) method. The basic idea of both CP and IP is to construct particle proposals at the target level, incorporating the measurements so as to bias the proposal toward the optimal importance density. We show that each has benefits and drawbacks and propose an adaptive partition (AP) method which automatically switches between the two as appropriate.

The permutation symmetry of the JMPD must be carefully accounted for when using these sampling schemes. The CP method proposes particles in a permutation invariant manner, however it has the drawback of being computationally demanding. When used on all partitions individually, the IP method is not permutation invariant. Our solution is to perform an analysis of the particle set to determine which partitions require the CP algorithm because they are involved in partition swapping and which partitions may be proposed via the IP method. This analysis leads to the AP method of proposal which is permutation invariant.

3.4.1 Independent-Partition (IP) Method.

Summarized in Table 4.3, the independent partition (IP) method of Orton [187] is a convenient way to propose particles when part or all of the joint target posterior density factors. The IP method proposes a new partition independently as follows. For a partition t, each particle at time $k-1$ has its tth partition proposed via the kinematic prior and weighted by the measurements. From this set of N_{part} weighted estimates of the state of the tth target, we select N_{part} samples (with replacement) to form the tth partition of the particles at time k.

Note that the importance density q is no longer simply the model of target kinematics $p(\mathbf{X}^k|\mathbf{X}^{k-1})$ as in the SIR Multi-target particle filter. Therefore the weight given by the weight equation (4.26) does not simply become the likelihood $p(\mathbf{y}^k|\mathbf{X}^k)$. There is a bias added which emphasizes partitions with high likelihood. To account for this sampling scheme, the biases corresponding to each particle for each target, $b_{\text{p},t}$, are retained to use in conjunction with the likelihood $p(\mathbf{y}^k|\mathbf{X}^k)$ when computing particle weights.

The IP method is predicated on the assumption that partition t in each particle corresponds to the same target. Therefore, the partitions in each particle must be identically ordered before this method is applied. If IP is applied to particles that have different orderings of partitions, multiple targets will be grouped together and erroneously used to propose the location of a single target.

3.4.2 Coupled Partition (CP) Proposal Method.

When the marginal posterior target distributions for different targets begin to overlap, the corresponding partitions are coupled and the IP method is no longer applicable. This situation requires a Coupled Partition (CP) scheme. We proceed as follows (see Table 4.4). To propose partition t of particle p, CP generates R possible realizations of the future state using the kinematic prior. The R proposed future states are then given weights using the current measurements and a single representative is selected. This process is repeated for each particle until the tth partition for all particles has been formed. This is an auxiliary

Table 4.3. Independent Partition Particle Filter (Table III from [145] which is ©2005 IEEE - used with permission).

1. For each partition, $t = 1 \cdots T_{max}$,
   (a) **Partition Proposal:** Propose partition t via Independent Partition Subroutine
2. **Particle weighting:** Compute $w_p^k = w_p^{k-1} * \frac{p(\mathbf{y}|\mathbf{X}_p)}{\prod_{t=1}^{T_p} b_{p,t}}$
3. **Weight normalization:** Normalize w_p^k to sum to one over p.

Independent Partition Subroutine for Target t:

1. For each particle p = 1, ..., N_{part},
   (a) **Particle partition proposal:** Sample $\mathbf{X}_{p,t}^* \sim p(\mathbf{x}|\mathbf{X}_{p,t}^{k-1})$
   (b) **Particle partition weighting:** Compute $\omega_p = p(\mathbf{y}|\mathbf{X}_{p,t}^*)$
2. **Partition weight normalization:** Normalize ω to sum to one over p.
3. For each particle p = 1, ..., N_{part},
   (a) **Index selection:** Sample an index j from the distribution defined by ω
   (b) **Particle partition selection:** Set $\mathbf{X}_{p,t} = \mathbf{X}_{j,t}^*$
   (c) **Bias balancing:** Retain bias of sample, $b_{p,t} = \omega_j$

particle filter of the type suggested in [190] where the multiplicity R plays the role of the auxiliary variable. As in the IP method, the final particle weights must be adjusted for biased sampling.

Stated in the language of genetic algorithms, the difference between CP and IP is that CP "maintains pedigree," i.e., all of the partitions in a new proposed particle must have come from a common ancestor while IP permits crossbreeding from different ancestors. Target birth and death are included in both CP and IP algorithms by adding or deleting partitions as determined by the target birth and death rates.

3.4.3 Adaptive Particle Proposal Method.

IP and CP can be combined adaptively to provide a scheme that delivers the speed advantage of IP for partitions that correspond to widely separated targets (usually the majority of targets) together with improved tracking for coupled targets. The Adaptive-Partition (AP) method analyzes each partition separately. Partitions that are sufficiently well separated according to a given metric (see below) from all other partitions are treated as independent and proposed using the IP method. When targets are not well-separated, the CP method is used.

Table 4.4. Coupled Partition Particle Filter (Table IV from [145] which is ©2005 IEEE - used with permission).

1. For each partition, $t = 1 \cdots T_{max}$
   (a) **Partition proposal:** Propose partition t via Coupled Partition Subroutine
2. **Particle weighting:** Compute $w_p^k = w_p^{k-1} * \frac{p(\mathbf{y}|\mathbf{X}_p)}{\prod_{t=1}^{T_p} b_{p,t}}$
3. **Weight normalization:** Normalize w_p^k to sum to one over p

Coupled Partition Subroutine for Target t

1. For each particle p = 1, ..., N_{part},
   (a) **Particle partition proposals:** For each proposal $r = 1, ..., R$
      i. Sample $\mathbf{X}_{p,t}^*(r) \sim p(\mathbf{x}|\mathbf{X}_{p,t}^{k-1})$
      ii. Compute $\omega_r = p(\mathbf{y}|\mathbf{X}_{p,t}^*(r))$
   (b) **Proposal weight normalization:** Normalize ω to sum to one.
   (c) **Index selection:** Sample an index j from the distribution defined by ω
   (d) **Partition selection:** Set $\mathbf{X}_{p,t} = \mathbf{X}_{p,t}^*(j)$
   (e) **Bias balancing:** Retain bias of sample, $b_{p,t} = \omega_j$

To provide a criterion for partition separation, we threshold based on distance in sensor space between the estimated state of the ith partition and the jth partition. Denote by $\hat{\mathbf{x}}_i$ the estimated x and y positions of the ith partition (4.42). Notice only the spatial states are used (i.e., velocities are neglected), as these are the states that measure distance in sensor space. We have computed the distance between two partitions using a Euclidean metric between the estimated centers, and the Mahalanobis metric (4.30), where $\hat{\Lambda}_j$ is the covariance associated with the estimate of the jth partition (4.43).

$$r^2 = (\hat{x}_i - \hat{x}_j)^\mathsf{T} \hat{\Lambda}_j^{-1} (\hat{x}_i - \hat{x}_j). \qquad (4.30)$$

We have studied the use of a nearest neighbor criterion, where partitions are considered coupled if any sample from partition i is closer to the center of partition j than any sample from partition j. In practice, we find that the Euclidean distance is less computationally burdensome and provides similar performance.

Table 4.5. Adaptive Proposal Method (Table V from [145] which is ©2005 IEEE - used with permission).

1. For each partition $t = 1 : T_{max}$
    (a) **Distance computation:** $d(t) = \min_{j \neq t} ||\hat{\mathbf{x}}_t - \hat{\mathbf{x}}_j||$
    (b) **Partition proposal:** if $d(t) > \tau$
          Propose partition t using IP method
        else
          Propose partition t using CP method
2. **Particle weighting:** For each particle p = 1, ..., N_{part}
   $w_{\text{p}}^k = w_{\text{p}}^{k-1} * \frac{p(\mathbf{y}|\mathbf{X}_{\text{p}})}{\prod_{t=1}^{T_{\text{p}}} b_{\text{p},t}}$
3. **Weight normalization:** Normalize w_{p}^k to sum to one.

3.5 Multi-target Particle Proposal Via Joint Sampling

In the IP, CP and AP methods as described above, samples are drawn independently for each target. Recall that this approach is motivated by the approximate factorization of the JMPD for well-separated targets. However, this approximate factorization does not hold for closely-spaced targets. As a result, if multiple targets are in close proximity drawing samples independently for each target leads to many particles being proposed in undesirable parts of the multi-target state space. Instead, particles for closely spaced targets should be drawn jointly, conditional on the previous target states and the current measurement.

Therefore a joint sampling refinement to the method promises to improve performance. In this method, those partitions that are deemed to be coupled are clustered according to the method of Section 3.3. This results in "partitions" that contain multiple targets – some with 2 targets, some with 3 targets, etc. Then instead of proposing each target individually, the clustered pairs (triplets, etc.) of targets are proposed all at once. This method is summarized in Table 4.6. Note that the idea of a partition containing multiple targets is present in the work of Orton [187], but adaptively deciding partition boundaries and partition clustering is a more recent innovation [145].

To understand how this joint sampling method works, Figure 4.1 illustrates a scenario with two targets moving along the real line in opposite directions with equal speed. The importance density should propose as many particles as possible with the correct arrangement of targets, in this case one target in each of the cells.

80 FOUNDATIONS AND APPLICATIONS OF SENSOR MANAGEMENT

Table 4.6. Modified Adaptive Proposal Method

1. **Partition clustering:** Construct non-overlapping target clusters $C_1, \ldots, C_s$, $s \leq r$ such that $\cup_{l=1}^{s} C_l = \{1, \ldots, T\}$, and if $i \in C_l$ then $\|\hat{x}_i - \hat{x}_j\| < \Gamma \Rightarrow j \in C_l$ where Γ is a threshold.
2. For each cluster, $l = 1 \cdots s$
   (a) **Cluster proposal:** if cluster C_l has one entry,
       Propose group c using IP method
   else
       Propose group c using CP method
3. **Particle weighting:** Compute the weights $w_p^k = w_p^{k-1} \frac{p(\mathbf{y}|\mathbf{X}_p)}{\prod_{l=1}^{s} b_{p,l}}$
4. **Weight normalization:** Normalize w_p^k to sum to one.

Let $\mathbf{x}_i^k = (\rho_i^k, v_k^i)$ denote the state of target i with ρ_i^k the position and v_i^k the velocity. The weight given to a particular arrangement of targets can be measured by the probability

$$I(V_1, V_2) = \int_{V_1} d\rho_1^k \int_{-\infty}^{\infty} dv_1^k \int_{V_2} d\rho_2^k \int_{-\infty}^{\infty} dv_2^k q(\mathbf{X}^k|\mathbf{X}^{k-1}, \mathbf{y}^k) \quad (4.31)$$

where the V_i are cell volumes. Ideally we would have $I([0,1], [-1,0]) = 1$, i.e., all samples for each target would be placed in the cell occupied by that target. If both targets are detected, the joint sampling importance density has $I([0,1], [-1,0]) = 0.977$ while the independent sampling importance density has $I([0,1], [-1,0]) = 0.754$. Thus, the joint sampling density places a higher proportion of particles in the correct cells compared to the independent sampling density. The benefits of joint sampling become even more apparent if only one target is detected. In this case $I([0,1], [-1,0]) = 0.692$ for joint sampling and $I([0,1], [-1,0]) = 0.003$ for independent sampling.

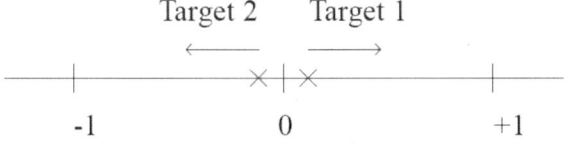

Figure 4.1. Crossing target scenario for demonstration of joint target proposals.

Joint sampling of particles is more computationally expensive than independent sampling. In particular, the computational expense of jointly drawing samples for a group of targets increases exponentially with the number of targets. Fortunately, this intractable increase in computational expense can be avoided by exploiting the approximate factorization of the JMPD for well-separated targets. This approximate factorization is illustrated in Figure 4.2 which shows the probability of correct placement of the particles plotted against target separation for joint and independent sampling. Results are given for the cases where both targets are detected and only one target is detected. It can be seen that for a sufficiently large separation the independent sampling density will almost certainly place particles in the correct location. In such cases the computational expense of joint sampling is unnecessary and should be avoided.

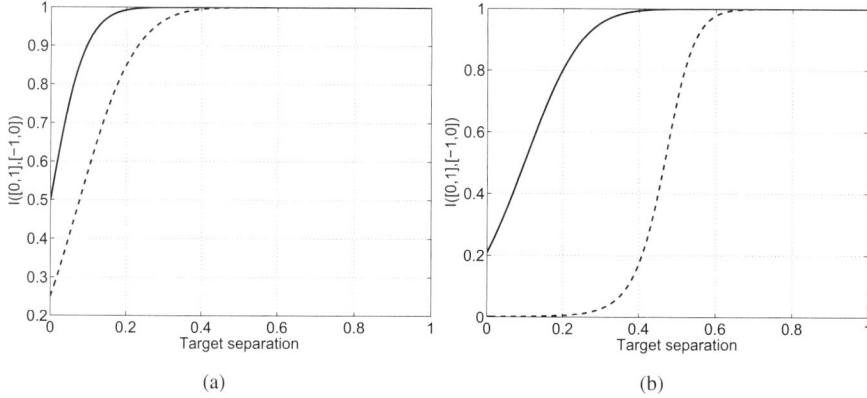

Figure 4.2. Probability of correct placement of particles for joint sampling (solid) and independent sampling (dashed) plotted against target separation when (a) both targets are detected and (b) only one target is detected.

In certain circumstances, the optimal importance density can be even more efficiently approximated than the sample based approach discussed here. In particular, if target dynamics are linear/Gaussian and measurements are made on a grid, the optimal proposal involves sampling from truncated normals [174]. In this case, a similar AP approach is used wherein partitions are first separated into groups that are uncoupled and then each group is treated by sampling from multidimensional truncated normals. This results in optimal sampling from each of the clusters giving even more efficient particle utilization.

3.6 Partition Ordering

As we have seen, the permutation symmetry associated with the JMPD discussed in Section 2 is inherited by the particle filter representation of the

JMPD. Each particle contains many partitions (as many as the number of targets it hypothesizes to exist) and the permutation symmetry of JMPD manifests through the fact that the relative ordering of targets may change from particle to particle.

The fact that partitions are in different orders from particle to particle is of no consequence when the object of interest is an estimate of the joint multi-target density. Each particle contributes the correct amount of mass in the correct location to the multi-target density irrespective of the ordering of its partitions.

However, the IP scheme gains efficiency by permitting hybridization across particles and so requires that particles be identically ordered. Furthermore, estimating the multi-target states from the particle filter representation of JMPD must also be done in a way that is invariant to permutations of the particles. Therefore, when estimating target states, we permute the particles so that each particle has the targets in the same order. We use a K-means [101] algorithm to cluster the partitions of each particle, where the optimization is done across permutations of the particles. In practice, this engenders a very light computational burden. The are two reasons for this. First, partitions corresponding to widely separated targets are not coupled and so remain ordered. Second, since ordering is applied at each time step, coupled partitions are always nearly ordered and so one iteration of the K-means algorithm is usually enough to find an optimal permutation.

As shown in Table 4.7, the K-means algorithm consists of the following. Under the permutation Π_p, the particle

$$\mathbf{X}_p = [\mathbf{x}_{p,1}, \mathbf{x}_{p,2}, \cdots, \mathbf{x}_{p,T_p}], \tag{4.32}$$

is reordered as

$$\mathbf{X}_p = [\mathbf{x}_{p,\Pi_p(1)}, \mathbf{x}_{p,\Pi_p(2)}, \cdots, \mathbf{x}_{p,\Pi_p(T_p)}]. \tag{4.33}$$

The mean of the tth partition under the permutation Π is

$$\bar{\mathbf{X}}_t(\Pi) = \sum_{p=1}^{N_{\text{part}}} w_p \mathbf{X}_{p,\Pi_p(t)}, \tag{4.34}$$

where it is understood that the summation is taken over only those particles that have partition t. Further, define the χ^2 statistic

$$\chi^2(\Pi) = \sum_{p=1}^{N_{\text{part}}} \sum_{t=1}^{T_p} w_p (\mathbf{X}_{p,\Pi_p(t)} - \bar{\mathbf{X}}_t(\Pi_p))^2. \tag{4.35}$$

Joint Multi-target Particle Filtering

To reorder the particles, the goal is to find the set of permutations Π that minimize χ^2, i.e.,

$$\hat{\Pi} = \min_{\Pi} \chi^2(\Pi). \quad (4.36)$$

The K-means algorithm is a well known method of approximately solving problems of this type. An initial permutation Π is assumed and perturbations about that value are made to descend and find the best (local) Π.

Table 4.7. K-means Algorithm Optimizing Over Partition Orderings (Table VI from [145] which is ©2005 IEEE - used with permission).

1. **Initialize ordering:** Initialize with Π = current ordering of partitions
2. **Compute means:** Compute $\bar{\mathbf{X}}_t(\Pi)$ for $t = 1 \cdots T_p$ using (4.34)
3. **Permute particles:** For each particle p, permute the particle (update Π_p) to yield

$$\Pi_p \leftarrow \arg\min_{\Pi_p} \sum_{t=1}^{T_p} (\mathbf{X}_{p,\Pi_p(t)} - \bar{\mathbf{X}}_t(\Pi_p))^2$$

4. **Attempt termination:** If no particles have changed permutation, quit. Otherwise set $\Pi = (\Pi_1, \cdots, \Pi_p, \cdots, \Pi_{N_{part}})$ and go to 2

3.7 Estimation

Particle estimates of various interesting quantities can be computed. The central quantity of interest for sensor management is the expected Renyi Divergence. Using the particle filter representation for the JMPD and inserting that into (4.16) yields the estimate of information gain

$$\hat{\mathrm{E}}[\mathcal{D}_\alpha(\mathbf{Y}^k)|\mathbf{Y}^{k-1}, a] = \frac{1}{\alpha - 1} \sum_{y=0}^{1} p(y) \log \frac{1}{p(y)^\alpha} \sum_{p=1}^{N_{part}} w_p p(y|\mathbf{X}_p)^\alpha \quad (4.37)$$

To compute the probability of exactly n targets in the surveillance volume, first define the indicator variable $I_p(n)$ for $p = 1...N_{part}$,

$$I_p(n) = \begin{cases} 1 & \text{if } T_p = n \\ 0 & \text{otherwise} \end{cases} \quad (4.38)$$

Then the probability of n targets in the surveillance volume, $p(n|\mathbf{Y})$, is given by

$$p(n|\mathbf{Y}) = \sum_{p=1}^{N_{\text{part}}} I_p(n) w_p \qquad (4.39)$$

The estimate of the probability that there are n targets in the surveillance volume is the sum of the weights of the particles that have n partitions. Note that the particle weights, w_p, are normalized to sum to unity for all equations given in this section.

To compute the estimated target state and covariance of target i, we first define a second indicator variable $\tilde{I}_p(i)$ which indicates if particle p has a partition corresponding to target i. This is necessary as each particle is a sample drawn from the JMPD and hence may have a different number of partitions (targets):

$$\tilde{I}_p(i) = \begin{cases} 1 & \text{if target } i \text{ exists in particle p} \\ 0 & \text{otherwise} \end{cases} \qquad (4.40)$$

Note that the sorting procedure of Section 3.3 has already identified an ordering of particles to allow $\tilde{I}_p(i)$ to be determined. Furthermore, we define the normalized weights to be

$$\hat{w}_p = \frac{w_p \tilde{I}_p(i)}{\sum_{l=1}^{N_{\text{part}}} \tilde{I}_l(i) w_l} \qquad (4.41)$$

So $\hat{w}_p$ is the relative weight of particle p, with respect to all particles tracking target i. Then the estimate of the state of target i is given by

$$\hat{\mathbf{X}}(i) = \mathbb{E}[\mathbf{X}(i)] = \sum_{p=1}^{N_{\text{part}}} \hat{w}_p \mathbf{X}_{p,i} \qquad (4.42)$$

which is simply the weighted summation of the position estimates from those particles that are tracking target i. The covariance estimate is

$$\hat{\mathbf{\Lambda}}(i) = \sum_{p=1}^{N_{\text{part}}} \hat{w}_p (\mathbf{X}_{p,i} - \hat{\mathbf{X}}(i))(\mathbf{X}_{p,i} - \hat{\mathbf{X}}(i))^{\mathsf{T}} \qquad (4.43)$$

3.8 Resampling

In the traditional method of resampling, after each measurement update, N_{part} particles are selected with replacement from $\mathbf{X}_p$ based upon the particle weights w_p. The result is a set of N_{part} particles that have uniform weight which approximate the multi-target density $p(\mathbf{X}|\mathbf{Y})$. The particular resampling that was used in this work is systematic resampling [10]. This resampling strategy is easily implemented, runs in order N_{part}, is unbiased, and minimizes the Monte Carlo variance. Many other resampling schemes and modifications are presented in the literature [73]. Of these methods, we have found that adaptively choosing at which time steps to resample [160] based on the number of effective particles leads to improved performance while reducing compute time. All results presented herein use the method of [160] to determine which times to resample and use systematic resampling [10] to perform resampling. We have also found that Markov Chain Monte Carlo (MCMC) moves using a Metropolis-Hastings scheme [73] leads to slightly improved performance in our application.

4. Multi-target Tracking Experiments

In this section we present performance results for the particle filtering techniques described above, focusing on tracking applications. Chapter 5 presents sensor management results obtained using these techniques. We begin with a set of experiments that shows the impact of the different proposal methods on tracking accuracy and numerical requirements. We then present a few results showing the effect of thresholding on tracker performance and the performance gain that can be achieved from using prethresholded measurements in our association-free approach.

We illustrate the performance of our multi-target tracking scheme by considering the following scenario. Targets move in a $5000m \times 5000m$ surveillance area. Targets are modeled using the four-dimensional state vector $\mathbf{x} = [x, \dot{x}, y, \dot{y}]^\mathsf{T}$. The target motion in the simulation is taken from a set of recorded data based on GPS vehicle measurements collected as part of a battle training exercise at the U.S. Army's National Training Center. This battle simulation provides a large number of real vehicles following prescribed trajectories over natural terrain. Based on an empirical fit to the data, we found that a nearly constant velocity model was adequate to model the behavior of the vehicles for these simulation studies and is therefore used in all experimental results presented here. (Estimation performance can be improved with a moderate increase in computational load using a multiple model particle filter with modes corresponding to nearly constant velocity, rapid acceleration, and stationarity [140].)

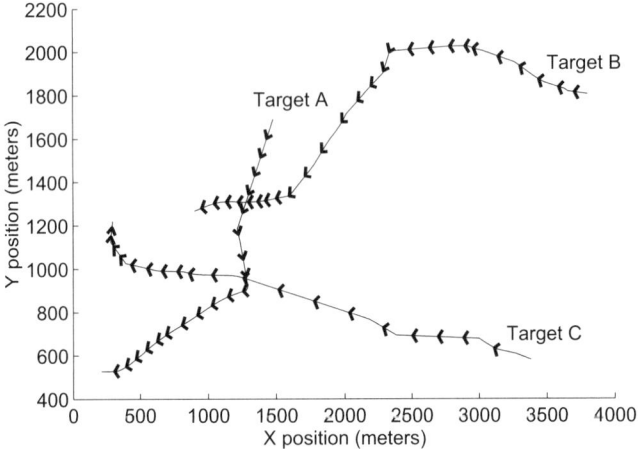

Figure 4.3. A schematic showing the motion of three of the ten targets in the simulation scenario. The target paths are indicated by the lines, and direction of travel by the arrows. There are two instances where the target paths cross (i.e., are at the same position at the same time). (Figure 1 from [145] - ©2005 IEEE used with permission).

We use the simple scalar sensor model described in Section 2.5. The sensor scans a fixed rectangular region of 50×50 pixels, where each pixel represents a $100m \times 100m$ area on the ground plane. The sensor returns Rayleigh-distributed measurements in each pixel, depending on the number of targets that occupy the pixel. Unthresholded measurements return energy according to (4.22) while thresholded measurements behave according to (4.24).

4.1 Adaptive Proposal Results

In Figure 4.4, we compare the performance of the Independent Partitions (Table 4.3), Coupled Partitions (Table 4.4), and Adaptive Partitions (Table 4.5) approaches with the traditional method of sampling from the kinematic prior (Table 4.2), in terms of RMS tracking error. In this example we use 3 targets with motion taken from recorded ground vehicle trajectories. The targets remain close in sensor space for about 50% of the time. Thresholded measurements with $P_d = 0.5$ are used and the SNR parameter λ is varied from 1 to 21. Each proposal scheme uses 100 particles to represent the JMPD. The filter is initialized with truth. Each point on the curve is an average of 100 runs. Table 4.8 shows the rough computational burden of the 4 methods, obtained using the "FLOPS" command of MATLAB®.

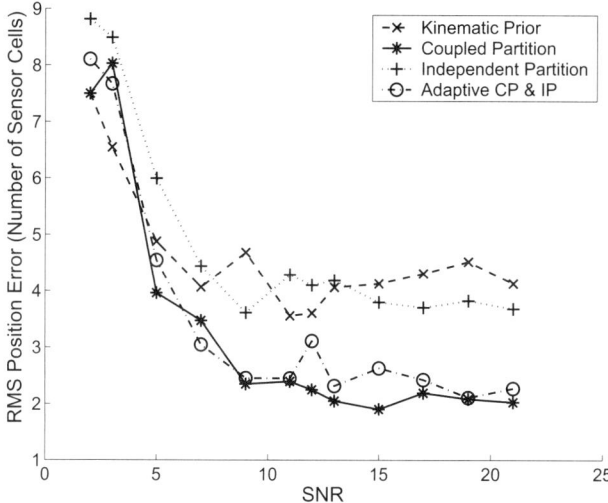

Figure 4.4. The Performance of the Coupled Partitions (CP), Independent Partitions (IP), and Adaptive Partitions (AP) schemes in comparison to simply using the kinematic prior. Performance is measured in terms of RMS position error. The units in the figure are sensor cells. For this simulation, we have extracted 3 targets from our large database of real recorded target trajectories. The targets were chosen so that they spent approximately one-half of the simulation in close proximity. The IP algorithm used alone is inappropriate during target crossings and so performs poorly here. The CP algorithm is always appropriate, but computationally demanding. The AP algorithm adaptively switches between IP and CP resulting in good performance at reduced computation (Figure 2 from [145] - ©2005 IEEE used with permission).

At low SNR all of the methods provide similar tracking performance. As SNR increases, the performance of all for methods improves (lower RMS error), but CP and AP provide consistently better estimation performance. The performance of KP is relatively low since it does not use the measurements in the proposal. IP adversely effected due to its failure to account for the correlation effects of nearby targets. The CP method makes no assumption about the independence of the targets and therefore performs very well, but at significantly higher computational cost. Most importantly, the adaptive method, which uses IP on partitions that are independent and CP otherwise, performs nearly as well as the CP method. AP achieves approximately a 50% reduction in computational burden (measured by floating point operations) as compared to the CP method alone.

To gain further insight into the relative performance of the IP, CP and KP methods, we simulate linear Gaussian target motion for five well-separated targets and examine the estimation performance as a function of the number of particles used in the different filters. The results are shown in Figure 4.5. In this model problem the Kalman filter is optimal and can therefore be used as

Table 4.8. FLOPS for KP, CP, IP, and AP Methods (Table VII from [145] which is ©2005 IEEE - used with permission).

Method	Flops
Coupled Partition	1.25e+8
Independent Partition	6.74e+6
Adaptive Partition	5.48e+7
Kinematic Prior	6.32e+6

a basis for comparison. We use the nearly constant velocity motion model for both the simulation of target motion and the filter. In each case, the filter is initialized with truth and run until it achieves steady state where the mean track error is measured. The interesting result is that IP and CP achieve the (optimum) Kalman filter performance bound with 200 and 1000 particles respectively while KP still has not quite met the bound with 20,000 particles. Thus even for this idealized problem for which the Kalman filter is optimal, KP performs poorly due to the high dimensionality of the multi-target state space.

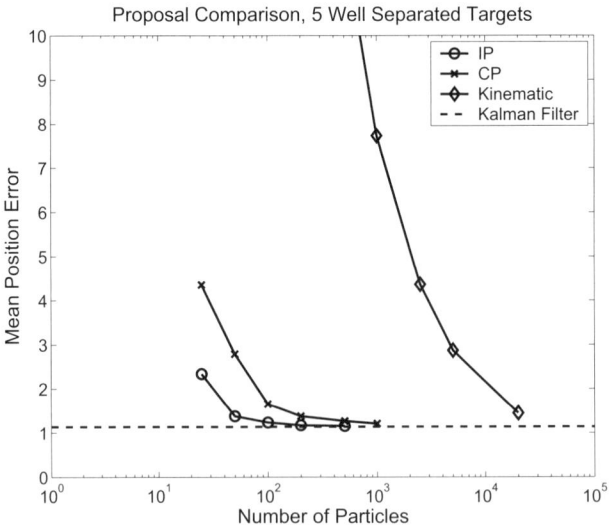

Figure 4.5. The performance of the CP and IP Proposal Schemes, in comparison to sampling from the Kinematic Prior. For the purposes of this example, we consider well separated targets with linear motion and linear state-to-measurement coupling. Therefore, in this simple example, the Kalman filter is optimal and is shown as a performance bound (Figure 3 from [145] - ©2005 IEEE used with permission).

4.2 Partition Swapping

Figures 4.6 and 4.7 illustrate how partition swapping occurs and how it is reduced by using the K-means partition sorting algorithm. Partition swapping leads to poor performance when it occurs while using the Independent Partition proposal algorithm. Figure 4.6 shows several snapshots of the particle distribution from a crossing target example without partition sorting (the number of particles has been reduced for illustration purposes here). Initially, the targets are well separated and identically ordered (e.g., Time=44) and the IP method is used for particle proposal. When the targets cross (Time=60), partition swapping occurs and the CP method must be used. The targets remain in the same detection cell for several time steps. Without partition sorting using the K-means algorithm of Section 3.3, this swapping persists even after the targets separated and the CP method must be used even at Time=84. This results in an inefficient algorithm, as the CP method is more computationally demanding.

Figure 4.7 is analogous to Figure 4.6, but this time we utilize the partition sorting algorithm outlined in Section 3.3 at each time step. While the CP method must still be used when the targets are occupying the same detection cell, when they separate (Time=72) the IP method may be used again. The partition sorting allows for the more computationally efficient IP method to be used for proposal by reordering the particles appropriately.

4.3 The Value of Not Thresholding

One of the strengths of association-free methods is their ability to use non-thresholded measurements without modification. Using non-thresholded measurements should improve performance because targets that are missed due to thresholding might be detected without thresholding. Here we quantify the relative performance of the tracker for a set of scenarios in which the target motion and the underlying measurement probability density functions are identical. The only difference is that in one case we threshold to obtain binary measurements with distribution (4.24) while in the other case we use unthresholded measurements and the envelope detected amplitude (4.22) is input into the tracker.

If one is using thresholded measurements, then the first step is to optimize the threshold. For Rayleigh measurements with a given SNR, setting the threshold is equivalent to setting the false alarm probability (see (4.23)). To assess performance, we again use the 3-crossing target data shown in Figure 4.3. The filters are initialized with ground truth and we assess the number

90 FOUNDATIONS AND APPLICATIONS OF SENSOR MANAGEMENT

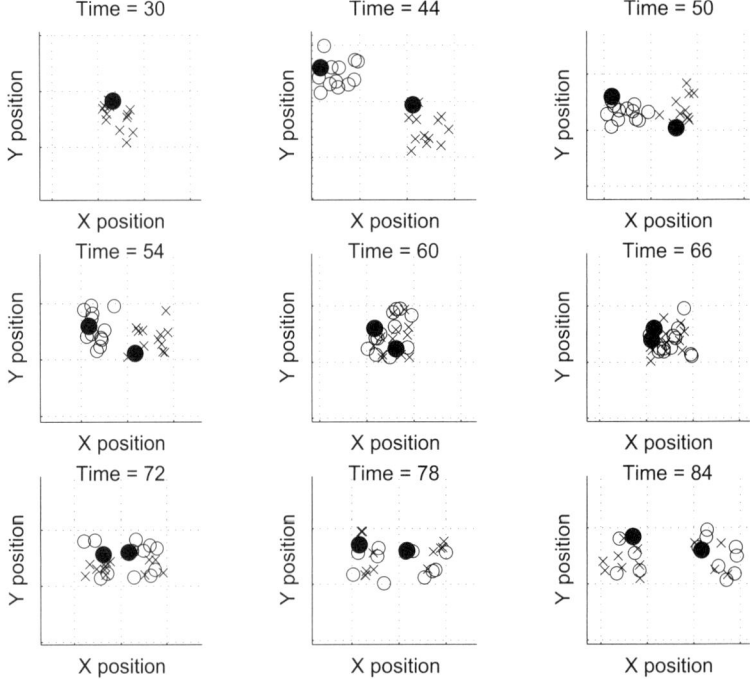

Figure 4.6. This figure illustrates the phenomenon of partition swapping that occurs in direct particle filter implementation of the SIR filter for JMPD. True target locations are indicated by a solid circle. The two partitions for each particle, plotted with × and ○, are well separated at time 44. From time 60 to 66, they occupy the same detection cell. At time 84, some partition swapping has occurred, indicated by the fact that there are mixtures of × and ○ corresponding to each target location (Figure 10 from [145] - ©2005 IEEE used with permission).

of targets in track at the conclusion of the vignette. Figure 4.8 shows a contour plot of the result as a function of P_d and SNR that increases with SNR and is broadly peaked around $P_d = 0.4$.

Figure 4.9 shows the performance of the algorithm using optimized thresholded measurements at $P_d = 0.4$ and the non-thresholded measurement algorithm. We see that non-thresholded measurements provide similar tracking performance at an SNR of 1 as the thresholded measurements provide at an SNR of 5, corresponding to a thresholding-induced loss of about 7dB.

4.4   Unknown Number of Targets

The ability of the JMPD joint particle filtering algorithm to determine the number of targets is illustrated in Figure 4.10 for the same data as in Figure 4.3.

Joint Multi-target Particle Filtering 91

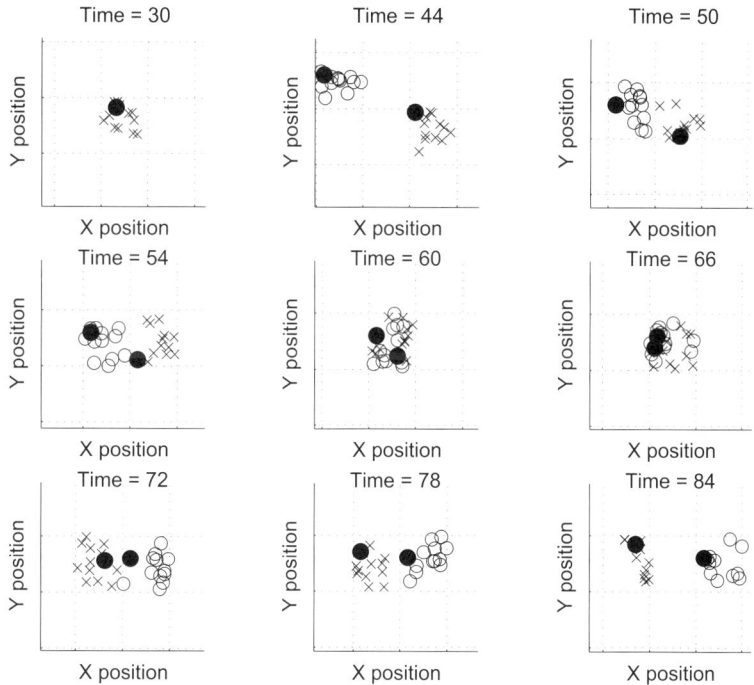

Figure 4.7. An example of the behavior of the particle filter based multi-target tracker during target crossing when partition sorting is employed. The nine time-sequential images focus on one of the ten targets that the filter is tracking. The ground truth location of the target (projected into the XY plane) is denoted by a solid circle. The partitions associated with the two targets are denoted by × and ○. The sensor cells are given by the gridlines. As discussed earlier, the sensor measures on a grid and receives energy from the target density if the cell is occupied or the false alarm density if the cell is empty. Initially (before time 50), this target is well separated from all of the others. At these points the IP algorithm is being used for proposal. During times 50 to 66, a second target is crossing (coming within sensor resolution) of the target of interest. Near time 72, the targets complete their crossing and again move apart (Figure 11 from [145] - ©2005 IEEE used with permission).

There are three targets in this simulation. We initialized the filter uniformly in target number space, allocating one-sixth the probability to $0, 1, \cdots, 5$ targets. Over time, the filter is able to accurately estimate the number of targets in the surveillance region. As the SNR improves, the time until correctly determining the number of targets decreases.

5. Conclusions

There is a need to develop sensor management techniques that are applicable across a broad range of Intelligence, Surveillance and Reconnaissance assets.

92 FOUNDATIONS AND APPLICATIONS OF SENSOR MANAGEMENT

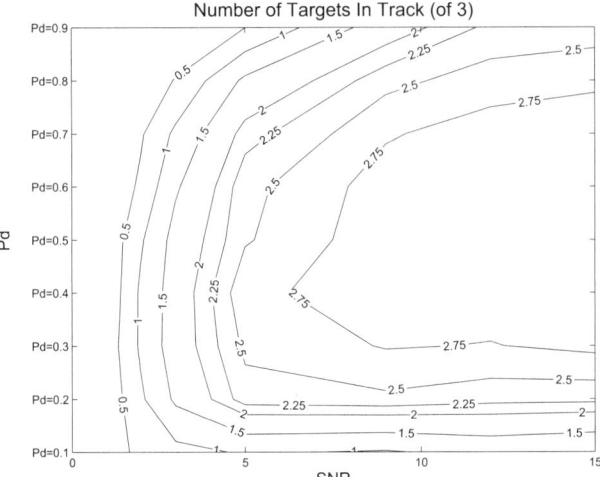

Figure 4.8. A contour plot showing the number of targets successfully tracked in this three target experiment versus P_d and SNR when using thresholded measurements (Figure 4 from [145] - ©2005 IEEE used with permission).

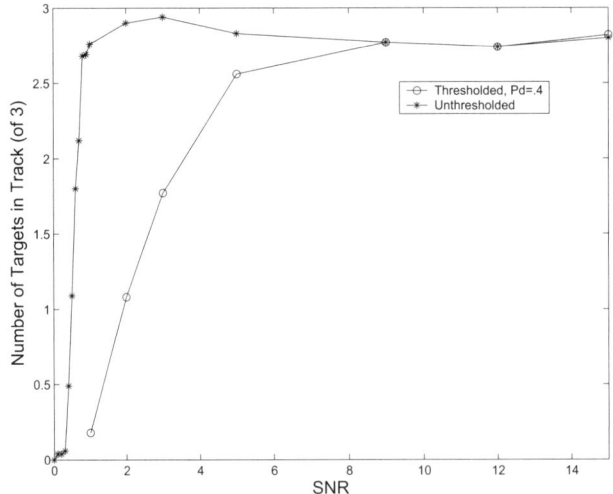

Figure 4.9. A plot of the number of targets successfully tracked in the three target experiment for thresholded measurements and unthresholded measurements as a function of SNR (Figure 5 from [145] -©2005 IEEE used with permission).

A critical component to sensor management is the accurate approximation of the joint posterior density of the targets, i.e., the belief state. The belief state is captured by the Joint Multi-target Probability Density and it quantifies the effect of uncertainty within the scene and permits the use of information-based

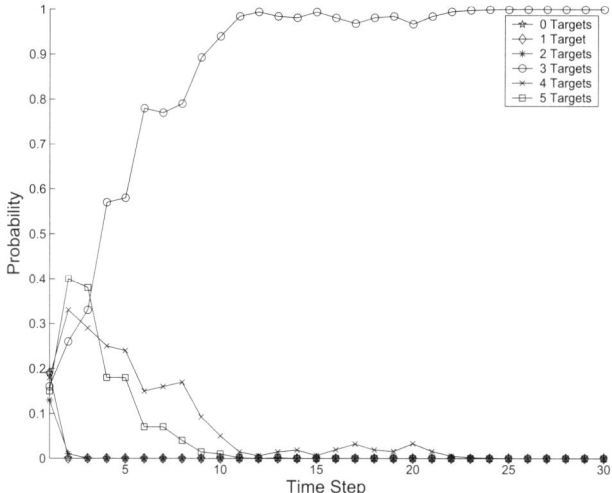

Figure 4.10. The estimate of number of targets in the surveillance region versus time with SNR=4. The filter is initialized with probability uniform for $0, 1, \cdots, 5$ targets. Measurements taken over time allow the filter to properly estimate the number of targets in the surveillance area (Figure 6 from [145] - ©2005 IEEE used with permission).

methods to quantify this uncertainty and select most informative sensing actions. The need to treat non-linear target motion and target birth and death motivates the use of non-linear techniques such as particle filtering. The inherent permutation invariance of the JMPD must be treated properly in developing proposal methods for the particle filter, leading to the IP, CP, AP and OID methods described in this chapter. These can be implemented with a moderate computational cost and can provide order-of-magnitude reductions in the amount of sensor resource required to achieve a given level of tracker performance. A side benefit is that these non-linear methods lead naturally to association free-trackers that can make use of unthresholded measurements, providing further improvement and effectively increasing the target SCNR by 3-5 dB.

The future directions for this work include expanding the information based sensor scheduling to long term planning discussed elsewhere in this volume (Chapters 2, 6). Another area requiring further work is in optimizing proposals during the track initiation phase as the work reported here focuses on optimizing proposals for firm tracks. The studies presented here have been performed in the context of air- and space-based sensors tracking ground moving targets. These results are readily extendable to other domains such as missile defense or tracking air targets.

In the next chapter the joint particle filtering methods introduced here will be applied to several sensor management scenarios.

Chapter 5

POMDP APPROXIMATION USING SIMULATION AND HEURISTICS

Edwin K. P. Chong

Colorado State University, Fort Collins, CO, USA

Christopher M. Kreucher

General Dynamics Michigan Research and Development Center, Ypsilanti, MI, USA

Alfred O. Hero III

University of Michigan, Ann Arbor, MI, USA

1. Introduction

This chapter discusses a class of approximation methods for sensor management under the partially observable Markov decision process (POMDP) model introduced in Chapter 2. In contrast to Chapter 2, which focuses on analytic methods for bounding and solving POMDPs, here we discuss methods based on heuristics and simulation. Our aim is to develop methods that are implementable on-line and find nearly optimal policies.

While the methods described herein are more generally applicable, the focus application will be sensor management for target tracking. Information theoretic measures discussed in Chapter 3 and particle filter approximations discussed in Chapter 4 will be implemented to illustrate these approximations.

It is informative to distinguish between *myopic* and *non-myopic* (also known as *dynamic* or *multistage*) resource management policies, a topic of much current interest (see, e.g., [139, 103, 104]). In myopic resource management, the objective is to optimize performance on a per-decision basis. For example, consider *sensor scheduling* for tracking a single target, where the problem is to select, at each decision epoch, a single sensor to activate. An example sensor-scheduling scheme is closest point of approach (CPA), which selects the sensor that is perceived to be the closest to the target. Another (more sophisticated) example is the method described in [141], where the authors present a sensor management method using alpha-divergence (or Rényi divergence) measures. Their approach is to make the decision that maximizes the expected information gain, which is measured in terms of the alpha-divergence.

Myopic sensor-management schemes may not be ideal when the performance is measured over a horizon of time. In such situations, we need to consider schemes that trade off short-term for long-term performance. We call such schemes non-myopic. Several factors motivate the consideration of non-myopic schemes:

Heterogeneous sensors. If we have sensors with different locations, characteristics, usage costs, and/or lifetimes, the decision of whether or not to use a sensor should consider what the overall performance will be, not whether or not its use maximizes the current performance.

Sensor motion. The future location of a mobile sensor affects how we should act now. To optimize a long-term performance measure, we need to be opportunistic in our choice of sensor decisions.

Target motion. If a target is moving, there is potential benefit in sensing the target before it becomes unresolvable (e.g., too close to other targets or to clutter, or shadowed by large objects). In some scenarios, we may need to identify multiple targets before they cross, to aid in data association.

The rest of this chapter is organized as follows. In Section 2, we give a concrete motivating example that advocates for the use of non-myopic scheduling. Next, in Section 3, we review the basic principles behind Q-value approximation. Then, in Section 4, we illustrate the basic lookahead control framework and describe the constituent components. In Section 5, we describe a host of Q-value approximation methods. Among others, this section includes descriptions of Monte Carlo sampling methods, heuristic approximations, and rollout methods. In Section 6, we provide a set of simulation results on a model problem that illustrate several of the approximate non-myopic scheduling methods described in this chapter. We conclude in Section 7 with some summary remarks.

2. Motivating Example

We now present a concrete motivating example that will be used to explain and justify the heuristics and approximations used in this chapter. This example involves a remote sensing application where the goal is to learn the contents of a surveillance region via repeated interrogation.

Consider a single airborne sensor which is able to image a portion of a ground surveillance region to determine the presence or absence of moving ground targets. At each time epoch, the sensor is able to direct an electrically scanned array so as to interrogate a small area on the ground. Each interrogation yields some (imperfect) information about the small area. The objective is to choose the sequence of pointing directions that lead to the best acquisition of information about the surveillance region.

A further complication is the fact that at each time epoch the sensor position causes portions of the ground to be unobservable due to obscuration. Obscuration is due to varying degrees of terrain elevation that can block line-of-sight from the sensor to the target on the ground. We assume that given the sensor position and the terrain elevation, the sensor can compute a visibility mask which specifies how well a particular spot on the ground can be seen by the sensor. As an example, in Figure 5.1 we give binary visibility masks that are computed from a sensor positioned (a) below and (b) to the left of the topographically nonhomogeneous surveillance region. As can be seen from the figures, sensor position causes "shadowing" of certain regions. These regions, if measured, would provide no information to the sensor.

This example illustrates a situation where non-myopic scheduling is highly beneficial. Using a known sensor trajectory and known topographical map, the sensor can predict locations that will be obscured in the future. This information can be used to prioritize resources so that they are used on targets that are predicted to become obscured in the future. Extra sensor dwells immediately before obscuration (at the expense of not interrogating other targets) will sharpen the estimate of target location. This sharpened estimate will allow better prediction of where and when the target will emerge from the obscured area. This is illustrated graphically with a six time-step vignette in Figure 5.2.

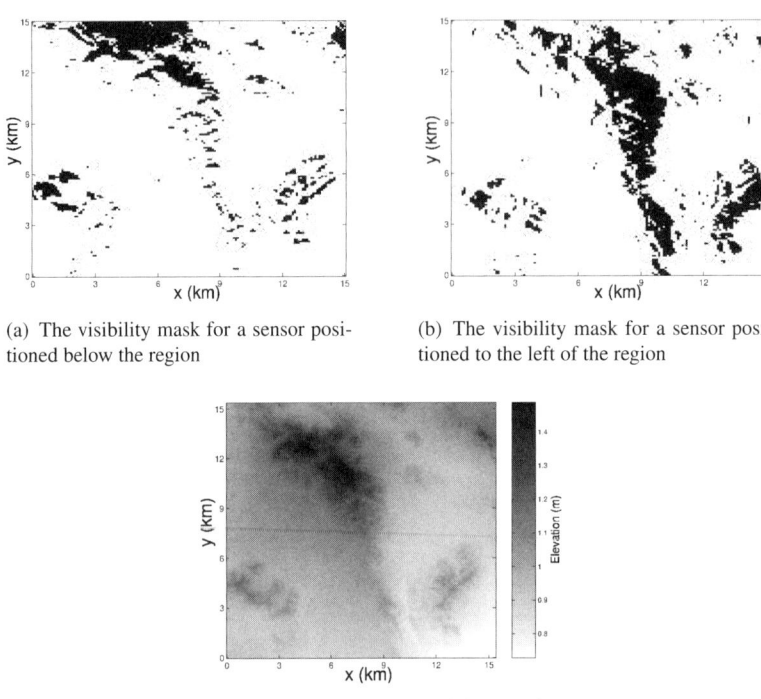

Figure 5.1. Visibility masks for a sensor positioned below and left of the surveillance region. We show binary visibility masks (non-visible areas are black and visible areas are white). In general, visibility may be between 0 and 1 indicating areas of reduced visibility, e.g., regions that are partially obscured by foliage (Figure 1 from [139] - ©2005 IEEE used with permission).

3. Basic Principle: Q-value Approximation

3.1 Optimal Policy

In general the action chosen at each time k should be allowed to depend on the entire history up to time k (i.e., the action at time k is a random variable that is a function of all observable quantities up to time k). However, it turns out that if an optimal choice of such a sequence of actions exists, then there is an optimal choice of actions that depends only on "belief-state feedback." In other words, it suffices for the action at time k to depend only on the belief state (or information state) π_k at time k (See Appendix, Section 2). Let π be the set of distributions over the underlying state space $\mathcal{X}$ (we call π the *belief-state space*). So what we seek is, at each time k, a mapping $\gamma_k^* : \pi \to \mathcal{A}$ such

POMDP Approximation Using Simulation and Heuristics 99

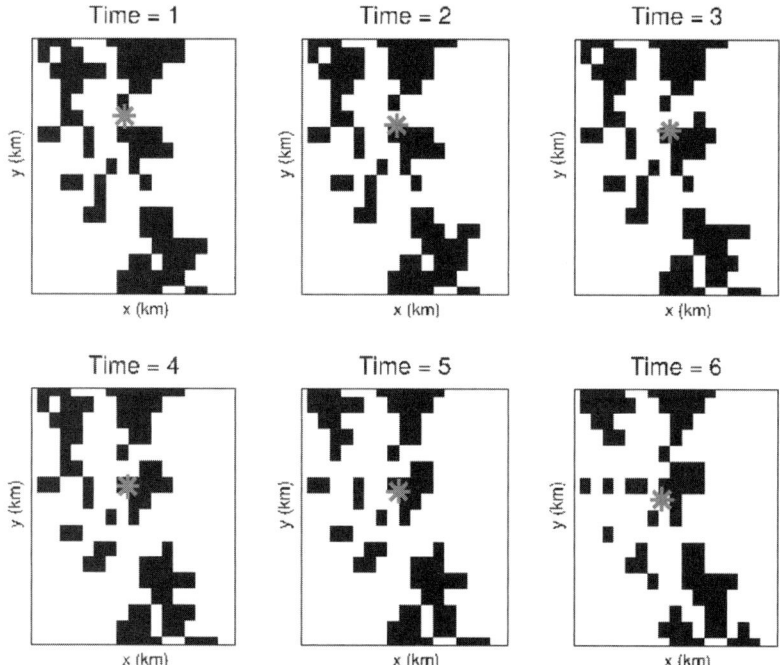

Figure 5.2. A six time step vignette where a target moves through an obscured area. Other targets are present elsewhere in the surveillance region but outside the area shown in the figures. The target is depicted by an asterisk. Areas that are obscured from the sensor point of view are black and areas that are visible are white. Extra dwells just before becoming obscured (time = 1) aid in localization after the target emerges (Figure 2 from [139] - ©2005 IEEE used with permission).

that if we perform action $a_k = \gamma_k^*(\pi_k)$, then the resulting objective function is maximized. As usual, we call such a mapping a *policy*.

3.2 Q-values

Let $V_H^*(\pi_0)$ be the optimal objective function value (over horizon H). Then, as discussed in Chapter 2, *Bellman's principle* states that

$$V_H^*(\pi_0) = \max_a (r(\pi_0, a) + \mathbb{E}_a[V_{H-1}^*(\pi_1)|\pi_0])$$

where $r(\pi_0, a)$ is the *reward* associated with taking action a at belief state π_0, and π_1 is the random next belief state, with distribution depending on a. Moreover,

$$\gamma_0^*(\pi_0) = \arg\max_a (r(\pi_0, a) + \mathbb{E}_a[V_{H-1}^*(\pi_1)|\pi_0]).$$

Define the *Q-value* of taking action a at state π_k as

$$Q_{H-k}(\pi_k, a) = r(\pi_k, a) + \mathbb{E}_a[V^*_{H-k-1}(\pi_{k+1})|\pi_k],$$

where π_{k+1} is the random next belief state. Then, Bellman's principle can be rewritten as

$$\gamma^*_k(\pi_k) = \arg\max_a Q_{H-k}(\pi_k, a)$$

In other words, the optimal action to take at belief-state π_k (at time k, with a horizon-to-go of $H-k$) is the one with largest Q-value at that belief state. This principle, called *lookahead*, forms the heart of approaches to solving POMDPs.

3.3 Stationary policies

In general, an optimal policy is a function of time k. It turns out that if H is sufficiently large, then the optimal policy is approximately *stationary* (independent of time k). This seems intuitively clear: if the end of the time horizon is a million years away, then how we should act today given a belief-state x is the same as how we should act tomorrow given the same belief state. To put it differently, if H is sufficiently large, the difference between Q_H and Q_{H-1} is negligible. Henceforth we will assume that there is a stationary optimal policy, and this is what we seek. We will use the notation γ for stationary policies (with no subscript k).

3.4 Receding horizon

Assuming that H is sufficiently large and that we seek a stationary optimal policy, at any time k we will write:

$$\gamma^*(\pi) = \arg\max_a Q_H(\pi, a).$$

Notice that the horizon is taken to be fixed at H, regardless of the current time k. This is justified by our assumption that H is so large that at any time k, the horizon is still approximately H time steps away. This approach of taking the horizon to be fixed at H is called *receding horizon control*. For convenience, we will also henceforth drop the subscript H from our notation unless it is needed for clarity.

3.5 Approximating Q-values

Recall that $Q(\pi, a)$ is simply the reward $r(\pi, a)$ of taken action a at belief-state π, plus the expected cumulative reward of applying the optimal policy

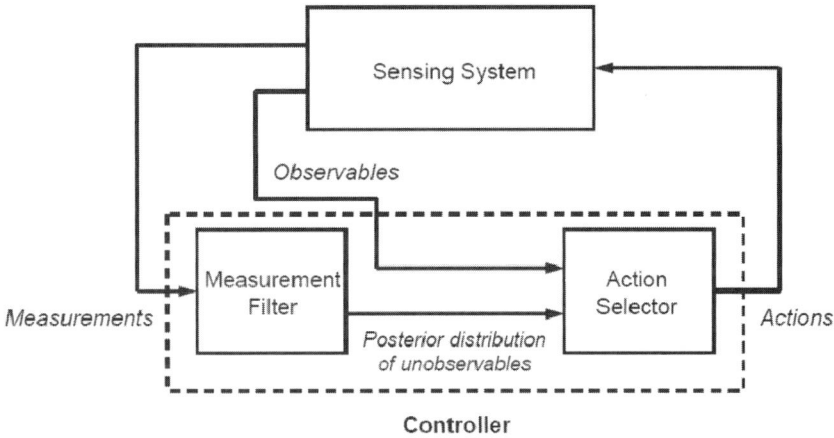

Figure 5.3. Basic lookahead framework.

for all future actions. This second term in the Q-value is in general difficult to obtain, especially for a problem with a large belief-state space. For this reason, approximation methods are necessary to obtain Q-values. Note that the quality of an approximation to the Q-value is not so much in the accuracy of the actual values obtained, but in the *ranking* of the actions reflected by their *relative* values.

In Section 5, we describe a variety of methods to approximate Q-values. But before discussing such methods, in the next section we first describe the basic control framework for using Q-values to inform control decisions.

4. Control Architecture

By Bellman's principle, knowing the Q-values allows us to make optimal control decisions. In particular, if we are currently at belief-state π, we need only find the action a with the largest $Q(\pi, a)$. This principle yields a basic control framework that is illustrated in Figure 5.3.

The top-most block represents the sensing system, which we treat as having an input and two forms of output. The input represents actions (external control commands) we can apply to control the sensing system. Actions usually include sensor-resource controls, such as which sensor(s) to activate, at what power level, where to point them, what waveforms to use, and what sensing modes to activate. Actions may also include communication-resource controls, such as the data rate for transmission from each sensor.

The two forms of outputs from the sensing system represent:

1 Fully observable aspects of the internal state of the sensing system (which we call *observables*), and

2 Measurements (observations) of those aspects of the internal state that are not directly observable (which we refer to simply as *measurements*).

We assume that the underlying state-space is the Cartesian product of two sets, one representing unobservables and the other representing observables. Target states are prime examples of unobservables. So, measurements are typically the outputs of sensors, representing observations of target states. Observables include things like sensor locations and orientations, which sensors are activated, battery status readings, etc. In the remainder of this section, we describe the components of our control framework. Our description starts from the architecture of Figure 5.3 and progressively fills in the details.

4.1 Controller

At each decision epoch, the *controller* takes the outputs (measurements and observables) from the tracking system and, in return, generates an action that is fed back to the tracking system. This basic closed-loop architecture is familiar to mainstream control system design approaches.

The controller consists of two main components. The first component is the *measurement filter*, which takes as its input the measurements, and provides as its output the posterior distribution of the unobservable internal states (which we henceforth simply call *unobservables*). In the typical situation where these unobservables are target states, the measurement filter outputs posterior distribution of the target states given the measurement history. We describe the measurement filter further below. The posterior distribution of the unobservables, together with the observables, form the belief state, the posterior distribution of the underlying state.

The second component of the controller is the *action selector*, which takes the belief state and computes an action (which is the output of the controller). The basis for action selection is Bellman's principle, using Q-values. We discuss this below.

4.2 Measurement filter

The measurement filter computes the posterior distribution given measurements. This component is present in virtually any target tracking system. It turns out that the posterior distribution can be computed iteratively: each time we obtain a new measurement, the posterior distribution can be obtained by updating the previous posterior distribution based on knowing the current action, the transition law, and the observation law. This update is based on Bayes' rule (see Chapter 2).

The measurement filter can be constructed in a number of ways. If we know beforehand that the posterior distribution always resides within a family of distributions that is conveniently parameterized, then all we need to do is keep track of the belief-state parameters. This is the case, for example, if the belief state is Gaussian. Indeed, if the unobservables evolve in a linear fashion, then these Gaussian parameters can be updated using a Kalman filter. See Chapter 10 for a detailed development of the Kalman filter for multiple target tracking with data association. In general, however, it is not practical to keep track of the exact belief state. A variety of options have been explored for belief-state representation and simplification (e.g., [203, 199, 258]). We will have more to say about belief-state simplification in Section 5.10.

Particle filtering is a Monte Carlo sampling method for updating posterior distributions. Instead of maintaining the exact posterior distribution, we maintain a set of representative samples from that distribution. It turns out that this method dovetails naturally with Monte Carlo sampling-based methods for Q-value approximation, as we will describe later in Section 5.7.

4.3   Action selector

As shown in Figure 5.4, the action selector consists of a search (optimization) algorithm that optimizes an objective function, the *Q-function*, with respect to an action. In other words, the Q-function is a function of the action—it maps each action, at a given belief state, to its Q-value. The action that we seek is one that maximizes the Q-function. So, we can think of the Q-function as a kind of "action-utility" function that we wish to maximize.

As is typical, the search algorithm iteratively generates a candidate action and evaluates the Q-function at this action (this numerical quantity is the Q-value), searching over the space of candidate actions for one with the largest Q-value. Methods for obtaining (approximating) the Q-values are described in the next section.

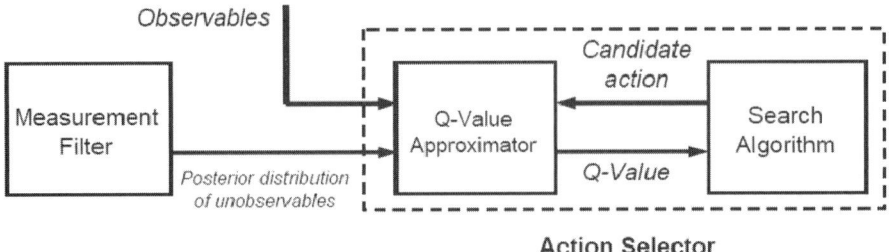

Figure 5.4. Basic components of the action selector.

5. Q-value Approximation Methods

5.1 Basic approach

Recall the definition of the Q-value,

$$Q(\pi, a) = r(\pi, a) + \mathbb{E}_a[V^*(\pi')|\pi], \qquad (5.1)$$

where π' is the random next belief state, with distribution depending on a. In all but very special problems, it is impossible to compute the Q-value exactly. In this section, we describe a variety of methods to approximate the Q-value. Because the first term on the right-hand side of (5.1) is usually easy to compute, most approximation methods focus on approximating the second term. As pointed out before, it is important to realize that the quality of an approximation to the Q-value is not so much in the accuracy of the actual values obtained, but in the *ranking* of the actions reflected by their *relative* values.

5.2 Monte Carlo sampling

In general, we can think of Monte Carlo methods simply as the use of computer generated random numbers in estimating expectations of random variables through averaging over many samples. With this in mind, it seems natural to consider using Monte Carlo methods to compute the value function directly based on Bellman's equation:

$$V_H^*(\pi_0) = \max_{a_0}(r(\pi_0, a_0) + \mathbb{E}_{a_0}[V_{H-1}^*(\pi_1)|\pi_0]).$$

Notice that the second term on the right-hand side involves expectations (one per action candidate a_0), which can be estimated using Monte Carlo sampling. However, the random variable inside each expectation is itself an objective function value (with horizon $H - 1$), and so it too involves a max of an expectation via Bellman's equation:

$$V_H^*(\pi_0) = \max_{a_0}\left(r(\pi_0, a_0) + \mathbb{E}_{a_0}\left[\max_{a_1}(r(\pi_1, a_1) + \mathbb{E}_{a_1}[V_{H-2}^*(\pi_2)|\pi_1])\bigg|\pi_0\right]\right).$$

Notice that we now have two "layers" of max and expectation, one "nested" within the other. Again, we see that the inside expectation involves the value function (with horizon $H-2$), which yet again can be written as a max of expectations. Proceeding this way, we can write $V_H^*(\pi_0)$ in terms of H layers of max and expectations. Each expectation that appears in this way of expressing the value function V_H^* can be computed using Monte Carlo sampling. The question that remains is, how computationally burdensome is this task?

Kearns, Mansour, and Ng [128] have provided a method to calculate the computational burden of approximating the value function using Monte Carlo sampling as described above, given some prescribed accuracy in the approximation of the value function. Unfortunately, it turns out that for practical POMDP problems this computational burden is prohibitive, even for modest degrees of accuracy. So, while Bellman's equation suggests a natural Monte Carlo method for approximating the value function, the method is not useful in practice. For this reason, we seek alternative approximation methods. In the next few subsections, we explore some of these methods.

5.3 Relaxation of optimization problem

Some problems that are difficult to solve become drastically easier if we *relax* certain aspects of the problem. For example, by removing a constraint in the problem, the "relaxed" problem may yield to well-known solution methods. This constraint relaxation enlarges the constraint set, and so the solution obtained may no longer be feasible in the original problem. However, the objective function value of the solution *bounds* the optimal objective function value of the original problem.

The Q-value involves the quantity $V^*(\pi')$, which can be viewed as the optimal objective function value corresponding to some optimization problem. The method of relaxation, if applicable, gives rise to a bound on $V^*(\pi')$, which then provides an approximation to the Q-value. For example, a relaxation of the original POMDP may result in a bandit problem (see Chapter 6), or may be solvable via linear programming (see Chapter 2). For further discussion on methods based on bandit approximation, see Chapter 7.

5.4 Heuristic approximation

In some applications, although we are unable to compute Q-values directly, we can use domain knowledge to develop some idea of how it behaves. If so, we can heuristically construct a Q-function based on this knowledge.

Recall from (5.1) that the Q-value is the sum of two terms, where the first term (the immediate reward) is usually easy to compute. Therefore, it often suffices to approximate only the second term in (5.1), which is the mean optimal objective function value starting at the next belief state, which we call the *expected value-to-go* (EVTG). Note that the EVTG is a function of both π and a, because the distribution of the next belief state is a function of π and a. In some problems, it is possible to construct a heuristic EVTG based on domain knowledge. If the constructed EVTG properly reflects tradeoffs in the selection of alternative actions, then the ranking of these actions via their Q-values will result in the desired "lookahead."

For example, consider the motivating example of tracking multiple targets with a single sensor. Suppose we can only measure the location of one target per decision epoch. The problem then is to decide which location to measure and the objective function is the aggregate (multi-target) tracking error. The terrain over which the targets are moving is such that the measurement errors are highly location dependent, for example because of the presence of topographical features which cause some areas to be invisible from a future sensor position. In this setting, it is intuitively clear that if we can predict sensor and target motion so that we expect a target is about to be obscured, then we should focus our measurements on that target immediately before the obscuration so that its track accuracy is improved and the overall tracking performance maximized in light of the impending obscuration.

The same reasoning applies in a variety of other situations, including those where targets are predicted to become unresolvable to the sensor (e.g., two targets that cross) or where the target and sensor motion is such that future measurements are predicted to be less reliable (e.g., a bearings-only sensor that is moving away from a target). In these situations, we advocate a heuristic method that replaces the EVTG by a function that captures the long-term benefit of an action in terms of an "opportunity cost" or "regret." That is, we approximate the Q-value as

$$Q(\pi, a) \approx r(\pi, a) + wN(\pi, a), \tag{5.2}$$

where $N(\pi, a)$ is an easily computed heuristic approximation of the long-term value, and w is a weighting term that allows us to trade the influence of the immediate value and the long-term value. As a concrete example of a useful heuristic, we have used the "gain in information for waiting" as a choice of $N(\pi, a)$ [140]. Specifically, let $\bar{g}_a^k$ denote the expected myopic gain when taking action a at time k. Furthermore, denote by $p_a^k(\cdot)$ the distribution of myopic gains when taking action a at time k. Then, in the notation of Chapter 4, a useful approximation of the long-term value of taking action a is the gain (loss) in information received by waiting until a future time step to take the action,

$$N(\pi, a) \approx \sum_{m=1}^{M} \gamma^m \operatorname{sgn}\left(\bar{g}_a^k - \bar{g}_a^{k+m}\right) \mathcal{D}_\alpha\left(p_a^k(\cdot) \| p_a^{k+m}(\cdot)\right) \quad (5.3)$$

where M is the number of time steps in the future that are considered.

Each term in the summand of $N(\pi, a)$ is made up of two components. First, $\operatorname{sgn}\left(\bar{g}_a^k - \bar{g}_a^{k+m}\right)$ signifies if the expected reward for taking action a in the future is more or less than the present. A negative value implies that the future is better and that the action ought to be discouraged at present. A positive value implies that the future is worse and that the action ought to be encouraged at present. This may happen, for example, when the visibility of a given target is getting worse with time. The second term, $\mathcal{D}_\alpha\left(p_a^k(\cdot) \| p_a^{k+m}(\cdot)\right)$, reflects the magnitude of the change in reward using the divergence between the density on myopic rewards at the current time step and at a future time step. A small number implies that the present and future rewards are very similar, and therefore the non-myopic term will have little impact on the decision making.

Therefore, $N(\pi, a)$ is positive if an action is less favorable in the future (e.g., the target is about to become obscured). This encourages taking actions that are beneficial in the long term, and not just taking actions based on their immediate reward. Likewise, the term is negative if the action is more favorable in the future (e.g., the target is about to emerge from an obscuration). This discourages taking actions now that will have more value in the future.

5.5 Parametric approximation

In situations where a heuristic Q-function is difficult to construct, we may consider methods where the Q-function is approximated by a parametric function (by this we mean that we have a function approximator parameterized by one or more parameters). Let us denote this approximation by $\tilde{Q}(\pi, \theta)$, where θ is a parameter (to be tuned appropriately). For this approach to be useful, the computation of $\tilde{Q}(\pi, \theta)$ has to be relatively simple, given π and θ. Typically, we seek approximations for which it is easy to set the value of the parameter θ appropriately, given some information of how the Q-values "should" behave (e.g., from expert knowledge, empirical results, simulation, or on-line observation). This adjustment or tuning of the parameter θ is called *training*.

As in the heuristic approximation approach, the approximation of the Q-function by the parametric function approximator is usually accomplished by approximating the EVTG, or even directly approximating the objective function V^*.[1] In the usual parametric approximation approach, the belief state π is first mapped to a set of *features*. The features are then passed through a parametric function to approximate $V^*(\pi)$. For example, in the problem of tracking multiple targets with a single sensor, we may extract from the belief state some information on the location of each target relative to the sensor, taking into account the topography. These constitute the features. For each target, we then assign a numerical value to these features, reflecting the measurement accuracy. Finally, we take a linear combination of these numerical values, where the coefficients of this linear combination serve the role of the parameters to be tuned.

The parametric approximation method has some advantages over methods based only on heuristic construction. First, the training process usually involves numerical optimization algorithms, and thus well-established methodology can be brought to bear on the problem. Second, even if we lack immediate expert knowledge on our problem, we may be able to experiment with the system (e.g., by using a simulation model, e.g., a generative model [128]). Such empirical output is useful for training the function approximator. Common training methods found in the literature go by the names of reinforcement learning, Q-learning, neurodynamic programming, and approximate dynamic programming.

The parametric approximation approach may be viewed as a systematic method to implement the heuristic-approximation approach. But note that even in the parametric approach, some heuristics are still needed in the choice of features and in the form of the function approximator. For further reading, see [28].

---

[1] In fact, given a POMDP, it turns out that the Q-value can be viewed as the objective function value for a related problem; see [28].

5.6 Action-sequence approximations

Let us write the value function (optimal objective function value as a function of belief state) as

$$V^*(\pi) = \max_{\gamma} \mathbb{E}_{\gamma}\left[\sum_{k=1}^{H} r(\pi_k, \gamma(\pi_k)) \,\bigg|\, \pi\right]$$

$$= \mathbb{E}\left[\max_{a_1,\ldots,a_H : a_k = \gamma(\pi_k)} \sum_{k=1}^{H} r(\pi_k, a_k) \,\bigg|\, \pi\right], \quad (5.4)$$

where the notation $\max_{a_1,\ldots,a_H : a_k = \gamma(\pi_k)}$ means maximization subject to the constraint that each action a_k is a (fixed) function of the belief state π_k. If we relax this constraint on the actions and allow them to be arbitrary random variables, then we have an upper bound on the value function:

$$\hat{V}_{\text{upper}}(\pi) = \mathbb{E}\left[\max_{a_1,\ldots,a_H} \sum_{k=1}^{H} r(\pi_k, a_k) \,\bigg|\, \pi\right].$$

In some applications, this upper bound provides a suitable approximation to the value function. The advantage of this approximation method is that in certain situations the computation of the "max" above involves solving a relatively easy optimization problem. This method is called *hindsight optimization* [58, 256].

One implementation of this idea involves averaging over many Monte Carlo simulation runs to compute the expectation above. In this case, the "max" is computed for each simulation run by first generating all the random numbers for that run, and then applying a static optimization algorithm to compute optimal actions $a_1, \ldots, a_H$. It is easy now to see why we call the method "hindsight" optimization: the optimization of the action sequence is done after knowing all uncertainties over time, as if making decisions in hindsight.

As an alternative to relaxing the constraint in (5.4) that each action a_k is a fixed function of the belief state π_k, suppose we further *restrict* each action to be simply fixed (not random). This restriction gives rise to a lower bound on the value function:

$$\hat{V}_{\text{lower}}(\pi) = \max_{a_1,\ldots,a_H} \mathbb{E}_{a_1,\ldots,a_H}[r(\pi_1, a_1) + \cdots + r(\pi_H, a_H) | \pi].$$

To use analogous terminology to "hindsight optimization," we may call this method *foresight optimization*—we make decisions before seeing what actually happens, based on our expectation of what will happen. For an application of this method to tracking, see [57].

5.7 Rollout

In this section, we describe the method of *policy rollout* The basic idea is simple. First let $V_\gamma(\pi_0)$ be the objective function value corresponding to policy γ. Recall that $V^* = \max_\gamma V_\gamma$. In the method of rollout, we assume that we have a candidate policy γ_{base} (called the *base policy*), and we simply replace V^* in (5.1) by $V_{\gamma_{\text{base}}}$. In other words, we use the following approximation to the Q-value:

$$Q_{\gamma_{\text{base}}}(\pi, a) = r(\pi, a) + \mathbb{E}_a[V_{\gamma_{\text{base}}}(\pi')|\pi].$$

We can think of $V_{\gamma_{\text{base}}}$ as the performance of applying policy γ_{base} in our system. In many situations of interest, $V_{\gamma_{\text{base}}}$ is relatively easy to compute, either analytically, numerically, or via Monte Carlo simulation.

It turns out that the policy γ defined by

$$\gamma(\pi) = \arg\max_a Q_{\gamma_{\text{base}}}(\pi, a) \qquad (5.5)$$

is at least as good as γ_{base} (in terms of the objective function); in other words, this step of using one policy to define another policy has the property of *policy improvement*. This policy-improvement result is the basis for a method known as *policy iteration*, where we iteratively apply the above policy-improvement step to generate a sequence of policies converging to the optimal policy. However, policy iteration is difficult to apply in problems with large belief-state spaces, because the approach entails explicitly representing a policy and iterating on it (remember that a policy is a mapping with the belief-state space π as its domain).

In the method of policy rollout, we do not explicitly construct the policy γ in (5.5). Instead, at each time step, we use (5.5) to compute the output of the policy at the current belief-state. For example, the term $\mathbb{E}_a[V_{\gamma_{\text{base}}}(\pi')|\pi]$ can be computed using Monte Carlo sampling. To see how this is done, observe that $V_{\gamma_{\text{base}}}(\pi')$ is simply the mean cumulative reward of applying policy γ_{base}, a quantity that can be obtained by Monte Carlo simulation. The term $\mathbb{E}_a[V_{\gamma_{\text{base}}}(\pi')|\pi]$ is the mean with respect to the random next belief-state π' (with distribution that depends on π and a), again obtainable via Monte Carlo simulation. We provide more details in Section 5.9. In our subsequent discussion of rollout, we will assume by default that the method is implemented using Monte Carlo simulation. For an application of the rollout method to sensor scheduling in target tracking, see [103, 104].

5.8 Parallel rollout

An immediate extension to the method of rollout is to use multiple base policies. So suppose that $\Gamma_B = \{\gamma^1, \ldots, \gamma^n\}$ is a set of base policies. Then replace V^* in (5.1) by

$$\hat{V}(\pi) = \max_{\gamma \in \Gamma_B} V_\gamma(\pi).$$

We call this method *parallel rollout* [54]. Notice that the larger the set Γ_B, the tighter $\hat{V}(\pi)$ becomes as a bound on $V^*(\pi)$. Of course, if Γ_B contains the optimal policy, then $\hat{V} = V^*$. It follows from our discussion of rollout that the policy improvement property also holds when using a lookahead policy based on parallel rollout. As with the rollout method, parallel rollout can be implemented using Monte Carlo sampling.

5.9 Control architecture in the Monte Carlo case

The method of rollout provides a convenient turnkey (systematic) procedure for Monte-Carlo-based decision making and control. Here, we specialize the general control architecture of Section 4 to the use of particle filtering for belief-state updating and a Monte Carlo method for Q-value approximation (e.g., rollout). Particle filtering, which, as discussed in Chapter 4, is a Monte Carlo sampling method for updating posterior distributions, dovetails naturally with Monte Carlo methods for Q-value approximation. An advantage of the Monte Carlo approach is that it does not rely on analytical tractability—it is straightforward in this approach to incorporate sophisticated models for sensor characteristics and target dynamics.

Figure 5.5 shows the control architecture specialized to the Monte Carlo setting. Notice that, in contrast to Figure 5.3, the a particle filter plays the role of the measurement filter, and its output consists of samples of the unobservables. Figure 5.6 shows the action selector in this setting. Contrasting this figure with Figure 5.4, we see that a Monte Carlo simulator plays the role of the Q-value approximator (e.g., through the use of rollout).

As a specific example, consider applying the method of rollout. In this case, the evaluation of the Q-value for any given candidate action relies on a simulation model of the sensing system operating some base policy. This simulation model is a "dynamic" model in the sense that it evaluates the behavior of the sensing system over some horizon of time (which is specified beforehand). The simulator requires as inputs the current observables together with samples of unobservables from the particle filter (to specify initial conditions) and a candidate action. The output of the simulator is a Q-value corresponding to

112 *FOUNDATIONS AND APPLICATIONS OF SENSOR MANAGEMENT*

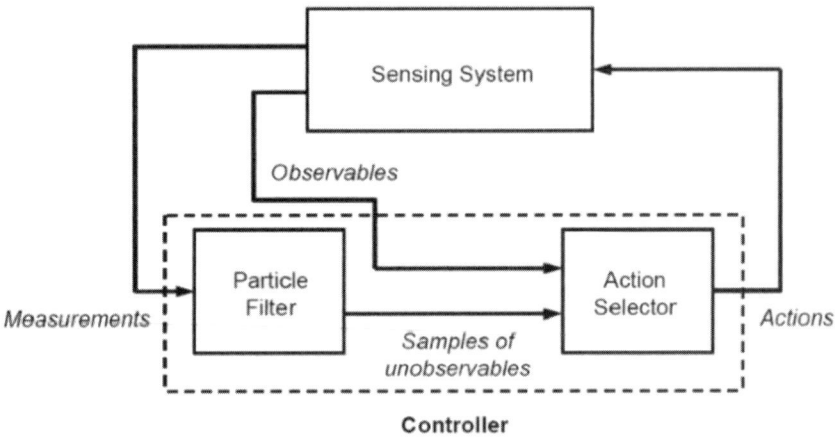

Figure 5.5.   Basic control architecture with particle filtering and rollout.

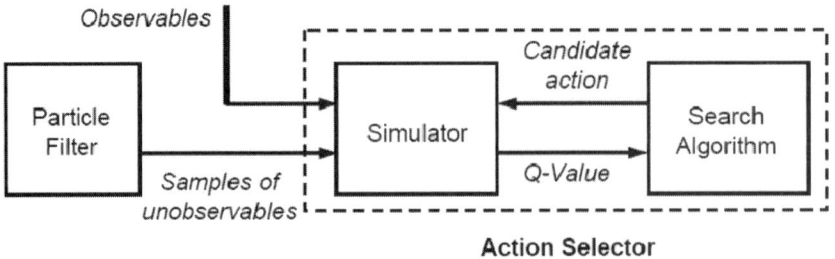

Figure 5.6.   Components of the action selector.

the current measurements and observables, for the given candidate action. The output of the simulator represents the mean performance of applying the base policy, depending on the nature of the objective function. For example, the performance measure of the system may be the negative mean of the sum of the cumulative tracking error and the sensor usage cost over a horizon of H time steps, given the current system state and candidate action.

To elaborate on exactly how the Q-value approximation using rollout is implemented, suppose we are given the current observables and a set of samples of the unobservables (from the particle filter). The current observables together with a single sample of unobservables represent a candidate current underlying state of the sensing system. Starting from this candidate current state, we simulate the application of the given candidate action (which then leads to a random next state), followed by application of the base policy for the remainder of the time horizon—during this time horizon, the system state evolves according to the dynamics of the sensing system as encoded within the simulation model. For this single simulation run, we compute the "action utility" of the system (e.g., the negative of the sum of the cumulative tracking error and sensor usage cost over that simulation run). We do this for each sample of the unobservables, and then average over the performance values from these multiple simulation runs. This average is what we output as the Q-value.

The samples of the unobservables from the particle filter that are fed to the simulator (as candidate initial conditions for unobservables) may include all the particles in the particle filter (so that there is one simulation run per particle), or may constitute only a subset of the particles. In principle, we may even run multiple simulation runs per particle.

The above Monte Carlo method for approximating POMDP solutions has some beneficial features. First, it is flexible in that a variety of sensor management scenarios can be tackled using the same framework. This is important because of the wide variety of sensors that may be encountered in practice. Second, the method does not require analytical tractability; in principle, it is sufficient to simulate a system component, whether or not its characteristics are amenable to analysis. Third, the framework is modular in the sense that models of individual system components (e.g., sensor types, target motion) may be treated as "plug-in" modules. Fourth, the approach integrates naturally with existing simulators (e.g., Umbra [94]). Finally, the approach is inherently non-myopic, allowing the trade-off of short-term gains for long-term rewards.

5.10 Belief-state simplification

If we apply the method of rollout to a POMDP, we need a base policy that maps belief states to actions. Moreover, we need to simulate the performance of this policy—in particular, we have to sample future belief states as the system evolves in response to actions resulting from this policy. Because belief states are probability distributions, keeping track of them in a simulation is burdensome.

A variety of methods are available to approximate the belief state. For example, we could simulate a particle filter to approximate the evolution of the belief state (as described in Chapter 4), but even this may be unduly burdensome. As a further simplification, as in Chapter 10 we could use a Gaussian approximation and keep track only of the mean and covariance of the belief state using a Kalman filter or any of its extensions, including *extended Kalman filters* and *unscented Kalman filters* [120]. Naturally, we would expect that the more accurate the approximation of the belief state, the more burdensome the computation.

An extreme special case of the above tradeoff is to use a delta distribution for belief states in our simulation of the future. In other words, in our lookahead simulation, we do away with keeping track of belief states altogether and instead simulate only a *completely observable* version of the system. In this case, we need only consider a base policy that maps underlying states to actions—we could simply apply rollout to this policy, and not have to maintain any belief states in our simulation. Call this method *completely observable (CO) rollout*. It turns out that in certain applications, such as in sensor scheduling for target tracking, a CO-rollout base policy is naturally available (see [103, 104]). Note that we will still need to keep track of (or estimate) the actual belief state of the system, even if we use CO rollout. The benefit of CO rollout is that it allows us to avoid keeping track of (simulated) belief states in our *simulation* of the future evolution of the system.

In designing lookahead methods with a simplified belief state, we must ensure that the simplification does not hide the good or bad effects of actions. In other words, we need to make sure that the resulting Q-value approximation properly ranks current actions. This requires a carefully designed simplification of the belief state together with a base policy that appropriately reflects the effects of taking specific current actions.

For example, suppose that a particular current action results in poor future rewards because it leads to belief states with large variances. Then, if we use the method of CO rollout, we have to be careful to ensure that this detrimental effect of the particular current action be reflected as a cost in the lookahead.

Otherwise, the effect would not be accounted for properly, because in CO rollout we do not keep track of belief states in our simulation of the future effect of current actions.

Another caveat in the use of simplified belief states in our lookahead is that the resulting rewards in the lookahead may also be affected (and this may have to be taken into account). For example, consider again the problem of sensor scheduling for target tracking, where the per-step reward is the negative mean of the sum of the tracking error and the sensor usage cost. Suppose that we use a particle filter for tracking (i.e., for keeping track of the actual belief state). However, for our lookahead, we use a Kalman filter to keep track of future belief states in our rollout simulation. In general, the tracking error associated with the Kalman filter is different from that of the particle filter. Therefore, when summed with the sensor usage cost, the relative contribution of the tracking error to the overall reward will be different for the Kalman filter compared to the particle filter. To account for this, we will need to scale the tracking error (or sensor usage cost) in our simulation so that the effect of current actions are properly reflected in the Q-value approximations from the rollout with the simplified belief state calculation.

5.11 Reward surrogation

In applying a POMDP approximation method, it is often useful to substitute a *surrogate* for the actual reward function. This can happen for for a number of reasons. First, we may have a surrogate reward that is much simpler (or more reliable) to calculate than the actual reward (see, e.g., the method of reduction to classification in Section 4 of Chapter 3). Second, it may be desirable to have a single surrogate reward for a range of different actual rewards. For example, information gain is often useful as a surrogate reward in sensing applications, taking the place of a variety of detection and tracking metrics (see Chapter 3). Third, reward surrogation may be necessitated by the use of a belief-state simplification technique. For example, under a Gaussian or Gaussian mixture model for the measurements a Kalman filter can be used to update the mean and covariance of the belief state (see Chapter 10).

The use of a surrogate reward can lead to many benefits. But some care must be taken in the design of a suitable surrogate reward. Most important is that the surrogate reward be sufficiently reflective of the true reward that the ranking of actions with respect to the approximate Q-values be preserved. A superficially benign substitution may in fact have unanticipated but significant impact on the ranking of actions. For example, recall the example raised in the previous section on belief-state simplification, where we substitute the tracking

116 *FOUNDATIONS AND APPLICATIONS OF SENSOR MANAGEMENT*

error of a particle filter for the tracking error of a Kalman filter. Superficially, this substitute appears to be hardly a "surrogate" at all. However, as pointed out before, the tracking error of the Kalman filter may be significantly different in magnitude from that of a particle filter.

6. Simulation Result

In this section, we illustrate the performance of several of the strategies discussed in this chapter on a common model problem. The model problem has been chosen to have the characteristics of the motivating example given earlier, while remaining simple enough so that the workings of each method are transparent.

In the model problem, there are two targets, each of which is described by a one-dimensional position x. The sensor may measure any one of the 16 cells, which span the possible target locations (see Figure 5.7). The sensor makes three (not necessarily distinct) measurements per time step, receiving binary returns independent from dwell to dwell. In occupied cells, a detection is received with probability P_d (set here at 0.9). In cells that are unoccupied a detection is received with probability P_f (set here at 0.01). At the onset, positions of the targets are known only probabilistically. The belief state for the first target is uniform across sensor cells $\{2 \cdots 6\}$ and for the second target is uniform across sensor cells $\{11 \cdots 15\}$.

The visibility of the cells changes with time to emulate the motivating example. At time 1, all cells are visible to the sensor. At times 2, 3, and 4, cells $\{11 \cdots 15\}$ become obscured. At time 5, all cells are visible again. This model problem reflects the situation where a target is initially visible to the sensor, becomes obscured, and then reemerges from the obscuration.

	0-1	1-2	2-3	3-4	4-5	5-6	6-7	7-8	8-9	9-10	10-11	11-12	12-13	13-14	14-15	15-16
	Cell 1	Cell 2	Cell 3	Cell 4	Cell 5	Cell 6	Cell 7	Cell 8	Cell 9	Cell 10	Cell 11	Cell 12	Cell 13	Cell 14	Cell 15	Cell 16
Time 1	X														X	
Time 2																
Time 3																
Time 4																
Time 5																

Figure 5.7. The model problem. At the onset, the belief state for target 1 is uniformly distributed across cells $\{2 \cdots 6\}$ and the belief state for target 2 is uniformly distributed across cells $\{11 \cdots 15\}$. At time 1 all cells are visible. At times 2, 3, and 4, cells $\{11 \cdots 15\}$ are obscured. This emulates the situation where one target is initially visible to the sensor, becomes obscured, and then reemerges (Figure 4 from [139] - ©2005 IEEE used with permission).

At time 1 a myopic strategy, having no information about the future visibility, will choose to measure cells uniformly from the set $\{2\cdots 6\} \cup \{11\cdots 15\}$ as they all have the same expected immediate reward. As a result, target 1 and target 2 will on the average be given equal attention. A non-myopic strategy, on the other hand, will choose to measure cells from $\{11\cdots 15\}$ as they are soon to become obscured. That is, the policy of looking for target 2 at time 1 followed by looking for target 1 is best.

Figure 5.8 shows the performance of several of the on-line strategies discussed in this chapter on this common model problem. The performance of each scheduling strategy is measured in terms of the mean squared tracking error at each time step. The curves represent averages over 10,000 realizations of the model problem. Each realization has randomly chosen initial positions of the targets and measurements corrupted by random mistakes as discussed above.

The performance of five different policies is given in Figure 5.8. These policies are described as follows.

- A **random** policy that simply chooses one of the 16 cells randomly for interrogation. This policy provides a worst-case performance and will bound the performance of the other policies.

- A **myopic** policy that takes the action expected to maximize immediate reward. Here the surrogate reward is information gain (see Chapter 3), so the value of an action is estimated by the amount of information it gains. The myopic policy is sub-optimal because it does not consider the long term ramifications of its choices. In particular, at time 1 the myopic strategy has no preference as to which target to measure because both are unobscured and have uncertain position. Therefore, half of the time, target 1 is measured, resulting in an opportunity cost because target 2 is about to disappear.

- The **heuristic EVTG approximation** described in Section 5.4. This policy gives weight to actions expected to be more valuable now than in the future. In particular, actions that correspond to measuring target 2 are given additional value because target 2 is predicted to be obscured in the future. This causes the relative ranking of actions corresponding to measuring target 2 higher than those corresponding to measuring target 1. For this reason, this policy (like the other non-myopic approximations described next) significantly outperforms the myopic strategy. This method has computational burden on the order of T times that of the myopic policy, where T is the horizon length.

- The **rollout** policy described in Section 5.7. The base policy used here is to point the sensor where the target is expected to be. This expectation is computed using the predicted future belief state, which requires the posterior (belief state) to be propagated in time. This is done using a particle filter to represent the posterior. We again use information gain as the surrogate metric to evaluate policies. The computational burden of this method is on the order of NT times that of the myopic policy, where T is the horizon length and N is the number of Monte Carlo trials used in the approximation (here $N = 25$).

- The **completely observable rollout** policy described in Section 5.10. The base policy here is also to point the sensor where the target is expected to be, but is modified to enforce the criterion that the sensor should alternate looking at the two targets. This slight policy modification is necessary due to the delta-function representation of the future belief state. Since the completely observable policy does not require predicting the posterior into the future, it is significantly faster than standard rollout (it is an order of magnitude faster in these simulations). However, completely observable rollout requires a different surrogate reward (one that does not require the posterior like the information gain surrogate metric does). Here we have chosen as a surrogate reward to count the number of detections received, discounting multiple detections of the same target.

7. Summary and Discussion

This chapter has presented approximation methods based on heuristics and simulation for partially observable Markov decision processes. We have highlighted via simulation on a simple model problem approaches based on rollout and a particular heuristic based on information gain. We have detailed some of the design choices that go into finding appropriate approximation, including choice of surrogate reward and belief-state representation.

Throughout this chapter we have taken special care to emphasize the limitations of the methods. Broadly speaking, all tractable methods require domain knowledge in the design process. Rollout methods require a base policy specially designed for the problem at hand; relaxation methods require one to identify the proper constraint(s) to remove; heuristic approximations require one to identify an appropriate approximation to the value-to-go function, and so on. That being said, when such domain knowledge is available it can often result in dramatic improvements in system performance over more traditional methods at a fixed computational cost.

POMDP Approximation Using Simulation and Heuristics

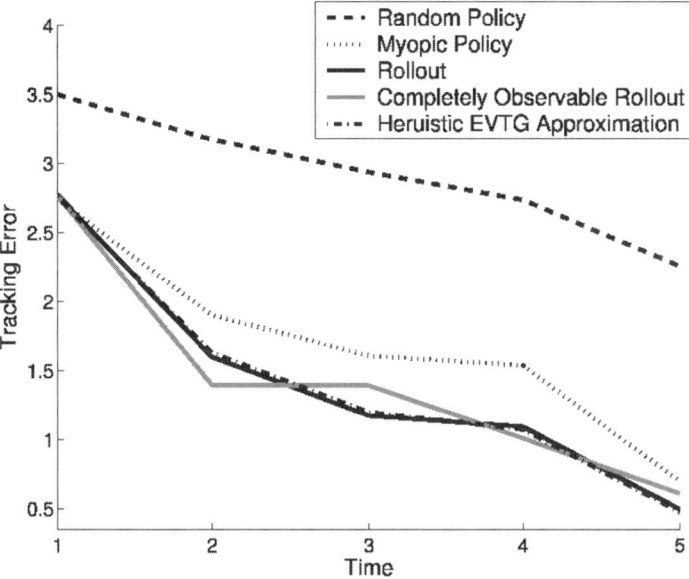

Figure 5.8. The performance of the five policies discussed above. Performance is measured in terms of mean squared tracking error at each time step, averaged over a large number of Monte Carlo trials.

Two aspects not discussed in this chapter are extensions for tracking reactive targets and to use of multiple platforms. The reactive target scenario falls outside of the scope of this book since sensor actions affect the target state in addition to the information state. A simple case of sensor management for reactive targets has been treated in [138]. In this work a single target behaves as a Markov chain with a hidden state that switches to low SNR mode (stealth) after the target detects that it is being illuminated by an active sensor. In [138] optimal strategies for scheduling between active and passive sensor modalities were developed using information gain approaches similar to those developed in this chapter. Multiple platforms complicate the implementation of the JMPD belief state approximation but fit otherwise fit into the framework of the previous three chapters. These techniques have been developed in [143]. Therein physico-mimetic particle fluid models are combined with embedded simulation to perform information-directed control of swarms of platforms for multiple target tracking in real time.

The next two chapters describe the multi-armed bandit (MAB) framework for sensor management. The MAB model can be viewed as an analytical approximation to the general POMDP model that has been the focus of this chapter. Multi-target tracking applications will be revisited in Chapters 7 and 10.

Chapter 6

MULTI-ARMED BANDIT PROBLEMS

Aditya Mahajan

University of Michigan, Ann Arbor, MI, USA

Demosthenis Teneketzis

University of Michigan, Ann Arbor, MI, USA

1. Introduction

Multi-armed bandit (MAB) problems are a class of sequential resource allocation problems concerned with allocating one or more resources among several alternative (competing) projects. Such problems are paradigms of a fundamental conflict between making decisions (allocating resources) that yield high current rewards, versus making decisions that sacrifice current gains with the prospect of better future rewards. The MAB formulation models resource allocation problems arising in several technological and scientific disciplines such as sensor management, manufacturing systems, economics, queueing and communication networks, clinical trials, control theory, search theory, etc. (see [88] and references therein).

In the classical MAB problem (discussed in Section 2) at each instant of time a single resource is allocated to one of many competing projects. The project to which the resource is allocated can change its state; the remaining projects remain frozen (do not change state). In the variants of the MAB problem (discussed in Section 3) one or more resources are dynamically allocated among several projects; new projects may arrive; all projects may change state; delays may be incurred by the reallocation of resources, etc.

In general, sequential resource allocation problems can be solved by dynamic programming, as discussed in Chapter 2. Dynamic programming, which is based on backwards induction, provides a powerful method for the solution of dynamic optimization problems, but suffers from the "curse of dimensionality." The special structure of the classical MAB problem has led to the discovery of optimal "index-type" allocation policies that can be computed by forward induction (see Section 2.2), which is computationally less intensive than backward induction. Researchers have also discovered conditions under which forward induction leads to the discovery of optimal allocation policies for variants of the classical MAB (see Section 3). Discovering conditions under which "index-type" allocation policies are optimal or nearly optimal for sequential resource allocation problems remains an active area of research.

In this chapter we present a qualitative description of the classical MAB problem and its variants. All problems are formulated in discrete time. MAB problems have also been studied in continuous time (see for example [167, 80]). The emphasis in this chapter is on describing the key features of MAB problems and explaining the nature of their solutions. A rigorous proof of the results can be found in the references. The chapter is organized as follows. In Section 2 we present the formulation and the nature of the solution of the classical MAB problem. In Section 3 we describe the formulation and the nature of the solution of several variants of the classical MAB problem. We briefly discuss the relations of the MAB problem and its variants to sensor management in Section 4. A presentation of σ-fields and stopping times, concepts that are important in understanding the nature of the solution of the classical MAB problem, appears in Section 3 of the Appendix.

Remark on notation. Throughout this Chapter, we use uppercase letters to represent random variables and the corresponding lowercase letters to represent their realizations.

2. The Classical Multi-armed Bandit

In this section we present a general formulation of the MAB problem, highlight its salient features and describe its optimal solution. We also discuss forward induction, which provides insight into the nature of an optimal solution.

Multi-armed Bandit Problems

2.1 Problem Formulation

2.1.1 A Bandit Process.
A (single-armed) bandit process is described by a machine/arm/project and is characterized by the pair of random sequences $\big(\{X(0), X(1), \ldots\}, \{R(X(0)), R(X(1)), \ldots\}\big)$, where $X(n)$ denotes the state of the machine[1] after it has been operated n times, and $R(X(n))$ denotes the reward obtained when the machine is operated for the nth time. The state $X(n)$ is a real-valued random variable and $R(X(n))$ is a random variable taking values in $\mathbb{R}_+$. In general when the machine is operated for the nth time its state changes according to

$$X(n) = f_{n-1}\big(X(0), \ldots, X(n-1), W(n-1)\big), \tag{6.1}$$

where $f_{n-1}(\cdot)$ is given and $\{W(n); n = 0, 1, \ldots\}$ is a sequence of independent real-valued random variables that are also independent of $X(0)$ and have known statistical description. Thus a (single-armed) bandit process is not necessarily described by a Markov process.

2.1.2 The Classical Multi-armed Bandit Problem.
A multi-armed (k-armed) bandit process is a collection of k independent single-armed bandit processes. The classical MAB problem consists a multi-armed bandit process and one controller (also called a processor). At each time, the controller can choose to operate *exactly* one machine; all other machines remain frozen. Each machine i, $i = 1, 2, \ldots, k$, is described by sequences $\{(X_i(N_i(t)), R_i(X_i(N_i(t)))); N_i(t) = 0, 1, 2, \ldots, t; t = 0, 1, 2, \ldots\}$, where $N_i(t)$ denotes the number of times machine i has been operated until time t. $N_i(t)$ is machine i's local time. Let $U(t) := \big(U_1(t), \ldots, U_k(t)\big)$ denote the control action[2] taken by the controller at time t. Since the controller can operate on exactly one machine at each time, the control action $U(t)$ takes values in $\{e_1, \ldots, e_k\}$, where $e_j = (0, \ldots, 0, 1, 0, \ldots, 0)$ is a unit k-vector with 1 at the jth position. Machines that are not operated remain frozen. Specifically, the system evolves according to

$$X_i\big(N_i(t+1)\big) = \begin{cases} f_{N_i(t)}\big(X_i(0), \ldots, X_i(N_i(t)), W_i(N_i(t))\big), & \text{if } U_i(t) = 1, \\ X_i\big(N_i(t)\big), & \text{if } U_i(t) = 0, \end{cases} \tag{6.2}$$

---
[1] The state $X(n)$ is similar to state s_n of Markov decision model adopted in Chapter 2, even though, as noted in this chapter, machines are not necessarily described by Markov processes.
[2] The control action $U(t)$ is similar to the control action a_t of the Markov decision model of Chapter 2.

and

$$N_i(t+1) = \begin{cases} N_i(t), & \text{if } U_i(t) = 0, \\ N_i(t)+1, & \text{if } U_i(t) = 1, \end{cases} \quad (6.3)$$

for all $i = 1, 2, \ldots, k$. Thus $N_i(t)$, the local time of machine i, is incremented only when the controller operates on machine i at t; i.e., only when $U_i(t) = 1$. $\{W_i(n);\ i = 1, \ldots, k;\ n = 0, 1, \ldots\}$ is a sequence of independent primitive random variables that are independent of $\{X_1(0), \ldots, X_k(0)\}$ and have known statistical description[3].

Machine i generates a reward only when it is operated. Let $R_i(t)$ denote the reward generated by machine i at time t; then

$$R_i(t) = R_i\big(X(N_i(t)), U_i(t)\big) = \begin{cases} R_i\big(X_i(N_i(t))\big), & \text{if } U_i(t) = 1, \\ 0, & \text{if } U_i(t) = 0, \end{cases} \quad (6.4)$$

A *scheduling policy* $\gamma := (\gamma_1, \gamma_2, \ldots)$ is a decision rule such that at each time instant t, the control action $U(t)$ takes values in $\{e_1, \ldots, e_k\}$ according to[4]

$$U(t) = \gamma_t\big(Z_1(t), \ldots, Z_k(t), U(0), \ldots, U(t-1)\big), \quad (6.5)$$

where

$$Z_i(t) = [X_i(0), \ldots, X_i(N_i(t))].$$

The MAB problem is the following: Determine a scheduling policy that maximizes

$$J^\gamma := \mathbb{E}\left[\sum_{t=0}^\infty \beta^t \sum_{i=1}^k R_i\big(X_i(N_i(t)), U_i(t)\big) \,\bigg|\, Z(0)\right]. \quad (6.6)$$

subject to (6.2)–(6.5).

This problem was originally formulated around 1940. It was known that it can be solved by a stochastic dynamic programming (SDP) approach, but no substantial progress in understanding the nature of its optimal solution was made until Gittins and Jones [89] showed that an optimal solution is of the *index type*. That is, for each bandit process one can compute what Gittins called a *dynamic allocation index* (DAI), which depends only on that process, and then at each time the controller operates on the bandit process with the

---

[3] As it is usually assumed, the processes $\{X_i(N_i(t)), W_i(N_i(t));\ N_i(t) = 0, 1, \ldots t;\ t = 0, 1, 2, \ldots;\ i = 1, \ldots, k\}$ are defined on the same probability space $(\Omega, \mathcal{F}, P)$.
[4] Formally $U(t)$ needs to be measurable with respect to appropriate σ-fields. However, in this exposition we adopt an informal approach and do not explicitly talk about measurability requirements.

highest index. Thus, finding an optimal scheduling policy, which originally requires the solution of a k-armed bandit problem, reduces to determining the DAI for k single-armed bandit problems, thereby reducing the complexity of the problem exponentially. The DAI was later referred to as the *Gittins index* in honor of Gittins's contribution.

Following Gittins several researchers provided alternative proofs of the optimality of the *Gittins index* rule (see [249, 240, 246, 85, 22, 29, 232, 166, 115, 116, 123, 129, 166, 167] and [88] and references therein). Another formulation of the MAB problem with a performance criterion described by a "learning loss" or "regret" has been investigated in the literature [156, 7, 8, 3, 5, 4]. We will not discuss this formulation as we believe it is not directly related to sensor management problems.

Before we present the solution of the MAB problem we briefly present the method of forward induction. Knowing when forward induction is optimal is critical in understanding the nature of the solution of the MAB problem.

2.2   On Forward Induction

For centralized stochastic control problems, SDP provides a methodology for sequentially determining optimal decision rules/control laws/decision strategies. The SDP algorithm proceeds backward in time and at every stage t determines an optimal decision rule by quantifying the effect of every decision on the current and future conditional expected rewards. This procedure is called *backward induction*. SDP provides a powerful methodology for stochastic optimization, but it is computationally hard in many instances.

A procedure (computationally) simpler than backward induction is to make, at each decision time t, a decision that maximizes the conditional expected reward acquired at t. This procedure concentrates only on the present and completely ignores the future; it is called a *myopic approach* and results in a *myopic policy*. In general, myopic policies are not optimal.

The notion of a myopic policy can be extended to form T-*step-look-ahead policies*: make, at each decision time t, a decision that maximizes the conditional expected reward acquired at t plus the conditional expected reward acquired over the next T stages. In general, T-step-look-ahead policies are suboptimal. As T increases their performance improves at the cost of an increase in computational complexity.

The notion of a T-step-look-ahead policy can be extending as follows: allow the number τ of steps over which we look ahead at each stage to depend on how the system evolves while these steps are taking place; thus the number τ

of steps is a stopping time[5] with respect to the increasing family of σ-fields that describe the information accumulated by the decision maker/controller during the evolution of the process. This extension of a T-step-look-ahead policy involves two maximizations. An inner maximization that assumes that decision rule for taking a sequence of decisions is given and chooses a stopping time τ to maximize the conditional expected reward rate. The outer maximization is to choose a decision rule to maximize the result of the inner maximization for that decision rule. This extension of the T-step-look-ahead policy works as follows. At $t = 0$, given the information about the initial state of the process, select a decision rule and a stopping time τ_1 and follow it for the next τ_1 steps. The process of finding a new decision rule and a corresponding stopping time τ_2 is then repeated by conditioning on the information accumulated during the first τ_1 steps. The new rule is followed for the next τ_2 steps, and this procedure is repeated indefinitely. This procedure determines a policy for the entire horizon and is called *forward induction*; the policy produced by this procedure is called a *forward induction policy*.

In general forward induction results in suboptimal policies for stochastic control/optimization problems. This is demonstrated by the the following example from Gittins [87, pg 152].

Consider the problem of choosing a route for a journey by car. Suppose there are several different possible routes all of the same length which intersect at various points, and the objective is to choose a route which minimizes the time taken for the journey. The problem may be modeled as a Markov decision process by interpreting the distance covered so far as the "time" variable, the time taken to cover each successive mile as negative reward, the position as the state, and by choosing a value just less than one for the discount factor β. The action space $\mathcal{U}(x)$ has more than one element when the state x corresponds to cross-roads, the different control actions representing the various possible exits.

For this problem the first stage in a forward induction policy is to find a route ζ_1, and a distance σ_1 along ζ_1 from the starting point, such that the average speed in traveling the distance σ_1 along ζ_1 is maximized. Thus a forward induction policy might start with a short stretch of highway, which is followed by a very slow section, in preference to a trunk road which permits a good steady average speed. The trouble is that irrevocable decisions have to be taken at each cross-roads in the sense that those exits which are not chosen are not available later on.

---

[5]For the definition of stopping time, we refer the reader to Section 3 of the Appendix.

The above example illustrates the reason why, in general, forward induction results in suboptimal policies. Irrevocable decisions have to be made at some stage of the decision process, that is, some alternatives that are available at that stage and are not chosen, do not remain available later.

Forward induction policies are optimal if the decisions made at any stage are not irrevocable; that is, any alternative that is available at any stage and is not chosen, may be chosen at a later stage and with exactly the same sequence of rewards (apart from a discount factor). Thus, there is no later advantage of not choosing a forward induction policy.

In the next section we explain why the MAB problem belongs to a class of stochastic controlled processes for which forward induction results in optimal policies.

2.3 Key Features of the Classical MAB Problem and the Nature of its Solution

Four features delimit the MAB problem within the general class of stochastic control problems:

(F1) only one machine is operated at each time instant. The evolution of the machine that is being operated is uncontrolled; that is, the processor chooses which machine to operate but not how to operate it;

(F2) machines that are not operated remain frozen;

(F3) machines are independent;

(F4) frozen machines contribute no reward.

Features (F1)–(F4)[6] imply that an optimal policy can be obtained by forward induction. Decisions made at any instant of time t are not irrevocable since any bandit process that is available for continuation at t but is not chosen at t, can be continued at any later time instant with exactly the same resulting sequence of rewards, apart from the discount factor. This means that there is no later compensation for any initial loss resulting from not choosing a forward induction policy. Consequently, without any loss of optimality, we can restrict attention to forward induction policies. The first stage of a forward induction policy must be such that the expected discounted reward per unit of expected

---

[6]In Section 3.6 we show that feature (F4) is not essential for obtaining an optimal policy by forward induction.

discounted time up to an arbitrary stopping time is maximized. Gittins [87] argued (and proved rigorously) that this maximum can be achieved by a policy under which only one bandit process is continued up to the stopping time in question. To determine the bandit process to be continued in the first stage and the corresponding stopping time the following arguments, due to Whittle [249], can be used. Consider arm i of the MAB process and let $x_i(0)$ be its state at $t = 0$ and let $N_i(t) = 0$. Suppose two options are available at $t = 0$: continue the process, or retire and receive a retirement reward v. Let $v_{X_i}(x_i(0))$ be the retirement reward that can be offered at $x_i(0)$ so that the controller is indifferent to both options. This reward is given by

$$v_{X_i}(x_i(0)) := \max_{\tau > 0} \frac{\mathbb{E}\left[\sum_{t=0}^{\tau-1} \beta^t R_i(X_i(t)) \Big| x_i(0)\right]}{\mathbb{E}\left[\sum_{t=0}^{\tau-1} \beta^t \Big| x_i(0)\right]}. \tag{6.7}$$

Then $v_{X_i}(x_i(0))$ is the maximum expected discounted reward per unit of expected discounted time that can be obtained at the first stage of a forward induction policy that continues the arm i with initial state $x_i(0)$. The corresponding stopping time $\tau_i(x_i(0))$ is the first time at which the expected discounted reward per unit of expected discounted time equals $v_{X_i}(x_i(0))$. Consequently, at $t = 0$ an optimal forward induction policy continues an arm j such that

$$v_{X_j}(x_j(0)) = \max_i v_{X_i}(x_i(0)).$$

Arm j is continued until $\tau_j(x_j(0)) - 1$. This constitutes the first stage of an optimal forward induction policy, and can be summarized as follows:

Step 1 Determine $v_{X_i}(x_i(0)), i = 1, \ldots, k$.

Step 2 Select an arm j such that

$$j = \arg\max_i v_{X_i}(x_i(0)).$$

Continue operating arm j until the minimum time that achieves the maximum in the right hand side of (6.7).

At the end of the first stage, the forward induction proceeds in the same way as at $t = 0$. Let τ_l be the end of the lth stage of an optimal forward induction policy for the MAB problem. At time τ_l we must decide which arm to operate next. The random time τ_l is a stopping time with respect to the family of σ-fields $\{\bigvee_{i=1}^{k} \mathcal{F}^i(t), t = 0, 1, 2, \ldots\}$, where $\mathcal{F}^i(t)$ is defined as the sigma field $\sigma(X_i(0), \ldots, X_i(N_i(t))), i = 1, 2, \ldots, k$. For any sample point ω in the sample space Ω of the MAB process (see footnote 3), let $\{x_i(0), x_i(1),$ $\ldots, x_i(N_i(\tau_l(\omega))), i = 1, 2, \ldots, k\}$ be the realization of the MAB process obtained under an optimal forward induction policy up to $\tau_l(\omega)$. The decision made by an optimal forward induction policy at $\tau_l(\omega)$ can be described by the following two-step process:

Step 1 For each $i = 1, \ldots, k$, let $x_i^l(\omega) := (x_i(0), \ldots, x_i(N_i(\tau_l(\omega))))$, and determine

$$v_{X_i}(x_i^l(\omega)) = \max_{\tau > \tau_l(\omega)} \frac{\mathbb{E}\left[\sum_{t=\tau_l(\omega)}^{\tau-1} \beta^t R_i(X_i(N_i(\tau_l) + t - \tau_l(\omega))) \Big| x_i^l(\omega)\right]}{\mathbb{E}\left[\sum_{t=\tau_l(\omega)}^{\tau-1} \beta^t \Big| x_i^l(\omega)\right]},$$

(6.8)

and the stopping time $\tau_i(x_i^l(\omega))$ achieves the maximum at the right hand side of (6.8).

Step 2 Select arm j such that

$$j = \arg\max_i v_{X_i}(x_i^l(\omega))$$

and operate it for $\tau_j(x_j^l(\omega)) - 1 - \tau_l(\omega)$ steps/time units.

The number $v_{X_i}(x_i^l(\omega))$ that denotes the maximum expected discounted reward per unit of expected discounted time obtained by arm i at $x_i^l(\omega)$ is the *Gittins index* of arm i at $x_i^l(\omega)$.

We examine now what happens between the stopping times τ_l and τ_{l+1}, $l = 0, 1, \ldots$. Suppose arm j is continued at τ_l. Then at $t = \tau_l + 1, \ldots, \tau_{l+1} - 1$ the *Gittins index* of arm j is higher than its index at τ_l (see [240]). The Gittins indices of all other arms remain the same as in τ_l since these arms are frozen. Thus an equivalent method to describe the above procedure is to consider the MAB problem at each instant of time t and continue the arm with the highest index. This observation allows us to understand the nature of the solution of some generalizations of the MAB problem (e.g., the arm-acquiring bandit).

In summary, an optimal scheduling policy for the MAB problem can be obtained by forward induction. Forward induction is computationally less intensive than backward induction; an optimal forward induction policy can be obtained by solving k one-dimensional problems (by computing the Gittins indices of each bandit process) instead of one k-dimensional problem.

In many formulations of the MAB problem it is assumed that the machines are Markovian[7], that is

$$f_{N_i(t)}\big(X_i(N_i(0)),\ldots,X_i(N_i(t)),W_i(N_i(t))\big)$$
$$= f_{N_i(t)}\big(X_i(N_i(t)),W_i(N_i(t))\big). \quad (6.9)$$

In this case (6.8) reduces to

$$v_{X_i}\big(x_i^l(\omega)\big) = v_{X_i}\big(x_i(N_i(\tau_l))\big)$$
$$= \max_{\tau > \tau_l(\omega)} \frac{\mathbb{E}\left[\sum_{t=\tau_l(\omega)}^{\tau-1} \beta^t R_i\big(X_i(N_i(\tau_l(\omega)) + t - \tau_l(\omega))\big) \,\big|\, x_i(N_i(\tau_l))\right]}{\mathbb{E}\left[\sum_{t=\tau_l(\omega)}^{\tau-1} \beta^t \,\big|\, x_i(N_i(\tau_l(\omega)))\right]}, \quad (6.10)$$

and such an index is considerably easier to compute (see [240, 127, 100, 126]).

2.4 Computational Issues

We concentrate on the classical MAB problem where the machines are time-homogeneous finite-state Markov chains (MCs)[8]. We assume that machine i, $i = 1, 2, \ldots, k$, has state space $\{1, 2, \ldots, \Delta_i\}$ and matrix of transition probabilities $P^{(i)} := \{P_{a,b}^{(i)}, a, b \in \{1, 2, \ldots, \Delta_i\}\}$.

In this case we do not need to keep track of the local time of the machines because of the Markovian property and the time-homogeneity of the Markov chains. The evolution of machine i, $i = 1, 2, \ldots, k$, can be described by the following set of equations. If $X_i(t) = a$, $a \in \{1, 2, \ldots, \Delta_i\}$, then

$$X_i(t+1) = a, \quad \text{if } U_i(t) = 0, \quad (6.11)$$

$$P\big(X_i(t+1) = b \mid X_i(t) = a\big) = P_{a,b}^{(i)}, \quad \text{if } U_i(t) = 1. \quad (6.12)$$

---

[7]Such formulations are considered in Section 2 of Chapter 7 where applications of MAB theory to sensor management is considered.
[8]Throughout this chapter we assume that the state of each machine is perfectly observable. For the case where state is imperfectly observable we refer the reader to [171].

Multi-armed Bandit Problems 131

Further, $\mathbf{X}(t) := \big(X_1(t), X_2(t), \ldots, X_k(t)\big)$ is an information state (sufficient statistic, see Chapter 3 and [153]) at time t. Thus (6.10) can be rewritten as

$$\nu_{X_i}\big(x_i(t)\big) = \max_{\tau > t} \frac{\mathbb{E}\left[\sum_{t'=t}^{\tau-1} \beta^{t'} R_i\big(X_i(t')\big) \Big| x_i(t)\right]}{\mathbb{E}\left[\sum_{t'=t}^{\tau-1} \beta^{t'} \Big| x_i(t)\right]} \quad (6.13)$$

Thus to implement the index policy we need to compute, for each machine i, $i = 1, \ldots, k$, the Gittins index of state $x_i(t)$. This can be done in either an off-line or an on-line manner. For an off-line implementation, we must compute the Gittins index corresponding to each state $x_i \in \{1, 2, \ldots, \Delta_i\}$ of each machine i, $i = 1, \ldots, k$. This computation can be done off-line and we only need to store the values of $\nu_{X_i}(x_i)$, $x_i \in \{1, \ldots, \Delta_i\}$ for each machine i, $i = 1, \ldots, k$. For an on-line implementation, at stage 0 we need to compute $\nu_{X_i}\big(x_i(0)\big)$ for each machine i, $i = 1, \ldots, k$ where $x_i(0)$ is given[9]. We operate machine $j = \arg\max_i \nu_{X_i}\big(x_i(0)\big)$ until the smallest time τ_1 at which machine j achieves its Gittins index. At any subsequent stage l, we only need to compute the Gittins index of the machine operated at stage $l - 1$. The computation of these Gittins indices has to be done on-line, but only for the stopping states that are reached during the evolution of the bandit processes. To achieve such a computation we need to store the reward vector and the matrix of transition probabilities for each machine.

We next describe the notions of continuation and stopping sets, which are key concepts for the off-line and on-line computation of the Gittins index rule (see [87]). Suppose we start playing machine i which is initially in state x_i. Then the state space $\{1, 2, \ldots, \Delta_i\}$ can be partitioned into two sets $C_i(x_i)$ (the continuation set of x_i) and $S_i(x_i)$ (the stopping set of x_i). When the state of machine i is in $C_i(x_i)$ we continue processing the machine. We stop processing machine i the first instant of time the state of the machine is in $S_i(x_i)$. Therefore, the Gittins index policy can be characterized by determining $C_i(x_i)$ and $S_i(x_i)$ for each $x_i \in \{1, 2, \ldots, \Delta_i\}$.

A computational technique for the off-line computation of the Gittins index rule, proposed by Varaiya, Walrand and Buyukkoc in [240], is based on the following observation. If for $a, b \in \{1, 2, \ldots, \Delta_i\}$ $\nu_{X_i}(a) > \nu_{X_i}(b)$, then $b \in S_i(a)$ and $a \in C_i(b)$. If $\nu_{X_i}(a) = \nu_{X_i}(b)$ then either $a \in C_i(b)$ and $b \in S_i(a)$, or $a \in S_i(b)$ and $b \in C_i(a)$. Thus, to determine $C_i(x_i)$ and $S_i(x_i)$ for each $x_i \in \{1, 2, \ldots, \Delta_i\}$ we must find first an ordering $l_1, l_2, \ldots, l_{\Delta_i}$ of

---

[9] It is the observed state of machine i at time 0.

the states of machine i such that

$$\nu_{X_i}(l_1) \geq \nu_{X_i}(l_2) \geq \cdots \geq \nu_{X_i}(l_{\Delta_i}), \tag{6.14}$$

and then set for all $l_j, j = 1, 2, \ldots, \Delta_i$,

$$\begin{aligned} C_i(l_j) &= \{l_1, l_2, \ldots, l_j\}, \\ S_i(l_j) &= \{l_{j+1}, l_{j+2}, \ldots, l_{\Delta_i}\}. \end{aligned} \tag{6.15}$$

To obtain such an ordering $l_1, l_2, \ldots, l_{\Delta_i}$ the following computational procedure was proposed in [240].

Given a machine i, $i = 1, 2, \ldots, k$, with state space $\{1, 2, \ldots, \Delta_i\}$, matrix of transition probabilities $P^{(i)} := \{P^{(i)}_{a,b},\ a, b \in \{1, 2, \ldots, \Delta_i\}\}$, and reward function $R_i(x_i)$, $x_i \in \{1, 2, \ldots, \Delta_i\}$ set

$$l_1 = \arg\max_{x_i} R_i(x_i). \tag{6.16}$$

Break ties by choosing the smallest x_i that satisfies (6.16). The Gittins index of state l_1 is

$$\nu_{X_i}(l_1) = R_i(l_1). \tag{6.17}$$

States $l_2, l_3, \ldots, l_{\Delta_i}$ can be recursively determined by the following procedure. Suppose $l_1, l_2, \ldots, l_{n-1}$ have been determined; then

$$\nu_{X_i}(l_1) \geq \nu_{X_i}(l_2) \geq \cdots \geq \nu_{X_i}(l_{n-1}). \tag{6.18}$$

Define

$$P^{(i,n)}_{a,b} := \begin{cases} P^{(i)}_{a,b}, & \text{if } b \in \{l_1, l_2, \ldots, l_{n-1}\} \\ 0, & \text{otherwise} \end{cases}, \tag{6.19}$$

and the vectors

$$\mathbf{R}_i := \big(R_i(1), \quad R_i(2), \quad \cdots, \quad R_i(\Delta_i)\big)^\mathsf{T}, \tag{6.20}$$

$$\mathbf{1} := \underbrace{(1, \quad 1, \quad \cdots, \quad 1)^\mathsf{T}}_{\Delta_i\,\text{times}}, \tag{6.21}$$

$$D^{(i,n)} := \beta\big[I - \beta P^{(i,n)}\big]^{-1}\mathbf{R}_i = \begin{pmatrix} D_1^{(i,n)} \\ D_2^{(i,n)} \\ \cdots \\ D_{\Delta_i}^{(i,n)} \end{pmatrix}, \tag{6.22}$$

$$B^{(i,n)} := \beta\big[I - \beta P^{((i,n))}\big]^{-1}\mathbf{1} = \begin{pmatrix} B_1^{(i,n)} \\ B_2^{(i,n)} \\ \cdots \\ B_{\Delta_i}^{(i,n)} \end{pmatrix}. \tag{6.23}$$

Then

$$l_n = \arg\max_{a \in \{1,2,\ldots,\Delta_i\} \setminus \{l_1,l_2,\ldots,l_{n-1}\}} \frac{D_a^{(i,n)}}{B_a^{(i,n)}}, \tag{6.24}$$

and

$$\nu_{X_i}(l_n) = \frac{D_{l_n}^{(i,n)}}{B_{l_n}^{(i,n)}}. \tag{6.25}$$

Another method for off-line computation of Gittins index, which has the same complexity as the algorithm of [240] presented above, appears in [29].

The following method for on-line implementation of the Gittins index was proposed by Katehakis and Veinott in [127]. As explained earlier, to obtain the Gittins index for state x_i only the sets $C_i(x_i)$ and $S_i(x_i)$ need to be determined. In [127], Katehakis and Veinott proposed the "restart in x_i" method to determine these sets. According to this method, we consider an alternative problem where in any state $a \in \{1,\ldots,\Delta_i\}$ we have the option either to continue operating machine i from state a or to instantaneously switch to state x_i and continue operating the machine from state x_i. The objective is to choose the option that results in the maximum expected discounted reward over an infinite horizon (see Chapter 2, Section 2.2). This approach results in a dynamic program

$$V(a) = \max\left\{ R_i(a) + \beta \sum_{b \in \{1,\ldots,\Delta_i\}} P^{(i)}_{a,b} V(b) \,,\, R(x_i) + \beta \sum_{b \in \{1,\ldots,\Delta_i\}} P^{(i)}_{x_i,b} V(b) \right\},$$
$$a \in \{1,\ldots,\Delta_i\} \quad (6.26)$$

that can be solved by various standard computational techniques for finite state Markov decision problems (see Chapter 2). The solution of this dynamic program determines the sets $C_i(x_i)$ and $S_i(x_i)$. These sets are given by

$$C_i(x_i) = \left\{ a \in \{1,\ldots,\Delta_i\} : R_i(a) + \beta \sum_{b \in \{1,\ldots,\Delta_i\}} P^{(i)}_{a,b} V(b) \geq V(x_i) \right\} \quad (6.27)$$

$$S_i(x_i) = \left\{ a \in \{1,\ldots,\Delta_i\} : R_i(a) + \beta \sum_{b \in \{1,\ldots,\Delta_i\}} P^{(i)}_{a,b} V(b) < V(x_i) \right\} \quad (6.28)$$

and the Gittins index is given by

$$\nu_{X_i}(x_i) = (1 - \beta) V(x_i) \quad (6.29)$$

Another method for on-line implementation similar in spirit to [240] appears in E. L. M. Beale's discussion in [87].

Several variations of the classical MAB problem have been considered in the literature. We briefly present them in Section 3.

3. Variants of the Multi-armed Bandit Problem

In this section we present various extensions of the classical MAB problem. In general, in these extensions, forward induction does not provide a methodology for determining an optimal scheduling policy. Index-type solutions are desirable because of their simplicity, but, in general, they are not optimal. We identify conditions under which optimal index-type solutions exist.

3.1 Superprocesses

A superprocess consists of k independent components and one controller/processor. At each time t each component $i = 1, 2, \ldots, k$ accepts control inputs $U_i(t) \in \mathcal{U}_i := \{0, 1, \ldots, M_i\}$. The control action $U_i(t) = 0$ is a freezing control; the action $U_i(t) = j, j = 1, 2, \ldots, M_i$ is a continuation control. Thus, each component of a superprocess is a generalization of an arm of a classical MAB problem (where at each t $U_i(t) \in \{0, 1\}$.) In fact, each component of a superprocess is a controlled stochastic process. For any fixed control law this stochastic process is a single-armed bandit process. Consequently, each component of a superprocess consists of a collection of bandit processes/machines, each corresponding to a distinct control law. Component $i = 1, 2, \ldots, k$ evolves as follows

$$X_i(N_i(t+1)) = X_i(N_i(t)) \quad \text{if } U_i(t) = 0, \tag{6.30}$$

and

$$X_i(N_i(t+1)) = f_{N_i(t)}(X_i(0), \ldots, X_i(N_i(t)), U_i(t), W_i(N_i(t)))$$
$$\text{if } U_i(t) \neq 0, \tag{6.31}$$

where $N_i(t)$ is the local time of component i at each t and $\{W_i(n), n = 1, 2, \ldots\}$ is a sequence of independent random variables that are also independent of $\{X_1(0), X_2(0), \ldots, X_k(0)\}$. Furthermore, the sequences $\{W_i(n); n = 1, 2, \ldots\}$, $\{W_j(n); n = 1, 2, \ldots\}$, $i \neq j, i, j = 1, 2, \ldots, k$, are independent.

A reward sequence $\{R_i(X_i(t), U_i(t)); t = 1, 2, \ldots\}$ is associated with each component $i = 1, 2, \ldots, k$, such that

$$R_i(t) = R_i(X_i(t), U_i(t)), \quad \text{if } U_i(t) \neq 0, \tag{6.32}$$

and

$$R_i(t) = 0, \quad \text{if } U_i(t) = 0. \tag{6.33}$$

At each time t the controller/processor can choose to operate/continue exactly one component. If the controller chooses component j at t, i.e., $U_i(t) = 0$ for all $i \neq j$, and $U_j(t)$ takes values in $\{1, 2, \ldots, M_j\}$, a reward $R_j(X_j(N_j(t)), U_j(t))$ is acquired according to (6.32).

A scheduling policy $\gamma := (\gamma_1, \gamma_2, \ldots)$ is a decision rule such that the action $U(t) = (U_1(t), U_2(t), \ldots, U_k(t))$ is a random variable taking values in $\bigcup_{i=1}^{k} \{0\}^{i-1} \times \{1, 2, \ldots, M_i\} \times \{0\}^{k-i}$, and

$$U(t) = \gamma_t(Z_1(t), Z_2(t), \ldots, Z_k(t), U(0), \ldots, U(t-1)), \tag{6.34}$$

where

$$Z_i(t) := [X_i(0), X_i(1), \ldots, X_i(N_i(t))]. \tag{6.35}$$

The objective in superprocesses is to determine a scheduling policy γ that maximizes

$$J^\gamma := \mathbb{E}^\gamma \left[\sum_{t=0}^{\infty} \beta^t \sum_{j=1}^{k} R_j(X_j(N_j(t)), U_j(t)) \,\bigg|\, Z(0) \right] \tag{6.36}$$

subject to (6.30)–(6.35) and the above constraints on $U(t)$, $t = 0, 1, 2, \ldots$ where
$$Z(0) := [X_1(0), X_2(0), \ldots, X_k(0)].$$

Even though features (F2)-(F4) of the MAB problem are present in the superprocess problem, (F1) is not, and as a result of this superprocesses do not in general admit an index-type of solution. Specifically: in the MAB problem, once a machine/process is selected for continuation, the evolution of this process and the accumulated rewards are uncontrolled; on the contrary, in superprocesses, once a component is selected for continuation, the evolution of this component and the accumulated rewards are controlled. Choosing the control law that maximizes the infinite horizon expected discounted reward for the component under consideration leads to a standard stochastic control problem which can only be solved optimally by backward induction.

Consequently, superprocesses are more complex problems than standard MAB problems. There is one situation where superprocesses admit an index form type of solution, namely, when each component has a *dominating machine*.

The concept of a dominating machine can be formally described as follows. Consider a machine $\{X(n), R(X(n)), n = 0, 1, 2, \ldots\}$ and let $\mu \in \mathbb{R}$. Define

$$\mathcal{L}(X, \mu) := \max_{\tau > 0} \mathbb{E}\left[\sum_{t=0}^{\tau-1} \beta^t \left[R(X(t))\right] - \mu\right], \quad (6.37)$$

where τ ranges over all stopping times of $\{\mathcal{F}_t := \sigma(X(0), X(1), \ldots, X(t)), t = 0, 1, 2, \ldots\}$ (see Appendix Section 3 for a discussion on σ-fields and stopping times). Notice that $\mathcal{L}(X, \mu) \geq 0$ since for $\tau = 1$ the right hand side of (6.37) is equal to zero with probability one.

DEFINITION 6.1 *We say that machine* $\{X(n), R(X(n)); n = 0, 1, 2, \ldots\}$ dominates *machine* $\{Y(n), R(Y(n)); n = 0, 1, 2, \ldots\}$ *if*

$$\mathcal{L}(X, \mu) \geq \mathcal{L}(Y, \mu), \quad \forall \mu \in \mathbb{R}. \quad (6.38)$$

This inequality can be interpreted as follows. Suppose one operates machines $\mathcal{M}(X) := \{X(n), R(X(n)); n = 0, 1, 2, \ldots\}$ and $\mathcal{M}(Y) := \{Y(n), R(Y(n)); n = 0, 1, 2, \ldots\}$ up to some random time after which one retires and receives the constant reward μ at each subsequent time. Then $\mathcal{M}(X)$ dominates $\mathcal{M}(Y)$ if and only if it is optimal to choose $\mathcal{M}(X)$ over $\mathcal{M}(Y)$ for any value of the retirement reward μ.

Multi-armed Bandit Problems 137

As mentioned above, a component of a superprocess is a collection of bandit processes $\{X^{\gamma_i}(n), R(X^{\gamma_i}(n), U^{\gamma_i}(n)); n = 0, 1, 2, \ldots\}$, $\gamma_i \in \Gamma_i$, where Γ_i is the set of all possible control laws for component i.

DEFINITION 6.2 *A component of a superprocess is said to have a dominating control law* γ^* *and a corresponding dominating machine* $\{X^{\gamma^*}(n), R(X^{\gamma^*}(n)); n = 1, 2, \ldots\}$ *if*

$$\mathcal{L}(X^{\gamma^*}, \mu) \geq \mathcal{L}(X^{\gamma}, \mu), \qquad \forall \gamma \in \Gamma, \forall \mu \in \mathbb{R}$$

where Γ is the set of all control laws for that component.

When each component of the superprocess has a dominating machine an index-type solution is optimal for the following reason. In every component of the superprocess one can restrict attention, without any loss of optimality, to its dominating machine. Each dominating machine is a single-armed bandit process. Thus, the superprocess problem reduces to a MAB problem for which an optimal solution is of the index type.

The condition that each component of a superprocess has a dominating machine is quite restrictive and difficult to satisfy in most problems.

3.2 Arm-acquiring Bandits

The arm-acquiring bandit problem is a variation of the MAB problem where one permits arrival of new machines. At time t, $K(t)$ independent machines are available. The machines available at t were either available at $t = 0$ or arrived during $1, \ldots, t-1$. Denote these machines by $\{(X_i(N_i(t)), R_i(X_i(N_i(t))));$ $N_i(t) = 0, 1, 2, \ldots, t; i = 1, 2, \ldots, K(t); t = 0, 1, 2, \ldots\}$. At each time instant, the controller decides to apply a continuation control to only one of the available machines and all other machines remain frozen. Define $U(t) := (U_1(t), \ldots, U_{K(t)}(t))$. Then $U(t) \in \{e_1(K(t)), \ldots, e_{K(t)}(K(t))\}$, where $e_i(j) = (0, \ldots, 0, 1, 0, \ldots, 0)$ is a j-dimensional unit vector with 1 at the ith position. The machines available at time t are independent and evolve in the same way as in the classical MAB problem.

At time t a set $A(t)$ of new machines arrive. These machines are available for operation from $(t+1)$ on and are independent of each other and of the $K(t)$ previous machines. Let $|A(t)|$ denote the number of machines in $A(t)$. Then,

$$K(t+1) = K(t) + |A(t)|$$

It is assumed that $\{|A(t)|; t = 1, 2, \ldots\}$ is a sequence of i.i.d. random variables. Further, $|A(t)|$ is independent of $U(0), \ldots, U(t)$.

In this context, a *scheduling policy* $\gamma := (\gamma_1, \gamma_2, \ldots)$ is a decision rule such that the action $U(t)$ at any time t is a random variable taking values in $\{ e_1(K(t)), \ldots, e_{K(t)}(K(t)) \}$ and

$$U(t) = \gamma_t\big(Z_1(t), \ldots, Z_{k(t)}(t), U(0), \ldots, U(t-1)\big), \quad (6.39)$$

where
$$Z_i(t) = [X_i(0), \ldots, X_i(N_i(t))];$$

that is, $U(t)$ denotes the machine operated at time t, and this decision depends on all past states of all machines.

The arm-acquiring bandit problem is to determine a scheduling policy that maximizes

$$J^\gamma := \mathbb{E}\left[\sum_{t=0}^{\infty} \beta^t \sum_{i=1}^{K(t)} R_i\big(X_i(N_i(t)), U_i(t)\big) \bigg| Z(0) \right], \quad (6.40)$$

subject to the aforementioned constraints on the evolution of the machines and the arrival of new machines.

Nash [179] first considered the arm-acquiring bandit problem using Hamiltonian and dynamic programming and he did not obtain an index-type of solution. Whittle [250] first showed that the *Gittins index* policy is optimal for the arm-acquiring bandit. Similar results on the optimality of the Gittins index rule for arm-acquiring bandits were later obtained by [240, 116]. Here we present briefly the arguments that lead to the optimality of the Gittins index rule.

Decisions are not irrevocable due to the following: bandit processes are independent; processes that are not operated on remain frozen; future arrivals are independent of past decisions; and the arrival process is a sequence of independent identically distributed random variables. Therefore, by the arguments presented in Section 2.3, forward induction obtains an optimal scheduling policy—at each instant of time continue the machine with the highest Gittins index. The expressions for the Gittins index of each machine are the same as in Equation (6.8). If the machines are described by Markov processes then their dynamics evolve as in (6.9) and the Gittins indices are given by (6.10).

3.3 Switching Penalties

In MAB problem with switching penalties we have the same model as in the classical MAB problem with one additional feature. Every time the processor switches from one machine to another, a switching penalty (switching cost c or switching delay d) is incurred. The inclusion of switching penalties is a realistic consideration. When the processor switches between different machines a new setup may be needed; during this setup time no bandit process is continued (therefore, no reward is acquired) and this lack of continuation can be modeled by a switching cost or switching delay.

The inclusion of switching penalties drastically alters the nature of the bandit problem. An index form of solution is no longer optimal. This has been shown in [12] and is illustrated by the following example from [11].

Consider a two-armed bandit problem with switching penalties. Each arm is described by a three-state Markov chain. The transition probabilities of the both are given by $P_{X_{t+1}|X_t}(2|1) = 1$, $P_{X_{t+1}|X_t}(3|2) = 1$, $P_{X_{t+1}|X_t}(3|3) = 1$, further, both Markov chains start in state 1. The rewards of the first arm are given by $R_1(1) = 20$, $R_1(2) = 18$, $R_1(3) = 0$; and of the second arm are given by $R_2(1) = 19$, $R_2(2) = 17$, $R_1(3) = 0$. Assume the switching cost $c = 3$ and the discount factor $\beta = 0.5$. If we operate the arms according to the Gittins index policy, the order of operation is 1,2,1,2 and the corresponding rewards are $(20-3) + (19-3)\beta + (18-3)\beta^2 + (17-3)\beta^3 = 30.5$, whereas a policy that operates in order 1,1,2,2 yields a reward $(20-3) + 18\beta + (19-3)\beta^2 + 17\beta^3 = 32.125$. Thus, the Gittins index policy is not optimal.

The nature of optimal scheduling/allocation strategies for the general MAB problem with switching penalties and an infinite horizon expected discounted reward including switching penalties is not currently known. Explicit solutions of special cases of the problem have been determined in Van Oyen et al. [234, 235]. Agrawal et al. [1] determined an optimal allocation strategy for the MAB problem with switching cost and the "learning loss" or "regret" criterion. Asawa and Teneketzis [11] determined qualitative properties of optimal allocation/scheduling strategies for the general MAB problem with an infinite horizon expected discounted reward minus switching penalties performance criterion. In this chapter we only consider switching costs. The main result in [11] states the following. Suppose that at $t = 0$, it is optimal to select machine j for continuation. If at $t = 0$ no switching penalty is incurred, it is optimal to continue the operation of machine j until its Gittins index corresponding to $x_j(0)$ is achieved. If at $t = 0$ a switching cost c is incurred, it is optimal to continue the operation of machine j until a switching index

$$\nu^s_{X_j}(x_j(0)) = \max_{\tau > 0} \frac{\mathbb{E}\left[\sum_{t=0}^{\tau-1} \beta^t R_j(t) - c \Big| x_j(0)\right]}{\mathbb{E}\left[\sum_{t=0}^{\tau-1} \beta^t \Big| x_j(0)\right]} \quad (6.41)$$

corresponding to $x_j(0)$ is achieved. In general, suppose that at decision epoch $\tau_l(\omega)$ it is optimal to select machine i for continuation. If machine i was operated at $\tau_l(\omega) - 1$ then it is optimal to continue the operation of machine i until its Gittins index corresponding to $(x_i^l(\omega))$ (and given by (6.8)) is achieved. If machine i was not operated at $\tau_l(\omega) - 1$, then it is optimal to continue its operation until a switching index

$$\nu_{X_i}^s(x_i^l) = \max_{\tau > \tau_l} \frac{\mathbb{E}\left[\sum_{t=\tau_l}^{\tau-1} \beta^t R_i(X_i(N_i(\tau_l) + t - \tau_l)) - \beta^{\tau_l} c \Big| x_i^l\right]}{\mathbb{E}\left[\sum_{t=\tau_l}^{\tau} \beta^t \Big| x_i^l\right]} \quad (6.42)$$

corresponding to $x_i^l(\omega)$ is achieved. (Recall that $x_i^l := x_i(0), \ldots, x_i(N_i(\tau_l))$).

The stopping time $\tau^s(x_i^l)$ that achieves the maximum on the RHS of (6.42) is related to the stopping time $\tau(x_i^l(\omega))$ that achieves the Gittins index as follows:

$$\tau^s(x_i^l(\omega)) \geq \tau(x_i^l(\omega)) \quad (6.43)$$

almost surely for all $x_i^l(\omega)$.

The main result in [11] does not describe which machine to select for continuation at each decision epoch. Such a selection must be determined by backward induction. Conditions under which it is possible to further simplify the search for an optimal allocation policy also appear in [11].

3.4 Multiple Plays

In MABs with multiple plays we have k independent processes/machines labeled $1, 2, \ldots, k$ and one controller that has m processors available ($m < k$). At each time instant the controller can allocate each processor to exactly one process. No process can be operated by more than one processor. Each bandit process and its evolution are modeled as in the classical MAB problem. A scheduling policy $\gamma := (\gamma_1, \gamma_2, \ldots)$ is a decision rule such that the action $U(t)$ at time t is a random variable taking values in $(d_1, d_2, \ldots, d_{\binom{k}{m}})$ where each d_i is a k-dimensional row vector consisting of m ones and $(k - m)$ zeros, and the positions of the ones indicate the machines/processes to which the processors are allocated. The objective in MABs with multiple plays is to determine a scheduling policy γ that maximizes

$$J^\gamma := \mathbb{E}^\gamma\left[\sum_{t=1}^{\infty} \beta^t \sum_{i=1}^{k} R_i(X_i(N_i(t)), U_i(t)) \Big| Z(0)\right], \quad (6.44)$$

subject to the constraints describing the evolution of the machines and the allocation of the processors, where

$$Z(0) := [X_1(0), X_2(0), \ldots, X_k(0)] \qquad (6.45)$$

and

$$R_i\big(X_i(N_i(t)-1), U_i(t)\big) = \begin{cases} R_i\big(X_i(N_i(t))\big), & \text{if } U_i(t) = 1, \\ 0, & \text{otherwise.} \end{cases}$$

In general operating machines with the m highest Gittins indices is not an optimal policy for MAB with multiple plays (see [115, 198]). Anantharam et al. [7, 8] determined an optimal scheduling policy for MABs with multiple plays and the "learning loss" or "regret" criterion. Furthermore, Agrawal et al. [2] determined an optimal scheduling policy for the MAB problem with multiple plays, a switching cost, and the "learning loss" criterion. Pandelis and Teneketzis [188] determined a condition sufficient to guarantee the optimality of the policy that operates the machines with the m highest Gittins indices at each instant of time. (We call this strategy the *Gittins index rule for MABs with multiple plays* or briefly the *Gittins index rule*.) The sufficient condition of [188] can be described as follows. For each machine i, $i = 1, 2, \ldots, k$, let τ_l^i denote the successive stopping times at which the Gittins indices of machine i are achieved, and let $\nu_{X_i}\big(X_i(0), \ldots, X_i(\tau_l^i)\big)$ denote the $(l+1)^{\text{th}}$ successive Gittins index of the process i. For every realization ω of the evolution of machine i we have the corresponding realizations $\tau_l^i(\omega)$ and $\nu_{X_i}\big(X_i(0,\omega), \ldots, X_i(\tau_l^i(\omega), \omega)\big)$, $l = 1, 2, \ldots$ of machine i, $i = 1, \ldots, k$. Consider the following condition.

(C1) For any realization ω of the problem, for any machines i, j such that $i \neq j$ and positive integers p, q such that

$$\nu_{X_i}\big(X_i(0,\omega), \ldots, X_i(\tau_p^i(\omega), \omega)\big) > \nu_{X_j}\big(X_j(0,\omega), \ldots, X_j(\tau_q^j(\omega), \omega)\big)$$

we have

$$\nu_{X_i}\big(X_i(0,\omega), \ldots, X_i(\tau_p^i(\omega), \omega)\big)(1 - \beta) \\ > \nu_{X_j}\big(X_j(0,\omega), \ldots, X_j(\tau_q^j(\omega), \omega)\big)$$

The main result in [188] states that if condition (C1) is satisfied then the Gittins index rule is optimal. The essence of the result of [188] is the following. Forward induction does not, in general, lead to optimal processor allocation decisions in MABs with multiple plays because at each stage of the allocation process the optimal scheduling policy jointly decides the m machines to be processed; thus, the forward induction arguments used in the classical MAB problem (and were discussed in Section 2.2) do not hold. Consequently, the full effect of future rewards has to be taken into account in determining an optimal scheduling policy. However, if the Gittins indices of different machines are sufficiently separated, the expected reward rate maximizing portions of each bandit process starting from its current history become the dominant factors in determining an optimal scheduling policy. In such situations, an optimal scheduling strategy can be determined by forward induction, and the Gittins index rule is optimal. Condition (C1) presents an instance where there is enough separation among the Gittins indices to guarantee the optimality of the Gittins index rule.

A search problem formulated as a MAB problem with multiple plays has been considered in [221]. Conditions under which the Gittins index rule is optimal for the above problem also appear in [221].

3.5 Restless Bandits

Restless bandits (RBs) consist of k independent machines and m identical processors, $m < k$. Each machine evolves over time even when it is not being processed, and hence is not a bandit process. Specifically, the evolution of machine i, $i = 1, 2, \ldots, k$, is described by

$$X_i(t+1) = f_{i,t}\left(X_i(0), \ldots, X_i(t), U_i(t), W_i(t)\right), \quad (6.46)$$

where $U_i(t) \in \{0, 1\}$, $U_i(t) = 0$ (respectively 1) means that machine i is not processed (respectively processed) at time t, and $\{W_i(t), t = 0, 1, 2, \ldots\}$ is a sequence of primitive random variables that are independent of $X_1(0), X_2(0), \ldots, X_k(0)$ and have known statistical description; furthermore, $\{W_i(t), t = 0, 1, 2, \ldots\}$ and $\{W_j(t), t = 0, 1, 2, \ldots\}$, $i \neq j$ are independent. (The reader is invited to contrast (6.46) with (6.2)). The reward received from machine i at time t is $R_i\left(X_i(t), U_i(t)\right)$[10]. At each instant of time each processor can process exactly one machine. Each machine can be processed by at most one processor. A scheduling policy is defined in exactly the same way as in the MAB problem with multiple plays. The performance criterion is defined by (6.44) and (6.45). The objective is to determine a scheduling policy to maximize an infinite horizon expected discounted reward criterion given by (6.44).

In general, forward induction does not result in an optimal allocation strategy for this problem. To see this, consider separately the cases where $m = 1$

---

[10] In [30] it was shown that without loss of generality we can assume that $R_i\left(X_i(t), 0\right) = 0$.

and $m > 1$. For $m = 1$, the model and optimization problem are a generalization of the classical MAB problem described in Section 2.1.2. In RBs any machine that is available for continuation at t but is not chosen at t changes its state after t, so decisions made at time t are irrevocable. Consequently, for $m = 1$ forward induction does not result in an optimal scheduling policy. For $m > 1$, the model and problem are a generalization of the MAB with multiple plays described earlier in this section. Forward induction does not, in general, result in an optimal scheduling policy even for MABs with multiple plays, and therefore does not, in general, result in an optimal scheduling policy for RBs.

Nevertheless, there are cases where RBs have an optimal solution that is of the index type. We describe two such cases below.

Case 1. Consider the situation where all machines are identical and each machine is describe by a finite-state controlled Markov chain that is irreducible under any stationary Markov policy. That is, (6.46) simplifies to

$$X_i(t+1) = f_{i,t}\left(X_i(t), U_i(t), W_i(t)\right), \qquad (6.47)$$

$i = 1, 2, \ldots, k$. Assume that the performance criterion is given by the infinite horizon average reward-per-unit-time-per-machine, that is

$$r^{\hat{\gamma}}(\alpha) = \frac{1}{k}\left[\lim_{T \to \infty} \frac{1}{T} \mathbb{E}^{\hat{\gamma}}\left[\sum_{t=1}^{T}\sum_{i=1}^{k} R_i\left(X_i(t-1), U_i(t)\right)\right]\right].$$

Such a performance criterion is not significantly different from the one considered so far, as infinite horizon expected discounted reward and infinite horizon expected reward-per-unit-time are related with one another [153, 251].

For the above model and problem assume that a subsidy Q is provided at time t to each machine that is not operated at t. Let $\nu(x_i)$ be the value of Q for which the expected reward-per-unit-time resulting from processing a machine (currently in state x_i) is equal to that resulting from not processing it plus the subsidy.

DEFINITION 6.3 *The value $\nu(x_i)$ is defined to be the index of a machine in state x_i.*

The notion of subsidy defined above can be used to introduce the concept of indexability that plays an important role in determining conditions sufficient to guarantee the optimality of an index-type solution for the above model.

DEFINITION 6.4 *Machine i is indexable if the following condition is satisfied.*

(C2) *Consider a value Q of the subsidy and suppose that when the machine is in state x, it is optimal not to operate it. Then, it is also optimal not to operate the machine in state x for any value of the subsidy higher than Q.*

A restless bandit is indexable if all its arms/machines are indexable.

The above notion of indexability may appear to be trivially true for any machine, but this it is not the case. In fact, indexability is a very strict requirement (see [252]). To proceed further we define:

1. an index policy $\hat{\gamma}$ according to which the machines with the m highest indices (specified by Definition 6.3) are operated at each time instant

2. for the index policy $\hat{\gamma}$

$$r^{\hat{\gamma}}(\alpha) := \lim_{\substack{k \to \infty \\ m \to \infty \\ \alpha = \frac{m}{k}}} \frac{1}{k} \left[\lim_{T \to \infty} \frac{1}{T} \mathbb{E}^{\hat{\gamma}} \left[\sum_{t=1}^{T} \sum_{i=1}^{k} R_i \left(X_i(t-1), U_i(t) \right) \right] \right] \quad (6.48)$$

Then the following result holds. If the RB process is indexable and certain technical conditions described in [247] hold,

$$r^{\hat{\gamma}}(\alpha) = \lim_{\substack{k \to \infty \\ m \to \infty \\ \alpha = \frac{m}{k}}} \frac{1}{k} \left[\sup_{\gamma \in \Gamma} \lim_{T \to \infty} \frac{1}{T} \mathbb{E}^{\gamma} \left[\sum_{t=1}^{T} \sum_{i=1}^{k} R_i \left(X_i(t-1), U_i(t) \right) \right] \right] \quad (6.49)$$

where Γ is the set of stationary Markov policies. That is, an optimal allocation strategy for the above class of RBs is an index policy. The above conditions are sufficient but not necessary to guarantee the optimality of an index policy $\hat{\gamma}$.

As pointed out above, indexability is a strict requirement and is often hard to check. These difficulties motivated the work in [30, 90, 183, 184]. In [30] Bertsimas and Niño-Mora provided a sufficient condition for indexability of a single restless bandit. In [183] Niño-Mora investigated an allocation policy for restless bandits and showed that if a set of conditions called Partial Conservation Laws (PCLs) are satisfied and the rewards associated with each machine belong to a certain "admissible region" (see [183]) then the allocation policy investigated in [183] is optimal. The ideas of [183] have been further refined in [184]. An approach to evaluating the sub-optimality of the index proposed

Multi-armed Bandit Problems

in [183] when the PCL conditions are satisfied but the rewards are not in the "admissible region" has been presented in [90]. An application of the results of [183] to queueing networks appears in [9].

Case 2. Consider k machines and one controller that has m processors available ($m < k$). Each machine $i, i = 1, \ldots, k$, is described by a controlled random process $\{X_i(t); t = 0, 1, 2, \ldots\}$ with a countable state-space $\{0, 1, 2, \ldots\}$ such that

$$X_i(t+1) = f_{i,t}(X_i(t), U_i(t), W_i(t)) \qquad (6.50)$$

where $U_i(t) \in \{0, 1\}$ and $U_i(t) = 0$ (respectively, $U_i(t) = 1$) means that the machine is not processed (respectively, processed) at time t. For each $i, i = 1, \ldots, k$, $\{W_i(s), s = 1, 2, \ldots\}$ is a sequence of random variables that take values in $\{0, 1, \ldots, m_i\}$, are not necessarily independent and have known statistics. The sequences $\{W_i(s), s = 1, 2, \ldots\}$ and $\{W_j(s), s = 1, 2, \ldots\}$ are independent for all $i, j, i \neq j$. Furthermore, each sequence $\{W_i(s), s = 1, 2, \ldots\}$ is perfectly observed; that is, for each $\omega \in \Omega$ and i, $W_i(0, \omega)$, $W_i(1, \omega), \ldots, W_i(t-1, \omega)$ are known by the controller at time t, before the allocation decision at t is made. The functions $f_{i,t}(\cdot), i = 1, 2 \ldots, k$, are

$$f_{i,t}(X_i(t), U_i(t), W_i(t)) = \begin{cases} X_i(t) + W_i(t), & \text{if } U_i(t) = 0, \\ X_i(t) - \Lambda_i + W_i(t), & \text{if } X_i(t) \neq 0, U_i(t) = 1, \\ W_i(t), & \text{if } X_i(t) = 0, \end{cases} \qquad (6.51)$$

where Λ_i is a random variable taking values in $\{0, 1\}$ with $P[\Lambda_i = 0] = q_i > 0$ and $W_i(t)$ is a random variable that is not necessarily independent of $W_i(0), \ldots, W_i(t-1)$.

At each instant of time t a machine is either available for processing (it is "connected"), or it is not (it is "not connected"). The probability that machine i is connected at t is $p_i, i = 1, 2, \ldots, k$, for all t. The reward received at time t from a connected machine is

$$R_i(X_i(t), U_i(t)) = \begin{cases} Y_i, & \text{if } X_i(t) \neq 0 \text{ and } U_i(t) = 1, \\ 0, & \text{if } U_i(t) = 0 \text{ or } X_i(t-1) = 0, \end{cases} \qquad (6.52)$$

where Y_i is a random variable taking values in $\{0, c_i\}$ with $P(Y_i = c_i) = q_i$. The reward received at time t from a machine that is not connected is zero. The performance criterion is given by (6.44)[11]. The objective is to determine a scheduling/processor allocation policy to maximize the performance criterion.

---

[11] In [161], where the model of Case 2 is proposed, the performance criterion is given by a holding cost instead of a reward. Using the transformation in [235] we can convert this performance criterion into (6.44).

The model described above arises in single-hop mobile radio systems (see [161] and references therein). The same model has also independent interest as a specific problem in queueing theory.

Consider a machine i, $i = 1, 2, \ldots, k$, which at time t is in state $x_i \neq 0$ and is connected. The *Gittins index* $\nu_i(x_i)$ of machine i is

$$\nu_i(x_i) = q_i c_i.$$

Define the *Gittins index policy* to be the allocation policy γ_{GI} that operates at each time t the connected machines that are not in the zero state and have the m highest Gittins indices. The following condition (C3) describes an instance where the allocation policy γ_{GI} is optimal.

(C3) $c_1 q_1 > c_2 q_2 > \cdots > c_k q_k$, and in addition

$$c_i q_i \left[\frac{1-\beta}{1-(1-q_i)\beta} \right] \geq c_j q_j$$

for all $i, j, 1 \leq i < j \leq k$.

The proof of optimality of the Gittins index policy γ_{GI} under (C3) can be found in [161].

The essence of the results of [161] is the following: If we were guaranteed that the system described above operated away from the $\mathbf{0} := (0, 0, \ldots, 0)$ (k times) state then it would be optimal to allocate the m processors to the connected machines with the m highest Gittins indices. Near the state $\mathbf{0}$, processor utilization becomes a critical issue in determining an optimal allocation policy. The Gittins index policy may result in processor under-utilization; thus, it may not be optimal in some neighborhood of the state $\mathbf{0}$. Therefore, if we require optimality of the Gittins index policy for the problem under consideration, we must identify conditions to ensure that the advantage gained by always allocating the processors to the highest index machines overcompensates potential losses resulting from processor under-utilization near the $\mathbf{0}$ state. Such a condition is expressed by (C3) which requires that the indices associated with the machines should be sufficiently separated from one another. Such a separation results in a priority ordering of the machines sufficient to guarantee the optimality of the Gittins index policy.

Variations of the model of Case 2 were considered by Ehshan and Liu in [78, 79, 76, 77, 75]. In [78, 77] the authors investigate RBs with imperfect (delayed) observations and a single processor; they identify conditions sufficient to guarantee the optimality of an index policy. In [76] the authors consider identical machines and a linear symmetric holding cost criterion that can be converted into (6.44) (see footnote 11). For this model they prove the optimality of the index rule. In [75, 79] the model is similar to that of [76] but the holding cost is convex. The authors identify conditions sufficient to guarantee the optimality of an index policy.

3.6 Discussion

Several of the extensions of the MAB problem presented in this chapter are related with one another. Specifically, the arm-acquiring bandit problem can be converted into a superprocess [240] and the MAB problem with switching cost can be converted into a RB problem [91].

Furthermore, there are other stochastic dynamic scheduling problems that are equivalent to the classical MAB problem. Two such problems are the tax problem [240, 253] and certain classes of Sensor Resource Management (SRM) problems [245].

In the tax problem there are k machines, each evolving according to (6.2) and (6.3). At each time instant exactly one machine is operated; the machines that are not operated remain frozen. If machine i is not operated at t a *tax* $T_i(X_i(t))$, depending on the state $X_i(t)$ of machine i at t, is charged. The objective is to determine a scheduling/processor allocation policy γ to minimize

$$\mathbb{E}^\gamma \left[\sum_{t=0}^{\infty} \beta^t \sum_{i=1}^{k} T_i(X_i(t), U_i(t)) \middle| Z(0) \right],$$

where
$$Z(0) := [X_1(0), X_2(0), \ldots, X_k(0)],$$

$U(t) := (U_1(t), \ldots, U_k(t))$, $t = 0, 1, 2, \ldots$, is a random variable taking values in $\{e_1, \ldots, e_k\}$, and

$$T_i(X_i(t), U_i(t)) = \begin{cases} T_i(X_i(N_i(t))), & \text{if } U_i(t) = 0, \\ 0, & \text{if } U_i(t) = 1. \end{cases}$$

Even though feature (F4) of the MAB problem is not present in the tax problem, the two problems are equivalent. For the details of transforming the tax problem into a MAB problem, we refer the reader to [240]. An example of a tax problem is Klimov's problem in queueing networks [132, 133, 240, 31].

Sensor management problems that are equivalent to the MAB problem are presented and discussed in Chapter 7.

4. Example

In this section we present a simple search problem and show how it can be viewed as a classical MAB problem. We also present briefly variations of the problem which lead to different variants of the classical MAB problem. The search model that we consider is not rich enough to lead to an arm-acquiring bandit variation. We refer the reader to Chapter 7 for more realistic sensor management scenarios.

We consider a search problem in which a stationary target is hidden in one of k cells. The *a priori* probability that the target is in cell i is denoted by $p_i(0)$, $i = 1, 2, \ldots, k$. One sensor capable of operating in only one mode is available to search the target; the sensor can search only one cell at every instant of time. The sensor is imperfect in the sense that if the target is in cell i and the sensor looks at cell i, it does not necessarily find the target, i.e.,

$$P(\text{sensor finds target in cell } i \mid \text{target is in cell } j) = \delta_{ij} q_j, \qquad (6.53)$$

where δ_{ij} is the Kronecker delta function. The search is completed when the target is found. Successful completion of the search at time t gives a reward β^t, where $0 < \beta < 1$. Such a reward captures the fact that the target must be identified as quickly as possible. The objective is to determine a sequential sensor allocation policy γ that maximizes the expected reward.

We show how the problem described above can be formulated as a classical MAB problem. Associate each cell with one machine/bandit process. Let $p_i(t)$ be the posterior probability that the target is in location i at time t given all previous search locations and the event that the target has not been found. The probability $p_i(t)$ can be considered as the state of machine i at time t; let $p(t) := \bigl(p_1(t), p_2(t), \ldots, p_k(t)\bigr)$ be the state of all machines. Denote by $U(t) := \bigl(U_1(t), U_2(t), \ldots, U_k(t)\bigr)$ the sensor allocation at t. $U(t)$ is a random variable taking values in $\{e_1, e_2, \ldots, e_k\}$ (see Section 2.1). The expected reward corresponding to any sequential sensor allocation policy γ is $\mathbb{E}^\gamma \left[\beta^\tau \mathbb{1}(\text{target is found at } \tau)\right]$, where $\mathbb{1}(E)$ is the indicator function of event E and τ is the time when the search process stops. For any arbitrary but fixed sensor allocation policy γ this expected reward can be alternatively written as,

$$\mathbb{E}^\gamma \left[\beta^\tau \mathbb{1}(\text{target is found at } \tau)\right] = \sum_{t=0}^{\infty} \beta^t \, \mathrm{P}^\gamma(\tau = t)$$

$$= \sum_{t=0}^{\infty} \beta^t \left[\sum_{i=1}^{k} \mathrm{P}^\gamma(\tau = t, U(t) = e_i)\right]$$

$$= \sum_{t=0}^{\infty} \beta^t \left[\sum_{i=1}^{k} p_i(t) q_i \, \mathbf{P}^\gamma \left(U(t) = e_i \right) \right]$$

$$= \sum_{t=0}^{\infty} \beta^t \left[\sum_{i=1}^{k} R_i \big(p_i(t), u_i(t) \big) \right],$$

where

$$R_i \big(p_i(t), u_i(t) \big) = \begin{cases} p_i(t) q_i, & \text{if } u_i(t) = 1, \\ 0, & \text{if } u_i(t) = 0, \end{cases}$$

and

$$u(t) = \big(u_1(t), u_2(t), \ldots, u_k(t) \big) = \gamma_t \big(p(t) \big).$$

By a careful examination of the above search problem, we find that features (F1), (F3), and (F4) of the MAB problem are present, but feature (F2) is not. This is because if we search location i at time t and do not find the target, then the state $p(t)$ evolves as follows:

$$p_i(t+1) = \frac{p_i(t)(1-q_i)}{c}, \tag{6.54}$$

$$p_j(t+1) = \frac{p_j(t)}{c}, \qquad j \neq i, \tag{6.55}$$

where $c = 1 - p_i(t) q_i$. Thus a particular allocation at t changes the state of all machines/cells.

The above problem can be converted into a classical MAB by considering an unnormalized probability $\hat{p}_i(t)$ as the state of machine i, $i = 1, 2, \ldots, k$ at time t and an appropriately modified reward function $\widehat{R}_i \big(\hat{p}_i(t), u_i(t) \big)$. Specifically the state $\hat{p}_i(t)$ of machine i, $i = 1, 2, \ldots, k$ evolves as follows:

$$\hat{p}_i(0), = p_i(0) \qquad \forall i, \tag{6.56}$$

$$\hat{p}_i(t+1) = \begin{cases} \hat{p}_i(t), & \text{if } u_i(t) = 0, \\ \hat{p}_i(t)(1-q_i), & \text{if } u_i(t) = 1, \end{cases} \tag{6.57}$$

and the modified reward function $\widehat{R}_i \big(\hat{p}_i(t), u_i(t) \big)$ is given by

$$\widehat{R}_i \big(\hat{p}_i(t), u_i(t) \big) = \begin{cases} \hat{p}_i(t) q_i, & \text{if } u_i(t) = 1, \\ 0, & \text{if } u_i(t) = 0. \end{cases} \tag{6.58}$$

The objective is to determine a sensor scheduling policy γ to maximize

$$\sum_{t=0}^{\infty} \beta^t \sum_{i=1}^{k} \widehat{R}_i \big(\hat{p}_i(t), u_i(t) \big),$$

where $u(t) = \gamma_t(\hat{p}(t))$. By using $\hat{p}_i(t)$ as the state of machine[12] i at time t, the modified problem has features (F1)–(F4) and is thus a classical MAB.

Even though the modified problem has a different reward structure and different cell/machine dynamics from the original problem, any optimal policy for one problem is also optimal for the other (for details, see [68, Sec 14.14]). The Gittins index of every machine is always achieved at $\tau = 1$ and is given by

$$\nu_{X_i}(\hat{p}_i(t)) = \hat{p}_i(t) q_i$$

(see [68]).

We call the above described model the *basic model*. Changes in the number and types of sensors or the target dynamics result in problems that can be transformed into one of the variants of the MAB problem, and are described below.

When the sensor can operate in one of M modes and everything else is the same as in the basic model, the resulting problem can be formulated as a superprocess where the state of cell/machine i is the unnormalized probability $\hat{p}_i(t)$.

When there is a setup cost or setup delay for switching the sensor from one location to another and everything else is the same as in the basic model, the resulting problem can be formulated as a MAB problem with switching penalties.

If there are m sensors $(1 < m < k)$ and everything else is the same as in the basic model, the resulting problem can be formulated as a MAB problem with multiple plays.

Finally, if the target is moving, there are m sensors $(1 < m < k)$, each with one mode, and everything else is the same as in the basic model, the resulting problem can be formulated as a restless bandit.

By considering various combinations of changes in target motion and the number and types of available sensors, we can obtain generalizations of the bandit problems. Some such generalizations appear in Chapter 7.

---

[12] $\hat{p}(t)$ is an information state for the modified problem (see Chapter 2).

5. Chapter Summary

We have presented the formulation of the classical MAB problem, and discussed its key features and the ideas that lead to the Gittins index solution. We also presented different variants of the classical MAB problem, explained their key differences from Gittins's original formulation, and highlighted conditions under which a Gittins-type index rule is optimal. We emphasized the qualitative features of MAB problem and gave an intuitive explanation of the key results. The technical details are available in the literature we cite in this chapter. We illustrated how different variants of a single search problem can be transformed to the classical MAB problem and its variants. More realistic sensor management problems and their relation to the MAB problem are discussed in Chapter 7.

Chapter 7

APPLICATION OF MULTI-ARMED BANDITS TO SENSOR MANAGEMENT

Robert B. Washburn

Parietal Systems, Inc., North Andover, MA, USA

1. Motivating Application and Overview

1.1 Introduction

Chapter 2 indicated how sensor management (SM) problems can be systematically formulated as stochastic control problems and solved by stochastic dynamic programming. However, although stochastic dynamic programming provides a systematic approach to solving SM problems, its practical application requires finding efficient algorithmic solutions to avoid the complexity of standard solutions of the dynamic programming equation (i.e., to break the curse of dimensionality). Consequently, in order to achieve necessary efficiency solutions of SM problems are often limited to suboptimal, myopic algorithms [175]. Thus, it is important to find classes of SM problems for which computationally efficient optimal or near-optimal solutions exist and which can improve the performance of myopic solutions.

In this chapter we study the application of the multi-armed bandit (MAB) theory discussed in Chapter 6 to SM problems. Under modeling assumptions to be specified later in this chapter, several SM problems of interest can be formulated as MAB problems that have optimal solutions that are priority index rules, as first proved by Gittins [89]. The significance of the MAB formulation for SM applications is that such priority index rules can be computed much more efficiently than the standard solution of a dynamic programming equation. In particular, the priority index rule has computational complexity that is linear in the number of targets instead of being exponential in the number of targets. This reduction in computational complexity is significant for real-time processing. Even when the assumptions required for strict formulation as an MAB problem are not satisfied, MAB solutions can form the basis for good approximations to more complicated SM problems (e.g., MAB can be used as the base policy of the rollout approximation discussed in Chapter 5). Furthermore, MAB research has investigated several extensions of the original Gittins work, which might prove fruitful as tools for developing optimal and near-optimal solutions of other, more realistic SM problems. Some of these extensions of the classical MAB were discussed in Chapter 6; the purpose of this chapter is to discuss how the MAB methodology applies to SM problems.

1.2 SM Example of Multi-armed Bandit

The MAB model has proved useful for modeling various decision problems such as machine scheduling, search problems, and drug and product testing problems [88]. One can also use it to model certain SM problems. For example, consider the search problem of detecting targets in a number N of locations or cells, which each contain either a target or noise alone. The sensor can observe only one cell at a time, and targets have a higher probability of giving a detection than noise does. The objective is to schedule the sequence of cells to look at in order to detect as many targets as quickly as possible. In this example, each cell corresponds to an arm of the MAB and $p_i(t)$ is the conditional probability that cell i contains a target based on observations up to time t. The incremental reward from cell i at time t is $R(p_i(t))$ where the function $R(p_i)$ is larger when the uncertainty is small. For example, $R(p_i) = p_i \log p_i$. The total reward to be maximized is the expected value of

$$\sum_{t=1}^{\infty} \beta^{t-1} \sum_{i=1}^{N} R(p_i(t))$$

where β is a discount factor. As we saw in the example of Chapter 6 (Section 4), MAB theory shows how to compute an optimal solution to this problem in terms of a priority index rule. That is, one can compute a function $m(p_i)$ such that the optimal choice of target i for the sensor to look at next is given by the value of i which maximizes $m(p_i(t))$. Thus, the optimal control is described in terms of the individual functions $m(p_i)$ instead of a much more complex function of the $p_1, ..., p_N$ jointly. This gain in efficiency for computing an optimal solution is what makes MAB methodology a useful tool for developing more efficient, near-optimal solutions of the SM problem.

1.3 Organization and Notation of This Chapter

In this chapter we discuss how the results of MAB research, as reviewed in Chapter 6, can be used to develop efficient, near-optimal solutions for SM problems such as occur in detecting, tracking, and classifying multiple targets with agile, controllable sensors. First in Section 2 we discuss the assumptions required to make the SM problem equivalent to a standard MAB problem for which optimal priority index rule solutions can be computed, and we review existing applications of this approach to SM problems. Next we discuss the correspondence between the general formulation of the SM problem and the various extensions of the MAB reviewed in Chapter 6. In theory these extensions might lead to efficient optimal or near-optimal solutions of more complex SM problems than the problems treated using standard MAB methodology. However, most of these MAB extensions have not yet been applied to SM problems, and we hope this book can provide a foundation and motivation for researchers interested in pursuing this approach. To illustrate the approach, Section 3 presents an example application of MAB index rules to obtain a non-myopic SM policy for a simplified tracking problem. Section 4 summarizes the current state-of-the-art in MAB approaches to SM problems and suggests directions for future work.

The notation of this chapter follows that of Chapter 6. Thus, $X(t), X_i(t)$ denote random variables which are the information states of targets at time t while x, x_i, ξ denote particular values (realizations) of the information state. Similarly, $\mathbf{U}(t), U_i(t), M(t)$ denote the random control (or action) taken at time t, and $\mathbf{e}_i, \mu$ denote particular control values. In addition, the notations $R, R_i, \tilde{R}$ denote reward functions, β denotes the discount factor, and V, V_i denote the optimal value function in the dynamic programming equations.

2. Application to Sensor Management

Having surveyed the current methodology available for solving multi-armed bandit (MAB) problems in Chapter 6, in this section we discuss the correspondence between SM problems and the MAB problems defined previously. In those cases that SM problems are equivalent to corresponding MAB problems the MAB methodology may offer new approaches to solving these SM problems. Even in cases where the MAB methodology does not yet provide effective solutions, it may offer good suboptimal approximations for SM problems.

2.1 Application of the Classical MAB

The general class of SM problems can be described as follows: allocate (possibly) multiple sensors to track multiple targets[1] so as to optimize a performance criterion that measures the accuracy of target tracking over a certain (possibly infinite) time horizon. The simplest SM problems is one where there is one sensor and multiple targets. Let k denote the number of targets and for each $i = 1, \ldots, k$, let $X_i(t)$ denote the information state[2] of target i at time t. The control action $\mathbf{U}(t) = [U_1(t), \ldots, U_k(t)]$ at time t is a random vector taking values in $\{\mathbf{e}_1, \ldots, \mathbf{e}_k\}$. The information state of the system evolves according to[3]

$$P(X_i(t+1) = x'_i | X_i(t) = x_i, \mathbf{U}(t)) = \begin{cases} p_i(x'_i | x_i), & \text{if } U_i(t) = 1 \\ q_i(x'_i | x_i), & \text{if } U_i(t) = 0 \end{cases} \quad (7.1)$$

The control action $\mathbf{U}(t)$ depends only on $X_i(0), \ldots X_i(t)$ for $i = 1, \ldots, k$. We assume that states $X_i(t)$ evolve independently of each other given the control sequence up to time t.

The performance criterion is given by

$$\mathbb{E}\left[\sum_{t=0}^{\infty} \beta^t \sum_{i=1}^{k} R_i\left(X_i(t), U_i(t)\right)\right] \quad (7.2)$$

where $R_i(X_i(t), U_i(t))$ denotes reward obtained by target i at time t when its state is $X_i(t)$ and an action $U_i(t)$ is taken. We allow the reward function $R_i(X_i(t), U_i(t))$ to be a general function of the state $X_i(t)$ and control action $U_i(t)$. For example, if the state is the error covariance, we can choose the reward function to be a monotonically decreasing function in $X_i(t)$ so that the smallest tracking errors have greatest rewards. A sensor allocation policy is defined in the same way as a scheduling policy in Chapter 6, Section 2.1. The objective is to determine a sensor allocation policy to maximize the performance criterion given by (7.2).

In order to apply the *classical* MAB results of Chapter 6, Section 2.1.2 we need to assume that unobserved states are frozen, namely that the transition probability $q_i(x'_i | x_i)$ in (7.1) is the trivial one for which $q_i(x'_i | x_i) = 1$

---

[1] Note that in the model "target" is an abstraction: it might correspond to individual objects such as vehicles being tracked, but it might also correspond to local regions of space being searched for the presence or absence of objects of interest.
[2] See Chapter 2 and Appendix, Section 2.
[3] The transition probability p accounts for changes in the information state of target i given that the target is observed (e.g., improvement in classification probabilities) while q accounts for changes in the information state of targets j which are not observed (e.g., increase in the tracking error covariance).

if $x'_i = x_i$ and is 0 otherwise. We also need to assume that the reward $R_i\left(X_i(t), U_i(t)\right)$ only accrues for observed states, namely when $U_i\left(t\right) = 1$ (i.e., i is the target being observed). In SM problems the assumption that unobserved states are frozen is only valid in some cases such as classification and some search problems problems (e.g., the examples of Chapter 2, section 1 and Chapter 6, Section 4). In those cases because the underlying physical state (e.g., the classification type of the target or the location of a stationary target) doesn't change with time, the information state (i.e., the conditional probability distribution of target class or target location) only changes when an observation is made. Contrast this situation with tracking problems where the information state (e.g., the tracking error covariance) is usually not frozen (e.g., prediction error increases the track error covariance even when the target is not observed). Nevertheless, even in SM problems where the frozen state assumption is not strictly valid, it may be approximately valid. This is particularly true if the information state changes very slowly when the target is not observed, as noted in [148], such as in the application of scheduling short duration MTI observations of an electronically scanned radar. Section 3 gives an example in which the classical MAB is a good approximation for problems that don't satisfy the frozen state assumption exactly.

As for the reward function, it is possible to transform (7.2) to a form consistent with the classical MAB formulation. This was first pointed out by Varaiya et al. in [240] in connection with the so-called tax problem, a variation of the classical MAB. In the tax problem the reward is accrued for all arms except the one being processed, but it is simple to extend the argument of [240] to the case in which reward is accrued for all arms. Gittins [88] refers to this MAB variation as an "ongoing bandit process" and describes the general transformation required for the reward function. For the SM problem with reward $R\left(X_i\left(t\right)\right)$ for each target i, the transformation of the reward function $R_i\left(x_i\right)$ is given by

$$\tilde{R}_i\left(x_i\right) = \sum_{\xi} R_i\left(\xi\right) p_i\left(\xi | x_i\right) - R_i\left(x_i\right), \qquad (7.3)$$

as shown in [245]. Under the transformed reward function $\tilde{R}_i\left(x_i\right)$ only the observed target will accrue a reward.

Using the transformed reward function $\tilde{R}_i$, one can solve for an index function $m_i\left(x_i\right)$ such that the control

$$U\left(t\right) = \arg\max_i m\left(X_i\left(t\right)\right)$$

is optimal. The index function can be computed by dynamic programming: solve

$$V_i(x_i, M) = \max\left\{M, \tilde{R}(x_i) + \beta \sum_{\xi} V_i(\xi, M) p(\xi|x_i)\right\},$$

for each i and then compute the index functions from

$$m_i(x_i) = \min\{M : V_i(x_i, M) = M\}.$$

Here M denotes a retirement reward, and the approach was first developed by Whittle [249] (see also Volume 2 of [25]). The dynamic program can be solved using general techniques as described in [25]. In Section 3 below we will solve these equations analytically for a tracking example.

An alternative dynamic programming approach is given by Katehakis and Veinott in [127] using a restart state formulation. For finite state problems (i.e., x_i takes values in a finite set), the stopping time method of Varaiya et al. (see [240] and [153]) is very efficient (i.e., the computational complexity is cubic in the number of values of x_i) and computations with hundreds of states are feasible[4]. These computational methods are described in more detail in Chapter 6, Section 2.1.2 which discusses the classical MAB.

The classical MAB has been applied to a few SM problems in the work of Krishnamurthy et al. [148], [147], [149], [155] and Washburn, Schneider, et al. [245], [181], [205]. The approach of [148], [147], [149], [155] is to approximate the original SM problem using a partially observed finite state Markov chain, and to compute index rules for this partially observed problem. The advantage of this approach is that it accounts directly for hidden tracking states[5], but it has the disadvantage of prohibitive computational requirements for more than a few discrete hidden states. On the other hand, the approach of [245], [181], [205] starts with a completely observed discrete state Markov chain approximation of the information state process[6], computes index rules for the discrete state problem, and uses the resulting MAB index rule with roll-out methods [26] to obtain a solution of the original SM problem. The advantage of this approach is that it allows computing index rules for more complicated models (e.g., tracking problems with many discrete states), but

---

[4]Indeed, since the index function can be computed offline, the remaining online computations are practical for real-time operation.
[5]For example, these might be the unknown target classification types in a SM problem seeking to classify k different targets.
[6]Here the information state is the totality of all statistical information about the target's state (e.g., about its kinematic state, identity, etc.) that is associated with the target's track (e.g., conditional mean, covariance, probability vector, etc.).

it has the disadvantage of requiring one to already have a discrete Markov chain approximation of the continuous state Markov chain corresponding to the original partially observed problem.

2.2 Single Sensor with Multiple Modes

The basic model of (7.1) and (7.2) can be generalized to include more complicated and realistic features. One generalization is to include, in addition to $\mathbf{U}(t)$, a control action $M(t)$ to account for the choice of other parameters (e.g., waveform parameters such as pulse repetition frequency (PRF), dwell time, and other types of sensor modes) that might affect the sensor's observations. In this case, the system evolves as follows

$$P\left(X_i(t+1) = x_i' | X_i(t) = x_i, M(t) = \mu, \mathbf{U}(t)\right) = \\ = \begin{cases} p_i(x_i'|x_i, \mu), & \text{if } U_i(t) = 1 \\ q_i(x_i'|x_i, \mu), & \text{if } U_i(t) = 0 \end{cases} . \quad (7.4)$$

The reward is given by $R_i\left(X_i(t), U_i(t), M(t)\right)$ and the optimization problem is to maximize

$$\mathbb{E} \sum_{t=0}^{\infty} \beta^t \sum_{i=1}^{k} R_i\left(X_i(t), U_i(t), M(t)\right)$$

Examples illustrating this generalization include:

1 $M(t)$ may determine the duration of observation of each target. In this case $M(t)$ takes values in $(0, T]$ and the system evolves according to the following

$$P\left(X_i\left(t + m(t)\right) = x_i' | X_i(t) = x_i, M(t) = \mu, \mathbf{U}(t)\right) = \\ = \begin{cases} p_i(x_i'|x_i, \mu), & \text{if } U_i(t) = 1 \\ q_i(x_i'|x_i, \mu), & \text{if } U_i(t) = 0 \end{cases}$$

2 A radar that is able to switch between fast detection/tracking modes and slow classification modes. In this case, $M(t)$ is a random variable taking values in $\{F, S\}$, where F is tracking mode and S is classification mode. The system evolution equations are given by (7.4).

These SM problems corresponds to a *superprocess MAB* (combined with a restless bandit if $q_i\left(x_i'|x_i, \mu\right)$ is not the trivial probability transition) as described in Section 3. Unfortunately, the superprocess does not admit an index type solution in general. Nevertheless, in the non-restless case[7], there is a sufficient condition that the superprocess have an optimal index policy if each component of the superprocess has a dominating machine. For the SM problem, this condition is interpreted to mean that for each target i, there is an optimal policy for controlling the sensor mode $M(t)$ (assuming the sensor is observing only i) which dominates any other mode control policy for any retirement reward μ (i.e., with respect to the reward criterion defined in (6.37) of Chapter 6 Section 3). A problem for future research is to determine conditions of the sensor management problem under which such mode domination is possible.

2.3 Detecting New Targets

A further generalization of the SM problem is to allow the number of targets $k(t)$ present at time t to vary with time, so as to model the discovery of new targets and disappearance of old ones. This is an important feature to include in SM problems that allocate sensor resources to simultaneously search and track (or search and identify) targets. When the search tasks generate new target detections, then new tracking tasks are added to the problem so that the total number $k(t)$ of tasks increases. Similarly, when tracks are dropped, $k(t)$ decreases.

This model corresponds to an *arm-acquiring bandit* described in Chapter 6, Section 3 and the solution is comparable to the classical MAB. In particular, for the non-restless case, the Gittins index policy is optimal and is computed the same as for the classical MAB, taking the highest Gittins index over $i = 1, ..., k(t)$ as shown in [250], [116], [247].

2.4 Sensor Switching Delays

The SM models discussed above assume that it is possible for the sensor to switch instantaneously from observing one target to observing another one, or from one sensor mode to another. This assumption may be a good approximation for electronically steered sensors such as radars, but may not be a good approximation for mechanically pointed sensors (e.g., video cameras that require time to pan, tilt, and zoom and require time to stabilize the camera after moving before effective video processing can resume). In this case, it is desirable

---

[7] Note that the SM problem of classifying the type of multiple targets with a single, multi-mode sensor corresponds to a superprocess which is not a restless bandit. Thus, non-restless superprocesses have significant application in SM.

to model switching times or switching costs as part of the sensor management problem and these correspond to *bandits with switching penalties* as described in Chapter 6 Section 3. The inclusion of switching penalties complicates the solution of the MAB (even in the non-restless case); the Gittins index policy is no longer optimal and the structure of the optimal policy is unknown except in special cases, which unfortunately don't correspond to sensor management problems. Nevertheless, Asawa and Teneketzis [11] showed that the optimal policy does have the property that the time at which to switch the sensor from one target to another (but not the choice of which target to switch to) can be determined by the Gittins index. As noted in [11], this result is based on a limited lookahead policy which might be extended further to obtain tighter conditions for optimality. The results also suggest evaluating limited lookahead policies as good suboptimal approximations to the optimal allocation.

2.5 Multiple Sensors

Another generalization of the basic model is to consider m sensors ($m > 1$). We assume that each sensor can look at only one of the k targets at a time, and each target can be observed by at most one sensor. The control action $\mathbf{U}(t) = [U_1(t), \ldots, U_k(t)]$ is a vector taking values in $\{d_1, \ldots, d_{\binom{k}{m}}\}$, where $d_i, i = 1, \ldots, \binom{k}{m}$, is a k dimensional vector consisting of m 1s and $(k-m)$ 0s, and the positions of 1s indicate the targets to which the sensors are allocated. The transition law, the performance criterion and the objective are same as in the basic model (7.1), (7.2). This model can be classified as an *MAB problem with multiple plays* as described in Chapter 6 Section 3. As noted previously, the Gittins index is generally not optimal for bandits with multiple plays and an infinite horizon expected discounted reward performance criterion. Nevertheless, Pandelis and Teneketzis [188] determined a sufficient condition (in the non-restless case) for the optimality of observing at each time the m targets with the highest Gittins index. The condition requires that Gittins indices of different targets are either equal or separated by a minimum distance defined by the discount factor of the performance criterion (see Section 3 for the precise condition). It remains to determine whether this condition is satisfied for significant sensor management problems. However, it is worth noting that the index function computed for a simple single sensor tracking example in Section 3 does satisfy this separation condition for sufficiently small discount factors. In that example, the index function is piecewise constant and the constant values decrease approximately exponentially by a factor that decreases as the discount rate decreases. See also [221], which considers a multi-sensor search problem and compares it to a corresponding multi-armed bandit problem and the results of [188] on multiple plays.

2.6 Application of Restless Bandits

As noted above, the assumption that unobserved states are frozen is true only for a few specialized SM problems. For SM problems involving kinematic tracking or joint kinematic tracking and identification, the transitions $q_i(x'_i|x_i)$ in (7.1) are not trivial and the unobserved states are not frozen. As a consequence, the basic SM problem with one sensor and multiple targets is equivalent to a *restless bandit* problem with one processor.

As for the classical MAB, it is necessary to transform the SM target reward $R_i(X_i(t), U_i(t))$ to a reward function which is zero for all unobserved targets (corresponding to inactive processes of the restless bandit). Although the transformation of [240] and [88] depends on the states being frozen for $U_i(t) = 0$, Bertsimas and Niño-Mora [30] were able to prove that such a transformation is possible in the general restless case also.

Restless bandit methods have not been applied to SM problems with the exception of current work by Kuklinski [151]. Part of the difficulty of applying restless bandit methods may lie in the difficulty of checking Whittle's indexability conditions described in Section 3 and the likelihood that these conditions are not satisfied for typical sensor management problems. Nevertheless, the preliminary report [151] describes an application of the linear programming method of [30] to sensor management for multiple target tracking and identification. In this application a finite state Markov chain is used to model the information state process, and the reward function accounts for sensor energy usage (e.g., for micro-sensor networks) as well as tracking and identification performance. The MAB method is used to compute an index policy value for possible tasks in a sensor network, and activate the tasks with the M largest values. However, it is unclear whether this sensor management problem satisfies conditions (e.g., the partial conservation laws of [183]) to guarantee that the index policy is optimal in this sensor management problem.

3. Example Application

To illustrate the application of MAB methodology to SM, this section defines a problem of maintaining track on multiple targets and shows how to find an approximate solution using MAB methods. It gives numerical results and compares the MAB method (numerically and qualitatively) to more conventional approximation techniques. The example is based on [245], which discusses further extension of the result to multidimensional target states.

3.1 MAB Formulation of SM Tracking Problem

Suppose that there are N targets being tracked, each of which has an estimated position and position error covariance. The sensor can look at one target at each time instant, and the objective is to schedule the sensor looks to keep as many targets, as often as possible within a specified root mean square error. To model this problem as an MAB, take as the information state the inverse error variance, denoted by $X_i(t)$. The information state dynamics are modeled by a Markov chain with a countable state space as follows. Let $\mathbf{U}(t) = [U_1(t), ..., U_n(t)]$ denote the control (i.e., $\mathbf{U}(t) = \mathbf{e}_i$ if the sensor looks at target i). If $\mathbf{U}(t) = \mathbf{e}_i$, then

$$X_i(t+1) = \begin{cases} X_i(t) + r^{-1} & \text{with probability } p \\ X_i(t) & \text{with probability } 1-p \end{cases},$$

where r denotes the variance of a single position measurement and p is the probability of detection for one measurement. If $\mathbf{U}(t) \neq \mathbf{e}_i$, then

$$X_i(t+1) = X_i(t).$$

Note that this second transition is necessary to satisfy the frozen state assumption for the standard MAB. It would be more realistic to assume that when $\mathbf{U}(t) \neq \mathbf{e}_i$, $X_i(t+1)$ decreases due to increased prediction error, namely

$$X_i(t+1) = \frac{X_i(t) q_i^2}{q_i^2 + X_i(t)}.$$

Unfortunately, this assumption makes the MAB restless and precludes a simple index rule solution. Consequently, we will assume $q_i = 0$ to design the index rule used for SM control. However, we will evaluate the control policy in cases of $q_i > 0$ in order to analyze the significance of ignoring restlessness.

The optimization goal is to maintain the root mean square error below a specified value, denoted $\delta > 0$. The individual reward function for target i is defined as

$$R_i(x_i) = \begin{cases} 0 & \text{if } x_i > \delta^{-2} \\ v_i & \text{if } x_i \leq \delta^{-2} \end{cases}.$$

Thus, there is a reward of v_i for keeping the error of target i below δ in one time period, but no greater reward for maintaining smaller errors, and there is no reward if the error exceeds δ. Varying v_i for different targets i allows assigning higher rewards to tracking some targets compared to others. We want to choose the stochastic control $\mathbf{U}(t)$ to maximize the total discounted reward

$$\mathbb{E}\left[\sum_{t=0}^{\infty} \alpha^t \sum_{i=1}^{N} R_i(X_i(t)) \middle| X_1(0), ..., X_N(0) \right].$$

So far, this model satisfies all the requirements for the standard MAB except the requirement that only target i for which $\mathbf{U}(t) = \mathbf{e}_i$ accrues reward in time period t and all other targets accrue no reward in that time period. As noted above, this deficiency can be remedied by transforming the reward function $R_i(x_i)$, which can be accomplished using (7.3) from the Section 2. In this particular case, the transformed reward function is

$$\tilde{R}_i(x_i) = R_i\left(x_i + r^{-1}\right)p - R_i(x_i)p,$$

or simply,

$$\tilde{R}_i(x_i) = \begin{cases} p & \text{if } \delta^{-2} - r^{-1} \leq x_i < \delta^{-2} \\ 0 & \text{if } x_i < \delta^{-2} - r^{-1} \text{ or } x_i \geq \delta^{-2}. \end{cases}$$

3.2   Index Rule Solution of MAB

For simplicity assume that $v_i = 1$. To find the index function corresponding to $\tilde{R}(x_i)$, we solve the dynamic programming equation

$$V_i(x_i, M) = \max\left\{M, \tilde{R}(x_i) + \beta \mathbb{E}\left[V_i(f(x_i, w_i), M)\right]\right\}$$

where $f(x_i, w_i)$ denotes the transition function for the information state. The expectation $\mathbb{E}[V_i(f(x_i, w_i), M)]$ is

$$\mathbb{E}\left[V_i(f(x_i, w_i), M)\right] = pV_i\left(x_i + r^{-1}, M\right) + (1-p)V_i(x_i, M),$$

so we have

$$V_i(x_i, M) = \max\left\{M, \begin{pmatrix} pI_{[\delta^{-2}-r^{-1},\delta^{-2})}(x_i) + \beta pV_i\left(x_i + r^{-1}, M\right) \\ +\beta(1-p)V_i(x_i, M) \end{pmatrix}\right\},$$

where $I_A(x)$ is the indicator function of the set A. The dynamic programming equation equation can be solved exactly by taking the limit of successive approximations of the value function (i.e., by using value iteration [25]). The resulting solution is

$$V_i(z_i, M) = \sum_{k > \frac{-a}{\Delta}}^{0} \Phi_k(M) I_{[a+k\Delta, a+(k+1)\Delta)}(z_i) + M I_{[a+\Delta, \infty)}(z_i).$$

where

$$\Phi_k(M) = \begin{cases} \kappa^{|k|+1}\beta^{-1} + \kappa^{|k|+1}M & \text{if } M \leq \frac{\kappa^{|k|+1}\beta^{-1}}{1-\kappa^{|k|+1}} \\ M & \text{if } M > \frac{\kappa^{|k|+1}\beta^{-1}}{1-\kappa^{|k|+1}} \end{cases}$$

with $a = \delta^{-2} - r^{-1}$, $\Delta = r^{-1}$, $\kappa = \frac{\beta p}{1-\beta(1-p)} < 1$.

Given V_i, the index function $m(x_i)$ is the solution of

$$m(x_i) = \min\{M : V_i(x_i, M) = M\}.$$

For $x_i \in [a + k\Delta, a + (k+1)\Delta)$, this is the smallest M such that $\Phi_k(M) = M$. Thus,

$$m(x_i) = \frac{\kappa^{|k|+1}\beta^{-1}}{1 - \kappa^{|k|+1}}$$

for $x_i \in [a + k\Delta, a + (k+1)\Delta)$. Similarly, for $x_i \in [a + \Delta, \infty)$, we have

$$m(x_i) = 0.$$

In terms of the parameters δ and r, we can express the index function as

$$m(x_i) = \begin{cases} \frac{\kappa^{|k|+1}\beta^{-1}}{1-\kappa^{|k|+1}} & \text{if } \delta^{-2} + (k-1)r^{-1} \leq x_i < \delta^{-2} + kr^{-1}, k \leq 0 \\ 0 & \text{if } x_i \geq \delta^{-2} \end{cases}.$$
(7.5)

The index rule control $\mathbf{U}(t) = \mathbf{e}_i$ where i maximizes $m(X_i(t))$. Equivalently, i maximizes $X_i(t)$ such that $X_i(t) < \delta^{-2}$. The latter expression is obvious, but at least confirms the general theorem. It's clear that the optimal strategy would not select $\mathbf{U}(t) = \mathbf{e}_i$ for any $X_i(t) \geq \delta^{-2}$ since these states already satisfy the goal and the state will not decrease. On the other hand, selecting the targets i with highest $X_i(t) < \delta^{-2}$ to look at first will maximize the expected discounted reward because then more $X_i(t)$ are likely to enter the goal set sooner than for any other strategy.

Nevertheless, if different rewards v_i are assigned to different targets, then it is easy to see that the corresponding index functions are $m_i(x_i)$ given by

$$m_i(x_i) = v_i \cdot m(x_i)$$

where $m(x_i)$ is given as in (7.5) above, and the control is not equivalent to selecting $\mathbf{U}(t) = \mathbf{e}_i$ to maximize $X_i(t) < \delta^{-2}$ if v_i are not all equal. Similarly, we can assign different error goals δ_i, different detection probabilities p_i, and different measurement variances r_i as well as different rewards v_i to each target. The corresponding index functions are $m_i(x_i)$ given by

$$m_i(z_i) = \begin{cases} v_i \frac{\kappa_i^{|k|+1}\beta^{-1}}{1-\kappa_i^{|k|+1}} & \text{if } \delta_i^{-2} + (k-1)r_i^{-1} \leq z_i < \delta_i^{-2} + kr_i^{-1}, k \leq 0 \\ 0 & \text{if } z_i \geq \delta_i^{-2} \end{cases}.$$

where

$$\kappa_i = \frac{\beta p_i}{1 - \beta(1 - p_i)}.$$

3.3 Numerical Results and Comparison to Other Solutions

To evaluate the index rule solution we compare its performance to three other SM control policies for a simple simulated tracking problem. The problem is to track N Brownian motion targets with a sensor that can look at only one target at a time. The objective is to track one target with high accuracy and the other targets with lower accuracy. The sensor measures position with a random error and a constant probability of missed detection. The SM control policies evaluated are an index rule, a myopic control based on predicted tracking error, a steady state control, and a "raster scan" that looks at each target, one after the other, without regard to the information state.

3.3.1 Tracking Model.
We can consider a simple problem of tracking N targets given position measurements. Each target satisfies a Brownian motion model

$$x_i(t+1) = x_i(t) + \sqrt{q} w_i(t),$$

for $i = 1, ..., N$ and $0 \leq t \leq T$ where $w_i(t)$ is a 0 mean, variance 1 Gaussian random variable. The parameter q is the process noise power and measures the maneuverability of target i. The sensor can look at one target at a time. We assume that there is a probability p_i of detecting target i if the sensor looks at target i. The measurement of target i has the model

$$y_i(t) = U_i(t) z_i(t) x_i(t) + \sqrt{r} n_i(t),$$

where $U_i(t) = 1$ if $\mathbf{U}(t) = \mathbf{e}_i$ and it selects which target to look at, $z_i(t)$ is a $0, 1$ random variable with

$$\mathbf{P}(z_i(t) = 1) = p_i,$$

and $n_i(t)$ is a 0 mean, variance 1 Gaussian random variable.

We will use the tracking error variance $\sigma_i^2(t)$ as an information state. The model for the evolution of the information state is simply

$$\sigma_i^2(t+\tau) = \left(\left(\sigma_i^2(t) + q \right)^{-1} + U_i(t) z_i(t) r^{-1} \right)^{-1}.$$

The initial value

$$\sigma_i^2(0) = \Sigma_i$$

corresponds to the size of the initial cue or the total search area for target i.

3.3.2 SM Measures of Performance.

Each target has a desired average mean square error (MSE) tracking accuracy δ_i^2. The basic question is how many targets can be tracked with the desired average MSE using a particular method of SM. Our basic measure of performance for a scenario is the distribution of the statistic

$$\chi_i = \frac{1}{T} \sum_{t=0}^{T} \frac{\sigma_i(t)^2}{\delta_i^2}.$$

In particular, we are interested in the case where there is one target ($i = 1$) which is a high value target with a small δ_1^2 and there are $N - 1$ lower value targets with larger $\delta_i^2 = \delta_2^2$ for $i \geq 2$. Thus, we will evaluate performance in terms of χ_1 (the time-averaged error variance for high-valued target 1) versus

$$\chi' = \frac{1}{N-1} \sum_{i=2}^{N} \chi_i,$$

the averaged error variance over the other $N - 1$ low value targets. It may happen that the SM can only track $m < N - 1$ of the less accurate targets. Thus, we compute

$$\rho = \frac{1}{N} \sum_{i=1}^{N} 1(\chi_i \leq 1),$$

the fraction of targets that meet the specified MSE.

3.3.3 Four Candidate SM Control Policies.

Index Rule SM. The index function for target i is given by

$$m_i(\sigma_i^2) = \begin{cases} v_i \frac{\kappa^{|k|+1}\beta-1}{1-\kappa^{|k|+1}} & \text{if } \delta_i^{-2} + (k-1)r^{-1} \leq \sigma_i^{-2} < \delta_i^{-2} + kr^{-1}, k \leq 0 \\ 0 & \text{if } \sigma_i^{-2} \geq \delta_i^{-2} \end{cases}$$

where

$$\kappa = \frac{\beta p_i}{1 - \beta(1 - p_i)}$$

and β is a discount factor. The corresponding index rule says to look at target i next based on

$$\arg\max_i \left\{ m_i\left(\sigma_i(t)^2\right) \right\}$$

where $\sigma_i(t)^2$ is the current estimated MSE error of track i. Note that the index rule doesn't explicitly specify q. This is because the index rule assumes that $q \approx 0$. This parameter enters the strategy indirectly in terms of the tracker update of $\sigma_i(t)^2$ after each measurement.

Myopic SM. The myopic SM selects which target $\mathbf{U}(t)$ to look at next based on *minimizing* the expected marginal cost. That is $\mathbf{U}(t) = \mathbf{e}_i$ such that $\mu_i\left(\sigma_i(t)^2\right)$ is minimized. The function $\mu_i\left(\sigma_i^2\right)$ is given by

$$\mu_i\left(\sigma_i^2\right) = p_i v_i \left(\left(\sigma_i^{-2} + r^{-1}\right)^{-1} - \left(\sigma_i^2 + q\right)\right).$$

Note that δ_i doesn't appear explicitly in the control strategy.

Steady State SM. Suppose that we look at target i every τ time units. Let $\sigma_i^2(t)$ denote the tracking error variance at time t immediately after a measurement update. Then

$$\sigma_i^2(t + \tau) = \left(\left(\sigma_i^2(t) + \tau q\right)^{-1} + r^{-1}\right)^{-1}.$$

The steady state error after update is

$$\sigma_i^2(\infty) = -\frac{1}{2}\tau q + \frac{1}{2}\sqrt{\tau^2 q^2 + 4r\tau q}.$$

However, the time-averaged MSE is

$$\delta^2 = \frac{\int_0^\tau \left(\sigma_i^2(\infty) + tq\right) dt}{\tau} = \sigma_i^2(\infty) + \frac{1}{2}\tau q = \frac{1}{2}\sqrt{\tau^2 q^2 + 4r\tau q}.$$

In terms of the dimensionless parameter

$$a = \frac{\tau q}{r},$$

we achieve the average mean square error δ^2 given by

$$\delta^2 = \frac{1}{2}r\sqrt{a^2 + 4a}.$$

If we wish to achieve accuracy δ, then we need to satisfy

$$\frac{\tau q}{r} = a = -2 + 2\sqrt{\left(1 + \left(\frac{\delta^2}{r}\right)^2\right)}.$$

Suppose that we wish to maintain at average steady state MSE δ_i^2 for each target i. For each target i, we require the revisit time

$$\tau_i = \frac{ra_i}{q} = \frac{r}{q}\left[-2 + 2\sqrt{\left(1 + \left(\frac{\delta_i^2}{r}\right)^2\right)}\right].$$

Thus, we need to revisit at the rate $\frac{1}{\tau_i}$ times per measurement. To account for missed detections, the revisit rate needs to be $\frac{1}{p_i \tau_i}$ in order to have an effective detection rate of $\frac{1}{\tau_i}$. Thus, to meet the steady state MSE δ_i^2 for each target i requires

$$\sum_{i=1}^{N} \frac{1}{p_i \tau_i} \leq 1.$$

The steady state analysis suggests a simple control policy defined as follows. For each target keep track of the time t_i since the last detection. At any time t use the control $\mathbf{U}(t) = \mathbf{e}_i$ such that v_i is the largest value with $t_i \geq \tau_i$ and where τ_i is defined as above. Note that

$$t_i(t+1) = \begin{cases} 0 & \text{if } \mathbf{U}(t) = \mathbf{e}_i \text{ and target } i \text{ is detected} \\ t_i(t) + 1 & \text{if } \mathbf{U}(t) \neq \mathbf{e}_i \text{ or target } i \text{ isn't detected} \end{cases}.$$

Raster Scan SM. Finally, define a raster scan SM which simply cycles through the N targets, one at a time. Thus, $\mathbf{U}(t)$ is the open loop control

$$\mathbf{U}(t) = \mathbf{e}_i \text{ where } i = t \bmod N + 1$$

for $t \geq 0$.

3.3.4 Evaluation.

We apply each SM control policy to the stochastic control problem

$$\sigma_i^2(t+\tau) = \left(\left(\sigma_i^2(t) + q \right)^{-1} + U_i(t) z_i(t) r^{-1} \right)^{-1}.$$

where $N, \delta_1, \delta_2, p_0, \Sigma_0$ are given and $n_1 = 1$, $n_2 = N - 1$, $\sigma_i(0)^2 = \Sigma_0$, $p_i = p_0$, $r = 1$. Select q to satisfy

$$q = p_0 \left(\left[-2 + 2\sqrt{(1+\delta_1^4)} \right]^{-1} + (N-1) \left[-2 + 2\sqrt{1+\delta_2^4} \right]^{-1} \right)^{-1} \quad (7.6)$$

so that $N - 1$ targets can be tracked with average MSE δ_2^2 and 1 target can be tracked with MSE δ_1^2.

The evaluation simulates the tracking and sensor control over T time periods, $0 \leq t < T$, and computes as performance measures the MSE of the high value target,

$$\chi_1 = \frac{1}{T} \sum_{t=1}^{T} \frac{\sigma_1(t)^2}{\delta_1^2},$$

the average MSE of the low value targets,

$$\chi' = \frac{1}{N-1} \sum_{i=2}^{N} \chi_i,$$

and the average probability that a target meets its MSE goal at any time,

$$\rho = \frac{1}{N} \sum_{i=1}^{N} \frac{1}{T} \sum_{t=1}^{T} 1\left(\frac{\sigma_i(t)^2}{\delta_i^2} \leq 1\right)$$

where

$$\chi_i = \frac{1}{T} \sum_{t=1}^{T} \frac{\sigma_i(t)^2}{\delta_i^2}$$

for each i.

In the following series of plots we show χ_1, χ', ρ versus the process noise q and the probability of detection p_0 for each SM. For these simulations we defined $\delta_1 = \frac{1}{2}$, $\delta_2 = 2$, $\Sigma_0 = 20$, $N = 10$. For all of the algorithms that use reward in their formulation (index rule, myopic, and steady state) the reward of the high value target is $r_1 = 100$ and the reward for all other targets is $r_2 = 1$. For the index rule SM, the discount factor is taken to be $\alpha = 0.9$. The time period is $T = 200$ for all evaluations. Figures 7.1, 7.2, 7.3 show χ_1, χ', ρ (MSE of high value target, average MSE of low value targets, and average probability of a target meeting its MSE goal) as a function of the probability p_0 of detection for each target. Figures 7.4, 7.5, 7.6 show χ_1, χ', ρ as a function of the process noise q for each target.

In Figures 7.1, 7.2, 7.3, $q = 0.04$. For $p_0 \leq 0.75$, the process noise is too high to maintain the MSE goals ($\delta_1 = \frac{1}{2}$ for the high value target and $\delta_2 = 2$ for 9 other targets). For $p_0 > 0.75$, the MSE goal is satisfied. Figure 7.1 shows χ_1 versus p_0. The three algorithms (index rule, myopic, and steady state) which use reward, have the same performance. The raster scan algorithm that looks at each target in turn, does much worse. In Figure 7.2 we see that the raster scan algorithm does the best on the low value targets (because it doesn't divert resources to the high value target). The index rule is next best with $\chi' \sim 1$ until $p_0 \sim 0.75$ (corresponding to the critical detection probability where resources are insufficient to meet the goal with a steady state solution). The myopic algorithm gives approximately a constant χ' which is worse than the index rule until $p_0 < 0.65$. The steady state algorithm performs much worse in this case. Figure 7.3 shows how often the algorithms meet the desired MSE goals for the high and low value targets. The raster scan algorithm has a uniformly high probability of meeting the goal. However, note that it never meets the goal for the high value target. The index rule is slightly better for $p_0 > 0.85$ and decreases rapidly for p_0 below the critical 0.75. The myopic algorithm is uniformly between 0.4 and 0.5, and the steady state algorithm performs worst, less than 0.25.

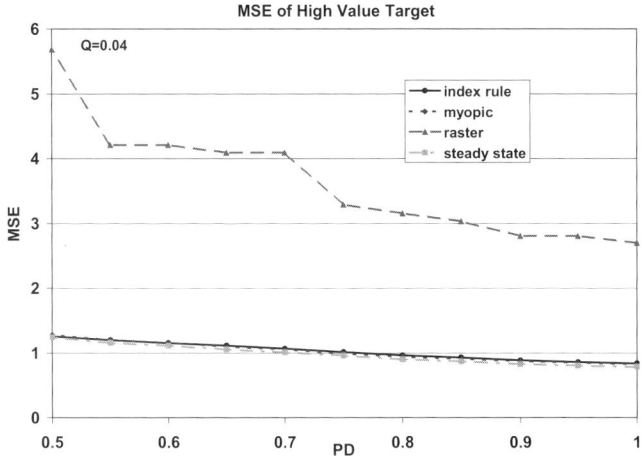

Figure 7.1. Mean square error χ_1 of high value target using four SM algorithms, plotted as a function of probability of detection p_0 with nominal process noise $q = 0.04$.

In Figures 7.4, 7.5, 7.6, $p_0 = 0.8$. For $q > 0.045$, the process noise is too high to maintain the MSE goals. Figure 7.4 shows χ_1 versus q. The three algorithms (index rule, myopic, and steady state) which use reward, have the same performance. The raster scan algorithm that looks at each target in turn, does much worse. In Figure 7.5 we see that the raster scan algorithm does the best on the low value targets. The index rule, closely followed by the myopic algorithm, is next best until it suddenly jumps up around $q \sim 0.065$ (above the critical point). Note that the index rule starts dropping low value targets (i.e., schedules no looks for them) to try to maintain the goal on the high value target and as many low value targets as possible. The myopic algorithm, by comparison, increases continuously as q increases, an it tries to allocate resources uniformly to all targets which have the same reward. The steady state algorithm performs much worse. Figure 7.6 shows how often the algorithms meet the desired MSE goals for the high and low value targets. The raster scan algorithm has a uniformly high probability of meeting the goal. However, note again that it doesn't meet the goal for the high value target unless $q = 0$. The index rule is best until $q \sim 0.035$ and decreases rapidly for q above the critical 0.065, for $q > 0.85$ it can no longer meet the MSE goal for the high value target (or any of the low value targets). The myopic algorithm decreases more

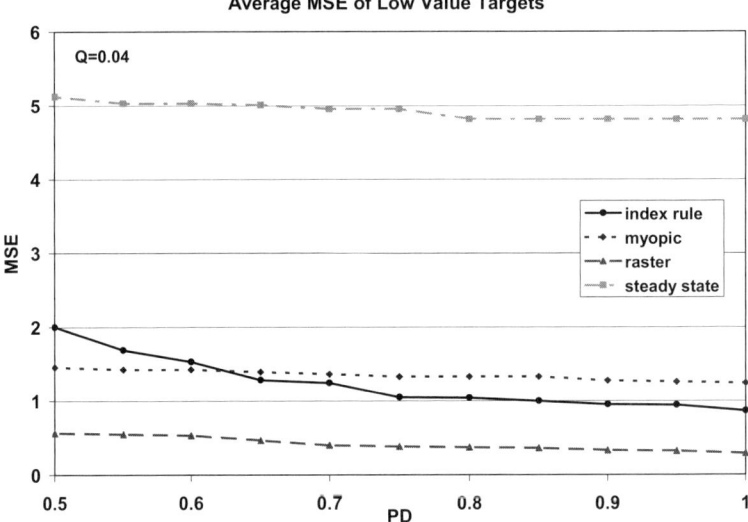

Figure 7.2. Average mean square error χ' of low value targets (targets $i > 1$) using four SM algorithms, plotted as a function of the probability of detection p_0 for nominal process noise $q = 0.04$.

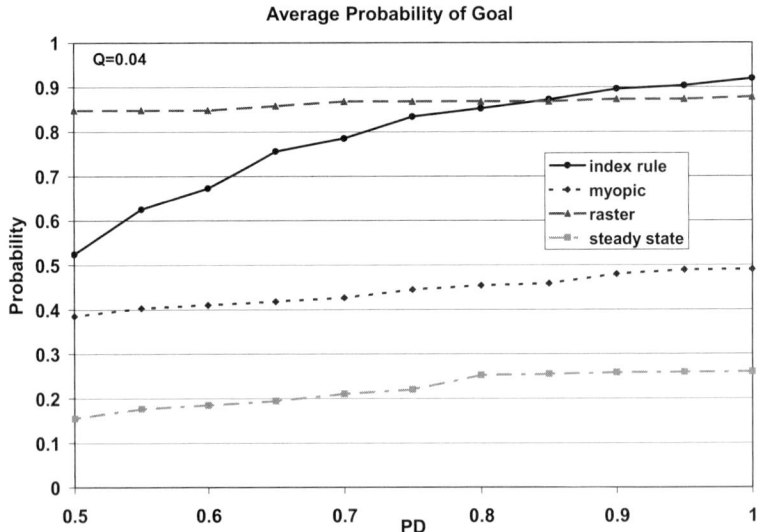

Figure 7.3. Average probability ρ of goal for all targets using four SM algorithms, plotted as a function of probability of detection p_0 for nominal process noise variance $q = 0.04$.

rapidly at first and then flattens out for q above about 0.08. The steady state algorithm performs worst except for q above about 0.08 where it does better than the index rule and myopic algorithms.

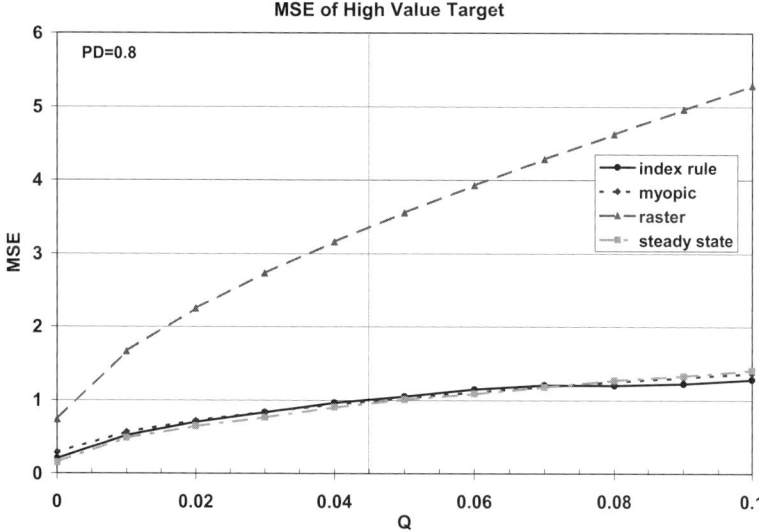

Figure 7.4. Mean square error χ_1 of high value target (target $i = 1$) using four SM algorithms, plotted as a function of process noise variance q with nominal probability of detection $p_0 = 0.8$.

4. Summary and Discussion

This chapter has illustrated the applicability and limitations of MAB methods for solving sensor management problems. The great value of the MAB approach is the possibility of computing priority index rule control strategies which are closed-loop feedback stochastic controls (and naturally non-myopic) and which can be computed very efficiently online. Although the classical MAB makes assumptions that restrict the corresponding SM problem to problems of static parameter estimation or classification (i.e., "restful" bandits), as the example to tracking in Section 3 illustrates, the classical MAB may be a very good suboptimal solution even when some of its assumptions are violated. Furthermore, it is possible to use the MAB solution as the basis for improved solutions using approximate dynamic programming, such as the rollout method described in Chapter 5.

In addition, as discussed in Chapter 6 MAB research has developed several generalizations of the original Gittins work, which correspond to more realistic

174 FOUNDATIONS AND APPLICATIONS OF SENSOR MANAGEMENT

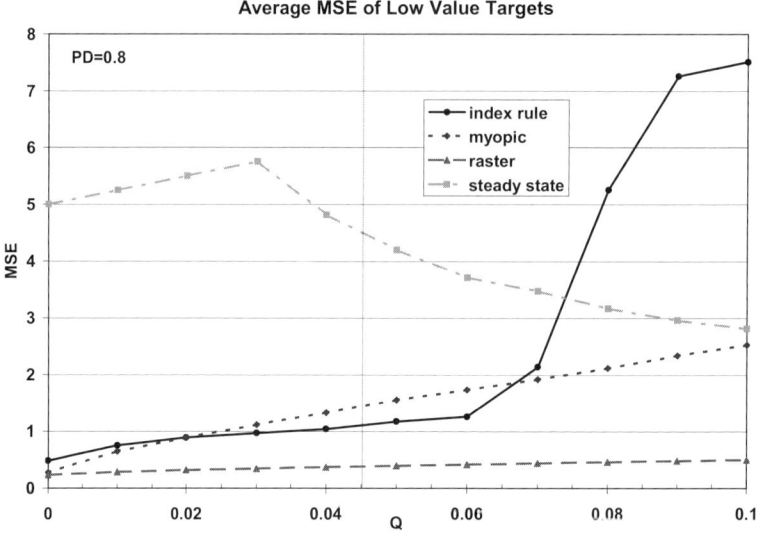

Figure 7.5. Average mean square error χ' of low value targets (targets $i > 1$) using four SM algorithms, plotted as a function of process noise variance q for nominal probability of detection $p_0 = 0.8$.

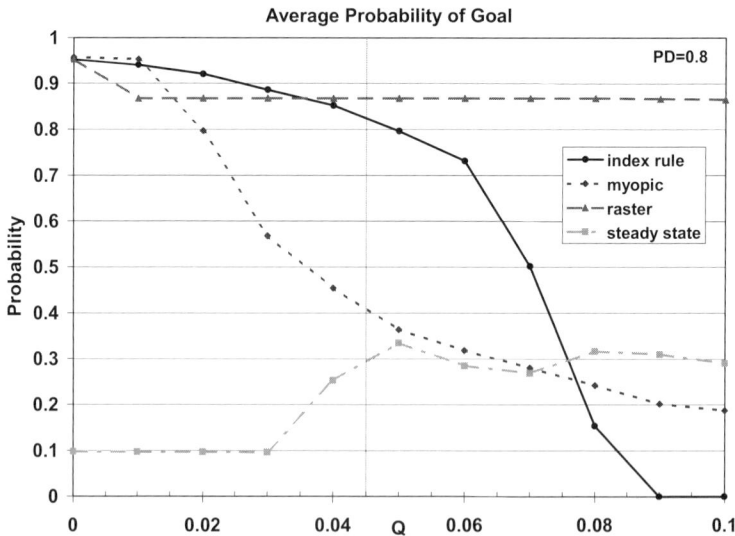

Figure 7.6. Average probability ρ of goal for all targets using four SM algorithms, plotted as a function of process noise variance q for nominal probability of detection $p_0 = 0.8$.

models of SM problems, such as multiple sensor modes (superprocesses), detecting new targets (arm-acquiring bandits), sensor switching delays (switching costs), multiple sensors (multiple plays), and prediction uncertainty (restless bandits).

So far these generalizations of MAB theory have not been applied widely to SM problems, and to be sure, these generalizations are themselves limited and require further work. For example, the different generalizations extend the classical MAB assumptions in different ways and there isn't one generalization that addresses all of these extensions simultaneously. In addition, the extended MAB problems often don't have an optimal index solution in general, and it is not yet known under what particular conditions they do have an optimal index solution, or whether there are good suboptimal solutions that are almost as efficient as index solutions. Nevertheless, the utility of the MAB model for many different applications continues to drive MAB research forward. We hope that the presentation of this chapter and the companion Chapter 6 will help SM researchers and practitioners to pick up and follow ongoing MAB research, and to use the MAB tools for developing real-time, near-optimal solutions of realistic SM problems.

Chapter 8

ACTIVE LEARNING AND SAMPLING

Rui Castro

University of Wisconsin, Madison, WI USA

Robert Nowak

University of Wisconsin, Madison, WI, USA

1. Introduction

Consider the problem of estimating a signal from noisy samples. The conventional approach (e.g., Shannon-Nyquist sampling) is to sample at many locations in a non-adaptive and more-or-less uniform manner. For example, in a digital camera we collect samples at locations on a square lattice (each sample being a pixel). Under certain scenarios though there is an extra flexibility, and an alternative and more versatile approach is possible: choose the sample locations 'on-the-fly', depending on the information collected up to that time. This is what we call *adaptive sampling*, as opposed to the conventional approach, referred to here as *passive sampling*. Although intuitively very appealing, adaptive sampling is seldom used because of the difficulty of design and analysis of such feedback techniques, especially in complex settings.

The topic of adaptive sampling, or *active learning* as it is sometimes called, has attracted significant attention from various research communities, in particular in the fields of computer science and statistics. A large body of work exists proposing algorithmic ideas and methods [63, 163, 84, 226, 37], but unfortunately there are few performance guarantees for many of those methods. Further most of those results take place in very special or restricted scenarios (*e.g.*, absence of noise or uncertainty, yielding perfect decisions). Under the adaptive sampling framework there are a few interesting theoretical results, some of which are presented here, namely the pioneering work of [43] regarding the estimation of step functions, that was later rediscovered in [98] using different algorithmic ideas and tools. Building on some of those ideas, the work in [136, 135, 92, 51] provides performance guarantees for function estimation under noisy conditions, for several function classes that are particularly relevant to signal processing and analysis.

In this chapter we provide an introduction to adaptive sampling techniques for signal estimation, both in parametric and non-parametric settings. The focus is on signals that are binary step-like functions over one or two dimensions. Even though we do not formulate the problem in terms of a Markov Decision Processes (MDP), it is worthwhile to put this adaptive sampling problem in the context of the previous chapters of this book. In the stochastic control terminology of Chapter 2, here the state will be the unknown location of the jump of a step function; the actions will be possible sampling locations, and the control policies for taking successive actions (samples) will be restricted to one-step-lookahead controls. Thus the state is static and the action space is uncountable.

We begin this chapter with some concrete applications that motivate active learning problems. Section 2 tackles a simple one-dimensional problem that sheds some light on the potential of the active learning framework. In Section 3 we consider higher dimensional function estimation problems that are more relevant in practical scenarios. Finally in Section 4 some final thoughts and open questions are presented.

1.1 Some Motivation Examples

When trying to learn a new concept, for example, asking someone to describe a scene or a painting, one usually asks questions in a sequential way, just as playing the twenty questions game. One can start by asking if the scene is outdoors. If the answer is affirmative one can then ask if there is sky depicted in the scene or if it is overcast or clear. Note that a key feature of this scheme is the feedback between the learner and the world. On the other hand, most imaging techniques pursue a completely different approach: all the 'questions' are asked in bulk. A digital scanner will give you the value of every pixel in the scene, regardless of the image, essentially by scanning the entire painting at the finest achievable resolution. If scanning time is a critical asset then this approach can be extremely inefficient. If it is known that the scene in question has some nice properties one can accelerate the scanning process by selectively choosing were to scan. To be more concrete consider an airborne laser scanning sensor. This kind of sensor is able to measure range (distance between sensor and observed object) and maybe some properties of the object, *e.g.*, the type of terrain: vegetation, sand, or rock. Suppose we want to use such a setup to construct a topographic map of a field, possibly an hostile environment, *e.g.*, a battlefield. For a variety of considerations, *e.g.*, safety concerns or fuel costs, we would like to limit the time of flight. Figure 8.1 illustrates the scenario. In this case the field is reasonably "flat", except for a ledge. Clearly, to estimate the topography of the "flat" regions low resolution scanning would suffice, but to accurately locate the ledge area higher resolution scanning is needed. If we are pursuing a non-adaptive sampling approach then we need to scan the entire field at the highest resolution, otherwise the ledge area might be inaccurately estimated. On the other hand, if an adaptive sampling technique is used then we can adapt our resolution to the field (based on our observations), thereby focusing the sampling procedure on the ledge area.

There are also other problems that share common characteristics with adaptive sampling, and that are in a sense dual problems. When the computational power is a critical asset (or the computation time is very expensive) one wants to focus the bulk of computation on tasks that are somehow more important (more rewarding). This is the case with most imaging systems (*e.g.*, satellite imaging): although we may face very few restrictions on the total amount of data collected, the subsequent processing of vast amounts of data can be excessively time consuming, therefore carefully choosing "where to look" during the processing phase can lead to significant computational savings. For example, in [50] some of the ideas/techniques of this chapter are applied to the construction of fast algorithms for image noise removal.

2. A Simple One-dimensional Problem

In this section we consider the adaptive sampling problem, essentially motivated by the laser scanning scenario. To gain some insight about the power of adaptive sampling we start with a simple, perhaps "toyish" problem. Consider estimating a step function from noisy samples. This problem boils down to locating the step location. Adaptively sampling aims to find this location with a minimal number of strategically placed samples.

Formally, we define the step function class

$$\mathcal{F} = \{f : [0,1] \to \mathbb{R} | f(x) = \mathbf{1}_{[0,\theta)}(x)\},$$

where $\theta \in [0,1]$. This is a parametric class, where each function is characterized by the transition point θ. Given an unknown function $f_\theta \in \mathcal{F}$ our goal is to estimate θ from n point samples. We will work under the following assumptions.

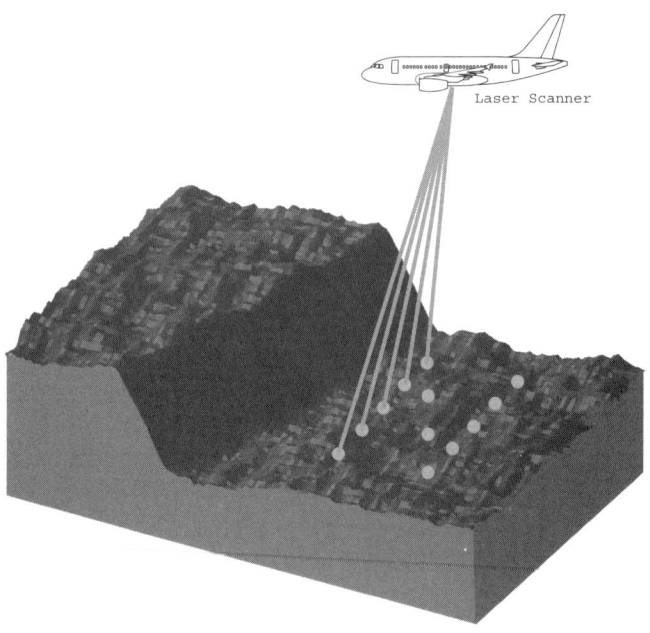

Figure 8.1. Airborne range sensor surveying a terrain.

A1.1 - The observations $\{Y_i\}_{i=1}^n$ are point samples of the *unknown* function f_θ, taken at sample locations $\{X_i\}_{i=1}^n$. These are corrupted by noise, that is with probability p we observe a one instead of a zero, and vice-versa. Formally

$$Y_i = \begin{cases} f_\theta(X_i) & \text{, with probability } 1-p \\ 1 - f_\theta(X_i) & \text{, with probability } p \end{cases} = f(X_i) \oplus U_i,$$

where $f_\theta \in \mathcal{F}$, $\oplus$ represents a sumation *modulo* 2 and $U_i \in \{0,1\}$ are Bernoulli random variables, with parameter $0 \leq p < 1/2$, independent and identically distributed (i.i.d.), and independent of $\{X_j\}_{j=1}^n$.

A2.1 - Non-Adaptive Sampling: The sample locations X_i are independent of $\{Y_j\}_{j \neq i}$.

A2.2 - Adaptive Sampling: The sampling location X_i depends only on $\{X_j, Y_j\}_{j<i}$. To be more specific let μ_i be a density defined as

$$\mu_i(X_1, \ldots, X_{i-1}, Y_1, \ldots, Y_{i-1}).$$

Finally let X_i be a sample taking according to this density. μ_i is called the *sampling strategy*, and completely defines our sampling schedule.

Under the non-adaptive sampling scenario (A2.1) the sample locations do not depend in any way on our observations, therefore the collection of sample points $\{X_i\}_{i=1}^n$ can be chosen before any observations are collected. On the other hand, the adaptive sampling scenario (A2.2) allows for the ith sample location to be chosen using all the information collected up to that point (the previous $i-1$ samples). In either case, our goal is to construct an estimate $\widehat{\theta}_n$ that is "close" to θ, using n samples, where close means that $\sup_{\theta \in [0,1]} |\widehat{\theta}_n - \theta|$ is small.

Consider first the case when there is no noise, that is, $p = 0$. Under the non-adaptive scenario (A2.1) the best we can hope to do is

$$\sup_{\theta \in [0,1]} |\widehat{\theta}_n - \theta| \leq \frac{1}{2(n+1)}.$$

This is achieved distributing the sample locations on a uniform grid over the interval $[0, 1]$,

$$\{X_i\}_{i=1}^n = \left\{ \frac{1}{n+1}, \frac{2}{n+1}, \ldots, \frac{n}{n+1} \right\}. \tag{8.1}$$

Any sample arrangement is going to induce a partition of the interval into $n+1$ intervals (unless there are overlapping samples), therefore we can only decide if the true parameter θ is inside one of these intervals. The performance is limited by the maximum size of these intervals (*i.e.*, the bias of any estimate is limited by the length of these intervals). Clearly the proposed sampling strategy is optimal for passive samples, since any other arrangement of the sample locations (even a randomized one) will lead to a possible degradation in performance.

Now suppose we are working under the adaptive scenario (A2.2). In this situation one can focus on θ much more effectively, using binary bisection: start by taking the first sample at $X_1 = 1/2$. Since there is no noise our observation is simply $Y_1 = f(X_1)$. If $Y_1 = 0$ then we know that $\theta \in [0, 1/2]$ and if $Y_1 = 1$ then $\theta \in (1/2, 1]$. We choose X_2 accordingly: If $Y_1 = 0$ then take $X_2 = 1/4$ and if $Y_1 = 1$ take $X_2 = 3/4$. We proceed according to this technique, always bisecting the set of possibilities. It is easy to see that

$$\sup_{\theta \in [0,1]} |\widehat{\theta}_n - \theta| \leq 2^{-(n+1)}.$$

This is clearly the best one can hope for with this measurement scheme: each measurement provides one bit of information, and with n bits we can encode the value of θ only up to the above accuracy.

If there is noise (*i.e.*, $p > 0$) the techniques one would use to estimate θ have to be modified appropriately. If we are working in the non-adaptive setup (A2.1) there is no reason to change the sampling scheme. We already know that our performance is going to be limited by $1/(2(n+1))$, because of our sampling strategy. To perform the estimation we can use the Maximum Likelihood Estimator (MLE). Define

$$S_n(\theta) = \sum_{i:X_i<\theta} Y_i + \sum_{i:X_i\geq\theta} 1 - Y_i.$$

The MLE estimator of θ is given by

$$\widehat{\theta}_n \equiv \arg\min_{\theta\in[0,1)}\left\{(1-p)^{n-S_n(\theta)} p^{S_n(\theta)}\right\}$$
$$= \arg\max_{\theta\in[0,1)} S_n(\theta).$$

Clearly this optimization has more than one solution, since the value of the likelihood is the same for all $\theta \in (X_i, X_{i+1}]$. For our purposes one can reduce the search to the midpoints of these intervals (i.e., $\widehat{\theta} \in \left\{\frac{1}{2(n+1)}, \frac{3}{2(n+1)}, \ldots, \frac{2n-1}{2(n+1)}\right\}$). It can be shown that this estimator performs optimally, in the sense that

$$\sup_{\theta\in[0,1]} E[|\widehat{\theta}_n - \theta|] \leq C(p)\frac{1}{n+1},$$

where $C(p)$ is an increasing function of p. The derivation of the above result is not at all trivial, and can be accomplished for example using the oracle bounds presented in [134]. Note that the expected error of this estimator behaves like $1/n$, the same behavior one has when there is no noise present, therefore the maximum likelihood estimator is optimal in this case (up to a constant factor).

If we are working under the adaptive framework, dealing with noise makes things significantly more complicated, in part because our decisions about the sampling depend on all the observations made in the past, which are noisy and therefore unreliable. Nevertheless there is a probabilistic bisection method, proposed in [110], that is suitable for this purpose. The key idea stems from Bayesian estimation. Suppose that we have a prior probability density function $\pi_0(x)$ on the unknown parameter θ, namely that θ is uniformly distributed over the interval $[0, 1]$ (that is $\pi_0(x) = 1$ for all $x \in [0, 1]$). To make the exposition clear assume a particular situation, namely that $\theta = 1/4$. Like before, we start by taking a measurement at $X_1 = 1/2$. With probability $1 - p$ we correctly observe a zero, and with probability p we incorrectly observe a one. Suppose a zero was observed. Given these facts we can compute the posterior density simply by applying Bayes rule. In this case we would get that

$$\pi_1(x|X_1, Y_1) = \begin{cases} 2(1-p) & \text{, if } x \leq 1/2, \\ 2p & \text{, if } x > 1/2, \end{cases}.$$

Initialization: Define the prior probability density function as $\pi_0 : [0,1] \to \mathbb{R}$, $\pi_0(x) = 1$ for all $x \in [0,1]$.

1. **Sample Selection after i samples were collected:** Define X_{i+1} to be the median of the posterior π_i. That is $X_{i+1} \in [0,1]$ satisfies
$$\int_0^{X_{i+1}} \pi_i(x) \mathrm{d}x = 1/2.$$

2. **Noisy Observation:** Observe $Y_{i+1} = f(X_{i+1}) \oplus U_{i+1}$.

3. **Update posterior:** Update the posterior function. This is simply the application of Bayes rule. If $Y_{i+1} = 0$ then
$$\pi_{i+1}(x) = \begin{cases} 2(1-p)\pi_i(x) & \text{if } x \leq X_{i+1} \\ 2p\pi_i(x) & \text{if } x > X_{i+1} \end{cases}.$$
If $Y_{i+1} = 1$ then
$$\pi_{i+1}(x) = \begin{cases} 2p\pi_i(x) & \text{if } x \leq X_{i+1} \\ 2(1-p)\pi_i(x) & \text{if } x > X_{i+1} \end{cases}.$$

4. **Final estimate:** Repeat steps 1,2 and 3 until n samples are collected. The estimate $\hat{\theta}_n$ is defined as the median of the final posterior distribution, that is, $\hat{\theta}_n$ is such that
$$\int_0^{\hat{\theta}_n} \pi_n(x) \mathrm{d}x = 1/2.$$

Figure 8.2. The probabilistic bisection algorithm.

The next step is to choose the sample location X_2. We choose X_2 so that it *bisects* the posterior distribution, that is, we take X_2 such that $\mathrm{P}_{\theta \sim \pi_1(\cdot)}(\theta > X_2 | X_1, Y_1) = \mathrm{P}_{\theta \sim \pi_1(\cdot)}(\theta < X_2 | X_1, Y_1)$. In other words X_2 is just the median of the posterior distribution. If our model is correct, the probability of the event $\{\theta < X_2\}$ is identical to the probability of the event $\{\theta > X_2\}$, and therefore sampling Y_2 at X_2 is most informative. We continue iterating this procedure until we have collected n samples. The estimate $\hat{\theta}_n$ is defined as the median of the final posterior distribution. Figure 8.3 illustrates the procedure and the algorithmic details are described in Figure 8.2. Note that if $p = 0$ then probabilistic bisection is simply the binary bisection described above.

The above algorithm seems to work extremely well in practice, but it is hard to analyze and there are few theoretical guarantees for it, especially pertaining error rates of convergence. In [43] a similar algorithm was proposed. Albeit its operation is slightly more complicated, it is easier to analyze. That algorithm (which we denote by BZ) uses essentially the same ideas, but enforces a parametric structure for the posterior. Also, in the application of the Bayes rule we use α instead of p, where $0 < p < \alpha$. The algorithm is detailed in Figure 8.4.

Pertaining to the BZ algorithm we have the following remarkable result.

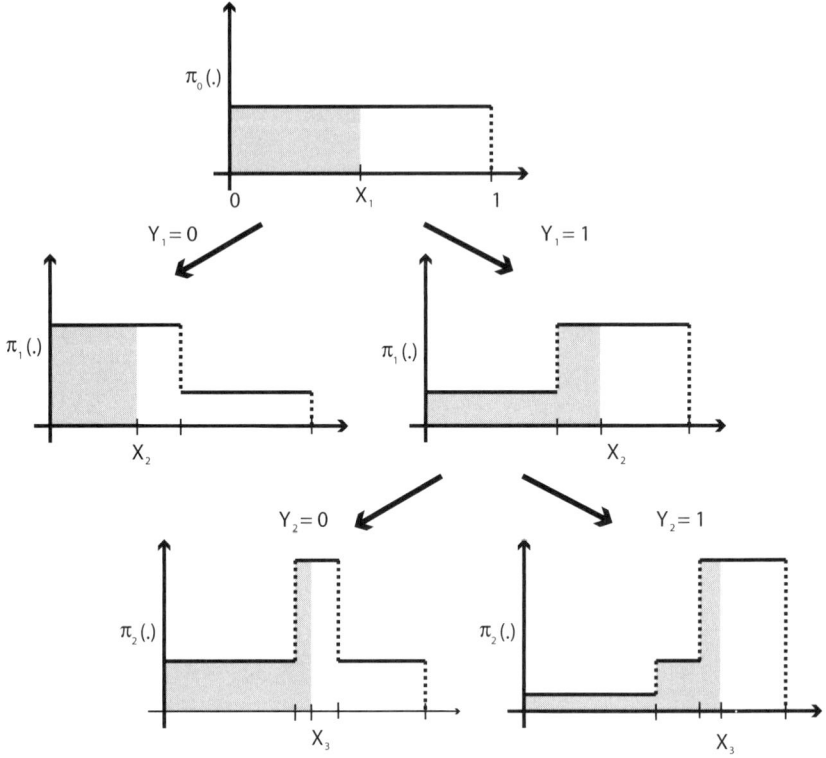

Figure 8.3. Illustration of the probabilistic bisection strategy. The shaded areas correspond to 1/2 of the probability mass of the posterior densities.

THEOREM 8.1 *Under the assumptions (A1.1) and (A2.2) the Burnashev-Zigangirov algorithm (Figure 8.2) satisfies*

$$\sup_{\theta \in [0,1]} P(|\widehat{\theta}_n - \theta| > \Delta) \leq \frac{1-\Delta}{\Delta} \left(\frac{1-p}{2(1-\alpha)} + \frac{p}{2\alpha} \right)^n.$$

Taking $\Delta^{-1} = \left(\frac{1-p}{2(1-\alpha)} + \frac{p}{2\alpha} \right)^{-n/2}$ *yields a bound on the expected error*

$$\sup_{\theta \in [0,1]} \mathbb{E}[|\widehat{\theta}_n - \theta|] \leq 2 \left(\frac{1-p}{2(1-\alpha)} + \frac{p}{2\alpha} \right)^{-n/2}.$$

Finally, taking $\alpha = \sqrt{p}/(\sqrt{p} + \sqrt{q})$ *minimizes the right hand side of these bounds, yielding*

$$\sup_{\theta \in [0,1]} \mathbb{E}[|\widehat{\theta}_n - \theta|] \leq 2 \left(\frac{1}{2} + \sqrt{p(1-p)} \right)^{n/2}.$$

Initialization: Let $\Delta > 0$ be such that $\Delta^{-1} \in \mathbb{N}$. Define the posterior after j measurements as $\pi_j : [0,1] \to \mathbb{R}$,

$$\pi_j(x) = \sum_{i=1}^{\Delta^{-1}} a_i(j) \mathbf{1}_{I_i}(x),$$

where $I_1 = [0, \Delta]$ and $I_i = (\Delta(i-1), \Delta i]$, for $i \in \{2, \ldots, \Delta^{-1}\}$. Notice that the collection $\{I_i\}$ is a partition of the interval $[0, 1]$. We initialize this posterior by taking $a_i(0) = \Delta$. Note that the posterior is completely characterized by the vector $\boldsymbol{a}(j) = [a_1(j), \ldots, a_{\Delta^{-1}}(j)]$, and that $\sum_{i=1}^{\Delta^{-1}} a_i(j) = 1$.

1 - **Sample Selection:** To preserve the parametric structure of the pseudo-posterior we need to take samples at the interval subdivision points. Define $k(j)$ such that

$$\sum_{i=1}^{k(j)-1} a_i(j) \leq 1/2, \quad \sum_{i=1}^{k(j)} a_i(j) > 1/2.$$

Note that $k(j) \in \{1, \ldots, \Delta^{-1}\}$. Select X_{j+1} among $\{\Delta(k(j)-1), \Delta k(j)\}$ by flipping a coin, choosing the first point with probability $P_1(j)$ and the second point with probability $P_2(j) = 1 - P_1(j)$, where $P_1(j) = \tau_2(j)/(\tau_1(j) + \tau_2(j))$ and

$$\tau_1(j) = \sum_{i=k(j)}^{\Delta^{-1}} a_i(j) - \sum_{i=1}^{k(j)-1} a_i(j),$$

and

$$\tau_2(j) = \sum_{i=1}^{k(j)} a_i(j) - \sum_{i=k(j)+1}^{\Delta^{-1}} a_i(j).$$

2 - **Noisy Observation:** Observe $Y_{j+1} = f(X_{j+1}) \oplus U_{j+1}$.

3 - **Update posterior:** Update the posterior function after collecting the measurement Y_{j+1}, through the application of Bayes rule, under the assumption that the Bernoulli random variables U_{j+1} have parameter $0 \leq \alpha < 1/2$. Let $\beta = 1 - \alpha$. Note that $X_{j+1} = \Delta k$, $k \in \mathbb{N}$ and define

$$\tau = \sum_{i=1}^{k} a_i(j) - \sum_{i=k+1}^{\Delta^{-1}} a_i(j).$$

If $i \leq k$ we have

$$a_i(j+1) = \begin{cases} \frac{2\beta}{1+\tau(\beta-\alpha)} & \text{if } Y_{j+1} = 0 \\ \frac{2\alpha}{1-\tau(\beta-\alpha)} & \text{if } Y_{j+1} = 1 \end{cases},$$

and if $i > k$ we have

$$a_i(j+1) = \begin{cases} \frac{2\alpha}{1+\tau(\beta-\alpha)} & \text{if } Y_{j+1} = 0 \\ \frac{2\beta}{1-\tau(\beta-\alpha)} & \text{if } Y_{j+1} = 1 \end{cases},$$

4 - **Final estimate:** Repeat steps 1,2 and 3 until n samples are collected. The estimate $\hat{\theta}_n$ is defined as the median of the final posterior distribution, that is, $\hat{\theta}_n$ is such that

$$\int_0^{\hat{\theta}_n} \pi_n(x) dx = 1/2.$$

Figure 8.4. The Burnashev-Zigangirov (BZ) algorithm.

Remarks: The above theorem shows that, even under noisy assumptions, there is a dramatic improvement in performance if one allows adaptive strategies (as opposed to passive strategies). Although the bounds display the exponential error decay behavior, also present in the noiseless scenario, the exponent depends on the noise parameter p, and it is clearly not optimal, since when $p \approx 0$ we would expect to obtain approximately the noiseless error bounds (i.e., $\mathbb{E}[|\widehat{\theta}_n - \theta|] \sim (1/2)^n$). Instead a weaker bound is attained, $\mathbb{E}[|\widehat{\theta}_n - \theta|] \sim (\sqrt{2}/2)^n$. It is possible to improve on this bounds by modifying the bounding strategy (as done in [43]). Finally, we note that although this result was derived for a particular noise model the result is applicable to other noise models. This can be done either by processing the observations, using a thresholding operator, or by modifying the likelihood structure (according to the noise model) in the proof of the Theorem.

The proof of Theorem 8.1 is extremely elegant and is presented below. The ideas in the proof can be used in various other contexts where feedback is present.

Proof of Theorem 8.1: For the proof we rely on the notation in the algorithm given in Figure 8.4. In particular recall that the unit interval is divided into subintervals of width Δ, $a_i(j)$ denotes the posterior probability that the changepoint θ is located in the i-th subinterval after the j-th sample, and $\widehat{\theta}_j$ denotes the median of the posterior after j samples.

Our first step is to construct an upper bound for the probability $P(|\widehat{\theta}_n - \theta| > \Delta)$. Let θ be fixed, but arbitrary, and define $k(\theta)$ to be the index of the bin I_i containing θ, that is $\theta \in I_{k(\theta)}$. Define

$$M_\theta(j) = \frac{1 - a_{k(\theta)}(j)}{a_{k(\theta)}(j)},$$

and

$$N_\theta(j+1) = \frac{M_\theta(j+1)}{M_\theta(j)} = \frac{a_{k(\theta)}(j)(1 - a_{k(\theta)}(j+1))}{a_{k(\theta)}(j+1)(1 - a_{k(\theta)}(j))}.$$

The reasoning behind these definitions is made clear later. For now, notice that $M_\theta(j)$ is a decreasing function of $a_{k(\theta)}(j)$.

After n observations our estimate of θ is the median of the posterior density π_n, which means that $\widehat{\theta}_n \in I_{k(n)}$. Taking that into account we conclude that

$$\begin{aligned} P(|\widehat{\theta}_n - \theta| > \Delta) &\leq P(a_{k(\theta)}(j) < 1/2) \\ &= P(M_\theta(n) > 1) \\ &\leq \mathbb{E}[M_\theta(n)], \end{aligned}$$

where the last step follows from Markov's inequality. The definition of $M_\theta(j)$ above is meant to get more leverage out of Markov's inequality, in a similar spirit of Chernoff bounding techniques. Using the definition of $N_\theta(j)$ and some conditioning we get

$$\begin{aligned}
\mathbb{E}[M_\theta(n)] &= \mathbb{E}[M_\theta(n-1)N_\theta(n)] \\
&= \mathbb{E}\left[\mathbb{E}[M_\theta(n-1)N_\theta(n)|\mathbf{a}(n-1)]\right] \\
&= \mathbb{E}\left[M_\theta(n-1)E[N_\theta(n)|\mathbf{a}(n-1)]\right] \\
&\vdots \\
&= M_\theta(0) E\left[E[N_\theta(1)|\mathbf{a}(0)]\cdots E[N_\theta(n)|\mathbf{a}(n-1)]\right] \\
&\leq M_\theta(0) \left\{ \max_{j\in\{0,\ldots,n-1\}} \max_{\mathbf{a}(j)} E[N_\theta(j+1)|\mathbf{a}(j)] \right\}^n. \quad (8.2)
\end{aligned}$$

The rest of the proof consists of showing that $E[N_\theta(j+1)|\mathbf{a_j}] \leq 1-\epsilon$, for some $\epsilon > 0$. Before proceeding we make some remarks about the above technique. Note that $M_\theta(j)$ measures how much mass is on the bin containing θ (if $M_\theta(j) = 0$ all the mass in our posterior is in the bin containing θ, the least error scenario). The ratio $N_\theta(j)$ is a measure of the improvement (in terms of concentrating the posterior around the bin containing θ) by sampling at X_j and observing Y_j. This is strictly less than one when an improvement is made. The bound (8.2) above is therefore only useful if, no matter what happened in the past, a measurement made with the proposed algorithm always leads on average to a performance improvement. This is the case with a variety of other useful myopic algorithms.

To study $E[N_\theta(j+1)|\mathbf{a}(j)]$ we are going to consider three particular cases: (i) $k(j) = k(\theta)$; (ii) $k(j) > k(\theta)$; and (iii) $k(j) < k(\theta)$. Let $\beta = 1-\alpha$ and $q = 1-p$. After tedious but straightforward algebra we conclude that

$$N_\theta(j+1) = \begin{cases} \frac{1+(\beta-\alpha)x}{2\beta} & \text{, with probability } q \\ \frac{1-(\beta-\alpha)x}{2\alpha} & \text{, with probability } p \end{cases},$$

where we have for the three different cases

1

$$x = \begin{cases} \frac{\tau_1(j)-a_{k(\theta)}(j)}{1-a_{k(\theta)}(j)} & \text{, if } X_{j+1} = \Delta(k(j)-1) \\ \frac{\tau_2(j)-a_{k(\theta)}(j)}{1-a_{k(\theta)}(j)} & \text{, if } X_{j+1} = \Delta k(j) \end{cases}$$

2

$$x = \begin{cases} -\frac{\tau_1(j)+a_{k(\theta)}(j)}{1-a_{k(\theta)}(j)} & \text{, if } X_{j+1} = \Delta(k(j)-1) \\ \frac{\tau_2(j)-a_{k(\theta)}(j)}{1-a_{k(\theta)}(j)} & \text{, if } X_{j+1} = \Delta k(j) \end{cases}$$

3

$$x = \begin{cases} \frac{\tau_1(j) - a_{k(\theta)}(j)}{1 - a_{k(\theta)}(j)} & \text{, if } X_{j+1} = \Delta(k(j) - 1) \\ -\frac{\tau_2(j) + a_{k(\theta)}(j)}{1 - a_{k(\theta)}(j)} & \text{, if } X_{j+1} = \Delta k(j) \end{cases}$$

Note that $0 \leq \tau_1(j) \leq 1$ and $0 < \tau_2(j) \leq 1$, therefore $|x| \leq 1$. To ease the notation define

$$\begin{aligned} g(x) &= \frac{q}{2\beta}(1 + (\beta - \alpha)x) + \frac{p}{2\alpha}(1 - (\beta - \alpha)x) \\ &= \frac{q}{2\beta} + \frac{p}{2\alpha} + \left(\frac{q}{2\beta} - \frac{p}{2\alpha}\right)(\beta - \alpha)x. \end{aligned}$$

It can be easily checked that $g(x)$ is an increasing function as long as $0 < p < \alpha$. Using this definition we have

1

$$\begin{aligned} &E[N_\theta(j+1)|\mathbf{a}(j)] \\ &= P_1(j)g\left(\frac{\tau_1(j) - a_{k(\theta)}(j)}{1 - a_{k(\theta)}(j)}\right) + P_2(j)g\left(\frac{\tau_2(j) - a_{k(\theta)}(j)}{1 - a_{k(\theta)}(j)}\right) \end{aligned}$$

2

$$\begin{aligned} &E[N_\theta(j+1)|\mathbf{a}(j)] \\ &= P_1(j)g\left(-\frac{\tau_1(j) + a_{k(\theta)}(j)}{1 - a_{k(\theta)}(j)}\right) + P_2(j)g\left(\frac{\tau_2(j) - a_{k(\theta)}(j)}{1 - a_{k(\theta)}(j)}\right) \end{aligned}$$

3

$$\begin{aligned} &E[N_\theta(j+1)|\mathbf{a}(j)] \\ &= P_1(j)g\left(\frac{\tau_1(j) - a_{k(\theta)}(j)}{1 - a_{k(\theta)}(j)}\right) + P_2(j)g\left(-\frac{\tau_2(j) + a_{k(\theta)}(j)}{1 - a_{k(\theta)}(j)}\right) \end{aligned}$$

Consider first cases (ii) and (iii). Note that $(\tau - a)/(1 - a) \leq \tau$ and $-(\tau + a)/(1 - a) < -\tau$ for all $0 < a < 1$. Therefore, for case (ii) we have

$$\begin{aligned} E[N_\theta(j+1)|\mathbf{a}(j)] &\leq P_1(j)g(-\tau_1(j)) + P_2(j)g(\tau_2(j)) \\ &= \frac{q}{2\beta} + \frac{p}{2\alpha} + \left(\frac{q}{2\beta} - \frac{p}{2\alpha}\right)(\beta - \alpha)(-P_1(j)\tau_1 + P_2(j)\tau_2) \\ &= \frac{q}{2\beta} + \frac{p}{2\alpha}. \end{aligned}$$

Analogously, for case (iii)

$$E[N_\theta(j+1)|\boldsymbol{a}(j)] \leq P_1(j)g(\tau_1(j)) + P_2(j)g(-\tau_2(j))$$
$$= \frac{q}{2\beta} + \frac{p}{2\alpha} + \left(\frac{q}{2\beta} - \frac{p}{2\alpha}\right)(\beta - \alpha)(P_1(j)\tau_1 - P_2(j)\tau_2)$$
$$= \frac{q}{2\beta} + \frac{p}{2\alpha}.$$

Finally, for case (i) a we need to proceed in a slightly different way. Begin by noticing that $\tau_1(j) + \tau_2(j) = 2a_{k(j)}(j) = 2a_{k(\theta)}(j)$. Then

$$E[N_\theta(j+1)|\boldsymbol{a}(j)]$$
$$= P_1(j)g\left(\frac{\tau_1(j) - a_{k(\theta)}(j)}{1 - a_{k(\theta)}(j)}\right) + P_2(j)g\left(-\frac{\tau_1(j) - a_{k(\theta)}(j)}{1 - a_{k(\theta)}(j)}\right)$$
$$= \frac{q}{2\beta} + \frac{p}{2\alpha} + \left(\frac{q}{2\beta} - \frac{p}{2\alpha}\right)(\beta - \alpha)\frac{\tau_1 - a_{k(\theta)}(j)}{1 - a_{k(\theta)}(j)}(P_1(j) - P_2(j))$$
$$= \frac{q}{2\beta} + \frac{p}{2\alpha} + \left(\frac{q}{2\beta} - \frac{p}{2\alpha}\right)(\beta - \alpha)\frac{\tau_1 - a_{k(\theta)}(j)}{1 - a_{k(\theta)}(j)}\frac{\tau_2(j) + \tau_1(j)}{\tau_1(j) + \tau_2(j)}$$
$$= \frac{q}{2\beta} + \frac{p}{2\alpha} + \left(\frac{q}{2\beta} - \frac{p}{2\alpha}\right)(\beta - \alpha)\frac{\tau_1 - a_{k(\theta)}(j)}{1 - a_{k(\theta)}(j)}\frac{2a_{k(\theta)}(j) - 2\tau_1(j)}{\tau_1(j) + \tau_2(j)}$$
$$\leq \frac{q}{2\beta} + \frac{p}{2\alpha}$$

Plugging in the above results into (8.2) yields

$$P(|\widehat{\theta}_n - \theta| > \Delta) \leq \frac{1 - \Delta}{\Delta}\left(\frac{q}{2\beta} + \frac{p}{2\alpha}\right)^n,$$

since $M_\theta(0) = (1 - \Delta)/\Delta$.

To get a bound on the expected error one proceeds by integration

$$\mathbb{E}[|\widehat{\theta}_n - \theta|] = \int_0^\infty P(|\widehat{\theta}_n - \theta| > t)dt$$
$$= \int_0^\Delta P(|\widehat{\theta}_n - \theta| > t)dt + \int_\Delta^1 P(|\widehat{\theta}_n - \theta| > t)dt$$
$$\leq \Delta + (1 - \Delta)P(|\widehat{\theta}_n - \theta| > \Delta)$$
$$\leq \Delta + \frac{(1 - \Delta)^2}{\Delta}\left(\frac{q}{2\beta} + \frac{p}{2\alpha}\right)^n.$$

Choosing Δ as in the statement of the theorem yields the desired result, concluding the proof. $\square$

3. Beyond 1d - Piecewise Constant Function Estimation

In this section we consider again the adaptive sampling scenario, but now in a higher dimensional setting. The one-dimensional setup in the previous section provided us with some insight about the possibilities of active learning, but it is quite restrictive: (i) the function is known up to the location of the step, that is, we know the function takes the values 0 and 1. (ii) One dimensional piecewise functions are extremely simple - they form a parametric class. Nevertheless even this simple type of problem can arise in some practical applications [211]. The kinds of functions we are going to consider next are generally higher dimensional, as in the case of laser field scanning, where the field can be described by a two dimensional function. Also the only prior knowledge we have about these functions is that they are piecewise "smooth", that is, these are composed of smooth regions (where the function varies slowly) separated by low dimensional boundary regions (where the function might change abruptly). One expects active learning to be advantageous for such function classes, since the complexity of such functions is concentrated around the boundary. Pin-pointing the boundary requires many more samples than estimation of the smooth regions so using the active learning paradigm it might be possible to focus most of the samples where they are needed: "near" the boundary. To make the description and discussion simpler we will consider solely piecewise constant functions, whose definition follows.

DEFINITION 1 *A function* $f : [0,1]^d \to \mathbb{R}$ *is* **piecewise constant** *if it is locally constant*[1] *at any point* $x \in [0,1]^d \setminus B(f)$, *where* $B(f) \subseteq [0,1]^d$ *is a set with upper box-counting dimension at most* $d - 1$. *Furthermore let* f *be uniformly bounded on* $[0,1]^d$ *(that is,* $|f(x)| \leq M$, $\forall x \in [0,1]^d$) *and let* $B(f)$ *satisfy* $N(r) \leq \beta r^{-(d-1)}$ *for all* $r > 0$, *where* $\beta > 0$ *is a constant and* $N(r)$ *is the minimal number of closed balls of diameter* r *covering* $B(f)$. *The set of all piecewise constant functions* f *satisfying the above conditions is denoted by* $\text{PC}(\beta, M)$.

The concept of box-counting dimension is closely related to topological dimension. The condition on the number of covering balls is essentially a measure of the $d - 1$-dimensional volume of the boundary set. Example of such a functions are depicted in Figures 8.5-8.7. Note that this class of functions is

---

[1] A function $f : [0,1]^d \to \mathbb{R}$ is locally constant at a point $x \in [0,1]^d$ if

$$\exists \epsilon > 0 : \forall y \in [0,1]^d : \quad \|x - y\| < \epsilon \Rightarrow f(y) = f(x).$$

non-parametric. These functions can provide a simple imaging model in various applications, for example in medical imaging, where one observes various homogeneous regions of tissue of differing densities.

In the rest of the chapter we are going to consider a slightly modified observation model, namely the observations are going to be samples of the function corrupted with additive white Gaussian noise.

A1.2 - The observations $\{Y_i\}_{i=1}^n$ are given by

$$Y_i = f(\boldsymbol{X}_i) + W_i,$$

where $f \in \text{PC}(\beta, M)$ and W_i are i.i.d. Gaussian random variables with zero mean and variance σ^2, and independent of $\{\boldsymbol{X}_j\}_{j=1}^n$.

Under this framework we are mainly interested in answering two questions:

Q1 - What are the limitations of active learning, that is, what is the best performance one can hope to achieve?

Q2 - Can a simple algorithm be devised such that the performance improves on the performance of the best passive learning algorithm?

Before attempting to answer these two questions it is important to know what are the limitations when the passive framework (A2.1) is considered. In [137] the following minimax lower bound is presented.

$$\inf_{\hat{f}_n, S_n} \sup_{f \in \text{PC}(\beta, M)} \mathbb{E}[\|\hat{f}_n - f\|^2] \geq c n^{-\frac{1}{d}}, \quad (8.3)$$

where $c > 0$ and the infimum is taken with respect to every possible estimator and sample distribution S_n.

There exist practical passive learning strategies that can nearly achieve the above performance bound. Tree-structured estimators based on *Recursive Dyadic Partitions* (RDPs) are an example of such a learning strategy [186]. These estimators are constructed as follows: (i) Divide $[0,1]^d$ into 2^d equal sized hypercubes. (ii) Repeat this process again on each hypercube. Repeating this process $\log_2 m$ times gives rise to a partition of the unit hypercube into m^d hypercubes of identical size. This process can be represented as a 2^d-ary tree structure (where a leaf of the tree corresponds to a partition cell). Pruning this tree gives rise to an RDP with non-uniform resolution. Let Ω denote the class of all possible pruned RDPs. The estimators we consider consist of a stair function supported over a RDP, that is, associated with each element in the partition there is a constant value. Let ω be an RDP; the estimators built over this RDP have the form $\tilde{f}^{(\omega)}(x) \equiv \sum_{A \in \omega} c_A \mathbf{1}\{x \in A\}$.

Since the location of the boundary is *a priori* unknown it is natural to distribute the sample points uniformly over the unit cube. There are various ways of doing this; for example, the points can be placed deterministically over a lattice, or randomly sampled from a uniform distribution. We will use the latter strategy. Let $\{X_i\}_{i=1}^n$ be i.i.d. uniform over $[0,1]^d$ and define the *complexity regularized estimator* as

$$\hat{f}_n \equiv \arg\min_{\tilde{f}^{(\omega)}:\omega \in \Omega} \left\{ \frac{1}{n} \sum_{i=1}^n \left(\tilde{f}^{(\omega)}(X_i) - Y_i \right)^2 + \lambda \frac{\log n}{n} |\omega| \right\}, \quad (8.4)$$

where $|\omega|$ denotes the number of cells of ω and $\lambda > 0$. The above optimization can be solved efficiently in $O(n)$ operations using a bottom-up tree pruning algorithm [42, 186].

The performance of the estimator in (8.4) can be assessed using bounding techniques in the spirit of [15, 186]. From that analysis we conclude that

$$\sup_{f \in \mathrm{PC}(\beta, M)} \mathbb{E}_f[\|\hat{f}_n - f\|^2] \leq C \left(\frac{n}{\log n} \right)^{-\frac{1}{d}}, \quad (8.5)$$

where the constant factor $C \equiv C(\beta, M, \sigma^2) > 0$. This shows that, up to a logarithmic factor, the rate in (8.3) is the optimal rate of convergence for passive strategies. A complete derivation of the above result is available in [52].

We now turn our attention to the active learning framework (A2.2). To address question (Q1) we will consider a subclass of the piecewise constant functions defined above, called the boundary fragments. Let $g : [0,1]^{d-1} \to [0,1]$ be a Lipshitz function, that is

$$|g(x) - g(z)| \leq \|x - z\|, \ \forall \ x, z \in [0,1]^{d-1}.$$

Define
$$G = \{(x, y) : 0 \leq y \leq g(x), (x, y) \in [0,1]^d\}, \quad (8.6)$$

and let $f : [0,1]^d \to \mathbb{R}$ be defined as $f(x) = M\mathbf{1}_G(x)$. The class of all functions of this form is called the *boundary fragment* class (usually taking $M = 1$), denoted by $\mathrm{BF}(M)$. An example of a boundary fragment function is depicted in Figure 8.5(a). It is straightforward to to show that $\mathrm{BF}(M) \subseteq \mathrm{PC}(\beta, M)$, for a suitable constant β. In [136] it was shown that under (A1.2) and A(2.2) we have the lower bound

$$\inf_{\hat{f}_n, S_n} \sup_{f \in \mathrm{BF}(M)} \mathbb{E}[\|\hat{f}_n - f\|^2] \geq cn^{-\frac{1}{d-1}},$$

Active Learning and Sampling 193

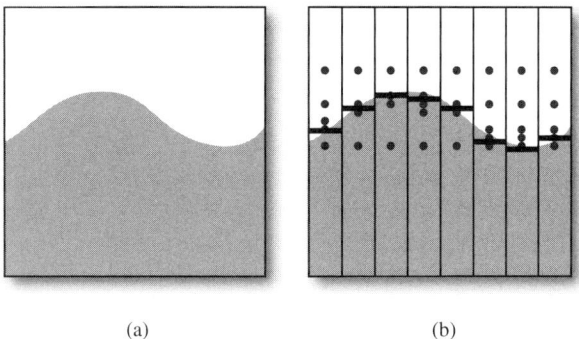

(a) (b)

Figure 8.5. (a) Illustration of a boundary fragment function for $d = 2$. (b) Adaptive sampling for boundary fragments. In each vertical stripe one uses the BZ algorithm to estimate a step function. The final estimate is a piecewise constant function whose boundary is a stair function.

for n large enough, where $c \equiv c(M, \sigma^2) > 0$, and the infimum is taken with respect to every possible estimator and sampling strategy. Since $\mathrm{BF}(M) \subseteq \mathrm{PC}(\beta, M)$ it follows that

$$\inf_{\hat{f}_n, S_n} \sup_{f \in \mathrm{PC}(\beta, M)} \mathbb{E}[\|\hat{f}_n - f\|^2] \geq cn^{-\frac{1}{d-1}}, \tag{8.7}$$

The above results hold for $d \geq 2$. Note that the error rates for the adaptive sampling framework display very significant improvement. For example, for $d = 2$ the passive learning error rate is $O(1/\sqrt{(n)})$, which is significantly slower that the active learning error rate of $O(1/n)$. Note that, for this two-dimensional case, the active learning rate coincides with the classical parametric rate, although this class of functions is non-parametric. In [136] an algorithm capable of achieving the above rate for the boundary fragment class is also presented. This algorithm takes advantage of the very special functional form of the boundary fragment functions. The algorithmic idea is very simply: begin by dividing the unit hypercube into $O(n/\log(n))$ "strips" and perform a one-dimensional change-point estimation in each of the strips (using the BZ algorithm with $\log(n)$ samples). This process is illustrated in Figure 8.5(b).

Unfortunately, the boundary fragment class is very restrictive and impractical for most applications. Recall that boundary fragments consist of only two regions, separated by a boundary that is a function of the first $d - 1$ coordinates. For a general piecewise constant function the boundaries oriented arbitrarily and generally are not aligned with any coordinate axis such that they can be described in a functional way. The class $\mathrm{PC}(\beta, M)$ is much larger and more general, so the algorithmic ideas that work for boundary fragments can no longer be used. A completely different approach is required, using radically different tools.

We now attempt to answer question (Q2), by proposing an active learning scheme for the piecewise constant class. The scheme is a two-step approach motivated by the tree-structured estimators for passive learning described above. Although the ideas and intuition behind the approach are quite simple, the formal analysis of the method is significantly difficult and cumbersome, therefore the focus of the presentation is on the algorithm and sketch of the proofs, deferring the details to the references. The main idea is to devise a strategy that uses the first sampling step to find advantageous locations for new samples, to be collected at the second step. More precisely in the first step, called the *preview step*, an estimator of f is constructed using $n/2$ samples (assume for simplicity that n is even), distributed uniformly over $[0, 1]^d$. In the second step, called the *refinement step*, we select $n/2$ samples near the perceived location of the boundaries (estimated in the preview step) separating constant regions. At the end of this process we will have half the samples concentrated in the perceived vicinity of the boundary set $B(f)$. Since accurately estimating f near $B(f)$ is key to obtaining faster rates, the strategy described seems quite sensible. However, it is *critical* that the preview step is able to detect the boundary with very high probability. If part of the boundary is missed, then the error incurred is going to propagate into the final estimate, ultimately degrading the performance. Conversely, if too many regions are (incorrectly) detected as boundary locations in the preview step, then the second step will distribute samples too liberally and no gains will be achieved. Therefore extreme care must be taken to accurately detect the boundary in the preview step, as described below.

Preview: The goal of this stage is to provide a coarse estimate of the location of $B(f)$. Specifically, collect $n' \equiv n/2$ samples at points distributed uniformly over $[0, 1]^d$. Next proceed by using the passive learning algorithm described before, but restrict the estimator to RDPs with leafs at a maximum depth of $J = \frac{d-1}{(d-1)^2+d} \log(n'/\log(n'))$. This ensures that, on average, every element of the RDP contains many sample points; therefore we obtain a low variance estimate, although the estimator bias is going to be large. In other words, we obtain a very "stable" coarse estimate of f, where stable means that the estimator does not change much for different realizations of the data. The justification for the particular value of J arises in the formal analysis of the method.

The above strategy ensures that most of the time, leafs that intersect the boundary are at the maximum allowed depth (because otherwise the estimator would incur too much empirical error) and leafs away from the boundary

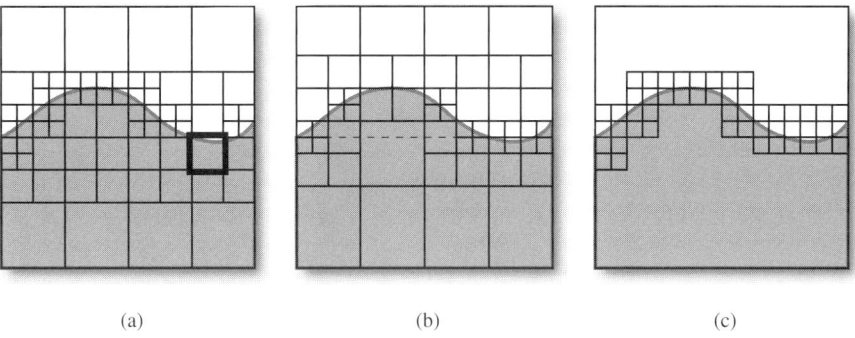

Figure 8.6. Illustration of the shifted RDP construction for $d = 2$: (a) RDP used obtained from the preview step with the regular RDPs. The highlighted cell intersects the boundary but it was pruned, since the pruning does not incur in severe error. (b) Shifted RDP, Obtained from the preview step over vertically shifted RDPs . In this case there is no pruning, since it would cause a large error. (c) These are the cells that are going to be refined in the refinement stage.

are at shallower depths. Therefore we can "detect" the rough location of the boundary just by looking at the deepest leafs. Unfortunately, if the set $B(f)$ is somewhat aligned with the dyadic splits of the RDP, leafs intersecting the boundary can be pruned without incurring a large error. This is illustrated in Figure 8.7(b); the cell with the arrow was pruned and contains a piece of the boundary, but the error incurred by pruning is small since that region is mostly a constant region. However, worst-case analysis reveals that the squared bias induced by these small volumes can add up, precluding the desired improved performance. A way of mitigating this issue is to consider multiple RDP-based estimators, each one using RDPs appropriately shifted. We use $d+1$ estimators in the preview step: one on the initial uniform partition, and d over partitions whose dyadic splits have been translated by 2^{-J} in each one of the d coordinates. Any leaf that is at the maximum depth on any of the $d + 1$ RDPs pruned in the preview step indicates the highly probable presence of a boundary, and will be refined in the next stage. This shifting strategy is illustrated in Figure 8.6

Refinement: With high probability, the boundary is contained in the leafs at the maximum depth. In the refinement step we collect additional $n/2$ samples on the corresponding partition cells, using these to obtain a refined estimate of the function f by applying again an RDP-based estimator. This produces a higher resolution estimate in the vicinity of the boundary set $B(f)$, yielding a better performance than the passive learning technique.

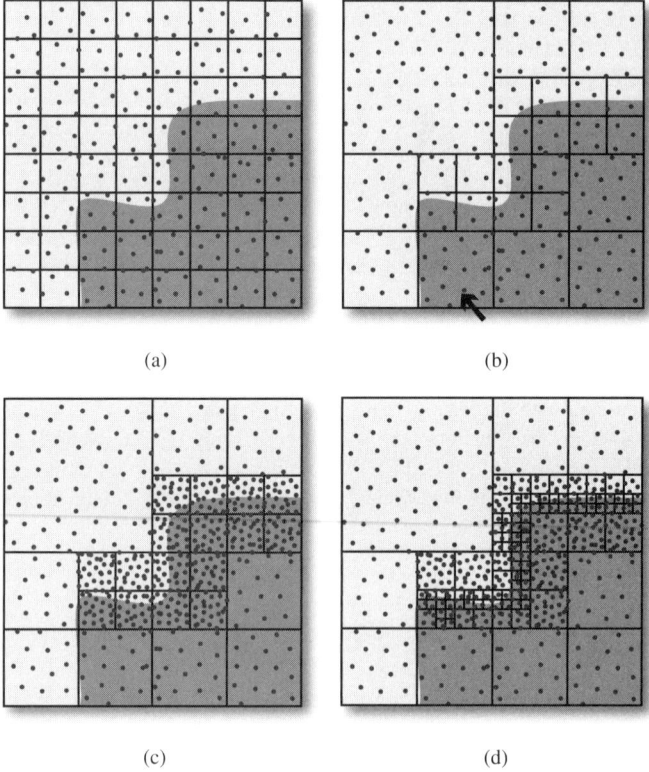

Figure 8.7. The two step procedure for $d = 2$: (a) Initial unpruned RDP and $n/2$ samples. (b) Preview step RDP. Note that the cell with the arrow was pruned, but it contains a part of the boundary. (c) Additional sampling for the refinement step. (d) Refinement step.

The final estimator is constructed assembling the estimate "away" from the boundary obtained in the preview step with the estimate in the vicinity of the boundary obtained in the refinement step.

To formally show that this algorithm attains the faster rates we desire we have to consider a further technical assumption, namely that the boundary set is "cusp-free"[2]. This condition is rather technical, but it is not very restrictive, and encompasses many interesting situations, including of course, boundary fragments. This condition seems to be necessary for the algorithm to perform well, and it is not simply an artifact of the proof. For a more detailed explanation see [52]. Under this condition we have the following theorem.

---

[2] A cusp-free boundary cannot have the behavior you observe in the graph of $|x|^{1/2}$ at the origin. Less "aggressive" kinks are allowed, such as in the graph of $|x|$.

THEOREM 8.2 *Under the active learning scenario (A1.2) and (A2.2) we have, for $d \geq 2$ and functions f whose boundary is cusp-free,*

$$\mathbb{E}\left[\|\hat{f}_n - f\|^2\right] \leq C \left(\frac{n}{\log n}\right)^{-\frac{1}{d-1+1/d}}, \tag{8.8}$$

where $C > 0$.

This bound improves on (8.5), demonstrating that this technique performs better than the best possible passive learning estimator. The proof of Theorem 8.2 is quite involved and is presented in detail in [52]. Below we present only a sketch of the proof.

A couple of remarks are important at this point. Instead of a two-step procedure one can iterate this algorithm, performing multiple steps (*e.g.*, for a three-step approach replace the refinement step with the two-step approach described above). Doing so can further improve the performance. One can show that the expected error will decay like $n^{-1/(d-1+\epsilon)}$, with $\epsilon > 0$, given a sufficiently large number of steps. Therefore we can get rates arbitrarily close to the lower bound rates in (8.7). These multi-step methods start to lack practical usefulness if the number of steps is too large, since the practical benefits appear only for a large number of samples. The two step procedure on the other hand displays excellent performance under practical scenarios as seen in [254].

Proof of Theorem 8.2: The main idea behind the proof is to decompose the error of the estimator for three different cases: (i) the error incurred during the preview stage in regions "away" from the boundary; (ii) the error incurred by not detecting a piece of the boundary (and therefore not performing the refinement step in that area); (iii) the error remaining in the refinement region at the end of the process. By restricting the maximum depth of the trees in the preview stage we can control the type-(i) error, ensuring that it does not exceed the error rate in (8.8). We start by defining f_J, a coarse approximation of f up to resolution J. Consider the partition of $[0, 1]^d$ into 2^{dJ} identical hypercubes and denote this partition by ω_J. Note that this partition can also be described by a RDP, where all the leafs are at depth J. Define $f_J : [0, 1]^d \to \mathbb{R}$ as

$$f_J(x) = \frac{1}{2^{dJ}} \sum_{A \in \omega_j} \left(\int_A f(t) dt\right) \mathbf{1}_A(x).$$

Note that f_J is identical to f "away" from the boundary, but in the vicinity of the boundary there is some averaging. We have the following key Lemma.

LEMMA 1 *Let $\hat{f}^p$ be the complexity regularized estimator of the preview step, using $n' = n/2$ samples. Then*

$$\mathbb{E}[\|\hat{f}^p - f_J\|] \leq C \frac{2^{(d-1)J} \log(n')}{n'},$$

for a suitable $C > 0$ and all $n' > 2$.

Lemma 1 characterizes the behavior of the final estimate "away" from the boundary, since f and f_J are identical in that region. So the above error bound controls the type-(i) error.

Type-(ii) error corresponds to the situations when a part of the boundary was not detected in the preview step. This can happen because of the inherent randomness of the noise and sampling distribution, or because the boundary is somewhat aligned with the dyadic splits, like in Figure 8.7(b). The latter can be a problem and this is why one needs to perform $d+1$ preview estimates over shifted partitions. If the boundary is cusp-free then it is guaranteed that one of those preview estimators is going to "feel" the boundary since it is not aligned with the corresponding partition. A piece of the boundary region is not refined if it is not detected in *all* the shifted partition estimators. The worst-case error can be shown not to exceed $C 2^{(d-1)J} \log(n')/n'$, for some $C > 0$, therefore failure to detect the boundary has the same contribution for the total error as the type-(i) error. The proof of this fact is detailed in [52].

Finally, analysis of type-(iii) error is relatively easy. Nonetheless one needs to make sure that the size of the region that needs to be refined is not too large, since in that case the density of samples in the refinement step might be not be sufficient to improve on passive methods. In other words, one needs to make sure that in the preview step not much more than the boundary region is detected. Let $\mathcal{R}$ be the set of cells that need to be re-sampled in the refinement step. One has to guarantee that, with high probability the number of boundary cells detected in the preview step (denoted by $|\mathcal{R}|$) is on the order of $2^{(d-1)J}$. The following Lemma [52] provides an affirmative answer.

LEMMA 2 *Let $|\mathcal{R}|$ be the number of cells detected in the preview step, that possibly contain the boundary (and therefore are going to be re-sampled in the refinement step). Then*

$$P(|\mathcal{R}| > C 2^{(d-1)J}) \leq 1/n,$$

for $C > 0$ and n sufficiently large.

In the regions that are going to be refined, that is, the regions in $\mathcal{R}$, we are going to collect further samples and apply the passive estimation strategy described in (8.4). As shown in the Lemma 2 we can assume that there are $O(2^{(d-1)J})$ elements in $\mathcal{R}$ (with high probability). We collect a total of $L \equiv n/(2|\mathcal{R}|)$ samples in each element of $\mathcal{R}$. The error incurred by $\hat{f}_r$, the refinement estimator, over each one of the elements of $\mathcal{R}$ is upper-bounded by

$$C\left(\frac{\log L}{L}\right)^{1/d} 2^{-dJ},$$

where $C > 0$ comes from (8.5), and 2^{-dJ} is just the volume of each element of $\mathcal{R}$. Therefore overall error contribution of the refinement step is upper-bounded by

$$C\left(\frac{\log L}{L}\right)^{1/d} 2^{-dJ}|\mathcal{R}|.$$

To compute the total error incurred by $\hat{f}_{\text{active}}$, our proposed active learning estimator, we just have to sum the contributions of (i), (ii) and (iii), and therefore we get

$$\mathbb{E}\left[\|\hat{f}_{\text{active}} - f\|^2\right] \leq C\left(\frac{\log L}{L}\right)^{1/d} 2^{-dJ}|\mathcal{R}| + C'\frac{2^{(d-1)J}\log n}{n},$$

with $C, C' > 0$. Assuming now that $|\mathcal{R}| = O(2^{(d-1)J})$ we can balance the two terms in the above expression by choosing

$$J = \left\lceil \frac{d-1}{(d-1)^2 + d}\log(n/\log(n)) \right\rceil,$$

yielding the desired result. $\square$

4. Final Remarks and Open Questions

The results presented in this chapter show that for certain scenarios active learning attains provable gains over the classical passive approaches, even if observation uncertainty is present. Active learning is intuitively appealing, and finds applications for many practical problems, for example in imaging techniques, some of them described here. Despite these benefits, the analysis of such active methods is quite challenging due to the existence of feedback in the measurement process. This creates statistical dependence in the observations (recall that now the sample locations are coupled with all the observations made in the past), precluding the use of the usual analysis tools, such as concentration inequalities and laws of large numbers, that require independence (or *quasi* independence) to be applicable. The piecewise constant function class studied provides a non-trivial canonical example that illustrates under what conditions one might expect the adaptive sampling framework to yield a performance improvement over more traditional passive sampling techniques. The algorithm presented here for actively learning members of the piecewise constant class demonstrates the possibilities of active learning in more general settings. A natural extension of this function class is the piecewise smooth class, whose element are Hölder smooth functions with smoothness parameter α. For this class of functions we conjecture that the best attainable performance under assumptions (A1.2) and (A2.2) is $O(\max\{n^{-1/(d-1)}, n^{-2\alpha/(2\alpha+d)}\})$. This is quite intuitive, since it is known that adaptive sampling is not effective for learning smooth functions [51]. Constructing a two-step algorithm to learn such functions is relatively simple and it has been done in the context of field estimation using wireless sensor networks [254]. The key modification is that now one needs to decorate the estimator RDPs with polynomials instead of constants. Despite its simplicity this algorithm is very hard to analyze (to be specific, it is difficult to generalize Lemma 1 in the proof).

Under the selective sampling framework there are even more open problems, that have recently spawned much interest. It is a known fact that most existing active learning algorithms tend to be too greedy: They work well when the number of collected samples/examples is small, but the performance quickly degrades as that number increases, leading to results that are worse than when using classical passive learning techniques. This creates some fertile ground for both practitioners and theoretical researchers, and an interesting interplay between the two.

Chapter 9

PLAN-IN-ADVANCE ACTIVE LEARNING OF CLASSIFIERS

Xuejun Liao

Duke University, Durham, NC, USA

Yan Zhang

Innovation Center of Humana, Inc., Louisville, KY, USA

Lawrence Carin

Duke University, Durham, NC, USA

1. Introduction

A standard classification problem can be defined by two underlying components: a data generator (sensor) and a true class labeler. The data generator generates data $\mathbf{x}$ according to a distribution $p(\mathbf{x})$ and the true class labeler labels $\mathbf{x}$ as class y according to a conditional distribution $p(y|\mathbf{x})$. One does not know $p(\mathbf{x})$ or $p(y|\mathbf{x})$, but only observes instances of $(\mathbf{x}, y)$, drawn from $p(\mathbf{x})p(y|\mathbf{x})$. The goal is to use these instances to build a classifier $f(\mathbf{x}; \mathbf{w})$ (parameterized by $\mathbf{w}$) such that for any $\mathbf{x}$ drawn from $p(\mathbf{x})$, $f(\mathbf{x}; \mathbf{w})$ predicts the true label y as accurately as possible.

Let $L(f(\mathbf{x}; \mathbf{w}), y)$ denote the loss due to prediction $f(\mathbf{x}; \mathbf{w})$ of the label of $\mathbf{x}$ when the true label is y. The optimality of f is measured by the expected loss $\psi(f) = \mathbb{E}[L(f(\mathbf{x}; \mathbf{w}), y)]$, where the expectation is taken with respect to $p(\mathbf{x})p(y|\mathbf{x})$ [238, 239]. In practice, both $p(\mathbf{x})$ and $p(y|\mathbf{x})$ are unknown and one computes the empirical loss $\widehat{\psi}(f) = \frac{1}{|\mathcal{X}_{\text{tr}}|}\sum_{\mathbf{x}\in\mathcal{X}_{\text{tr}}} L(f(\mathbf{x}; \mathbf{w}), y)$, where $\mathcal{X}_{\text{tr}} = \{\mathbf{x} : y \text{ is acquired}\}$ constitutes a set of training data with their labels drawn from $p(y|\mathbf{x})$, and $|\cdot|$ denotes set cardinality. Given $\mathcal{X}_{\text{tr}}$ and the associated set of labels denoted by $\mathcal{Y}_{\text{tr}}$, the best classifier is determined as the one that minimizes the empirical loss, with regularization (i.e., a prior) usually imposed on $\mathbf{w}$.

It is clear that, for given $p(y|\mathbf{x})$, a classifier depends on the training set $\mathcal{X}_{\text{tr}}$. The choice of $\mathcal{X}_{\text{tr}}$ can have great impact on the resulting classifier. Intuitively a good $\mathcal{X}_{\text{tr}}$ should be such that its collection of labels is representative of the between-class boundaries. Since labels are more uncertain near the boundaries than off the boundaries, this is equivalent to saying that $\mathcal{X}_{\text{tr}}$ should favor those $\mathbf{x}$ that give large $\mathcal{H}(y|\mathbf{x}) = -\sum_y p(y|\mathbf{x})\log p(y|\mathbf{x})$, the entropy of y given $\mathbf{x}$. In addition, labeling a datum usually involves expensive instrumental measurement and time-consuming analysis, therefore it is important to only label data that will improve the classifier. Toward this goal, we should let $\mathcal{X}_{\text{tr}}$ drawn selectively, instead of independently, from $p(\mathbf{x})$; i.e., we should construct $\mathcal{X}_{\text{tr}} \overset{\pi}{\sim} p(\mathbf{x})$ instead of $\mathcal{X}_{\text{tr}} \overset{i.i.d.}{\sim} p(\mathbf{x})$, where π is a selection criterion. This represents the notion called "active learning" in machine learning [62, 150].

The process of acquiring the label y of a given $\mathbf{x}$ can be viewed as an experiment, the nature of which may be physical, chemical, computational, etc., depending on the application. The experiment is controlled by $\mathbf{x}$, which determines a point in the data space at which the experiment is performed. The data space is determined by $p(\mathbf{x})$. In practice, it suffices to represent this space by $\mathcal{X}$, a set of $\mathbf{x}$ drawn independently from $p(\mathbf{x})$, provided that $p(\mathbf{x})$ can be reconstructed from $\mathcal{X}$ with high accuracy using standard density estimators. Thus selection of $\mathcal{X}_{\text{tr}}$ is to choose $\mathcal{X}_{\text{tr}} \subset \mathcal{X}$. After $\mathcal{X}_{\text{tr}}$ are selected and labeled, the remaining data, i.e., $\mathcal{X}_{\text{te}} = \mathcal{X} \setminus \mathcal{X}_{\text{tr}}$, constitute the testing set.

In this chapter, we introduce a plan-in-advance approach to active learning of classifiers (the meaning of "plan-in-advance" will be clear shortly). We emphasize that the plan-in-advance strategy is different from previous active learning methods ([62, 150, 231, 163, for example]) in that it decomposes the active learning into two disjoint stages: the pre-labeling stage and the labeling stage. In the pre-labeling stage one determines the analytic form of $f(\mathbf{x};\mathbf{w})$ and selects the training data $\mathcal{X}_{\text{tr}}$, without seeing the associated labels $\mathcal{Y}_{\text{tr}}$. In the labeling stage, one acquires labels of $\mathcal{X}_{\text{tr}}$ and estimates $\mathbf{w}$ to complete the classifier. It is important to note that through the entire pre-labeling stage, one does not need to know any labels. It is after completion of the pre-labeling stage that one actually goes out to acquire the labels of $\mathcal{X}_{\text{tr}}$. Therefore, this methodology addresses the problem for which there are no labeled data *a priori*.

The plan-in-advance active learning, with a strict separation of data selection from label acquisition as described above, has great significance in practice. In remote sensing applications, labeling of different data are often performed at different geological locations. Using the approach described above, one can *plan in advance* the visits to the selected data. A reasonable plan can avoid wasted travel. Another example is in biological applications, where labeling of multiple data can be done in parallel. The separation of data selection from label acquisition means that one does not have to wait for the labeling results of the previously selected data to proceed to selecting next datum. Instead one can *plan in advance* the labeling experiments for the selected data and save a tremendous amount of time.

We now give an overview of the subsequent sections. In Section 2, we formulate the possible analytic forms of the classifier and define the loss function and its variants. In Section 3 we develop a criterion, which is directly related to the loss function but which is independent of labels, for determining an analytic form of f from a given set of candidate basis functions. As discussed previously, an optimal $\mathcal{X}_{tr}$ should be constructed as $\mathcal{X}_{tr} \overset{\pi}{\sim} p(\mathbf{x})$ instead of $\mathcal{X}_{tr} \overset{i.i.d.}{\sim} p(\mathbf{x})$. With $\mathcal{X}$ a sufficient set of samples of $p(\mathbf{x})$, we select $\mathcal{X}_{tr}$ from $\mathcal{X}$ based on selection criterion π. In Section 4, we develop a selection criterion that is directly related to the loss function but independent of labels. The optimal $\mathcal{X}_{tr}$ thus selected from $\mathcal{X}$ yields a classifier $f(\mathbf{x}; \mathbf{w})$ that generalizes better to the remaining data $\mathcal{X}_{te} = \mathcal{X} \setminus \mathcal{X}_{tr}$ than a classifier trained on a non-optimal $\mathcal{X}_{tr}$.

Since we have viewed data labeling as an experiment, it is natural to connect our approach to the theory of optimal experiments [56, 81, 191]; this connection is addressed in Section 5. As discussed earlier, the plan-in-advance approach is particularly suitable to remote sensing. In Section 6 we demonstrate our approach on the problem of unexploded ordnance (UXO) detection, where $\mathcal{X}$ correspond to the data across the entire site under test. As the actively-learned classifier is specific to $\mathcal{X}$, its detection performance is much improved over that of its non-active counterpart.

2. Analytical Forms of the Classifier

We consider a classifier of the analytical form

$$f(\mathbf{x}; \mathbf{w}) = \mathbf{w}^\mathsf{T} \phi(\mathbf{x}) \tag{9.1}$$

where $\phi(\mathbf{x}) = [1, \phi^1(\mathbf{x}), \cdots, \phi^n(\mathbf{x})]^\mathsf{T}$ is a column containing n basis functions and $\mathbf{w} = [w_0, w_1 \cdots, w_n]^\mathsf{T}$ is a column of parameters weighting the basis functions, with w_0 accounting for a bias term.

We consider a loss function of the form $L(f(\mathbf{x}, \mathbf{w}), y) = [f(\mathbf{x}, \mathbf{w}) - y]^2$, a training data set $\mathcal{X}_{tr} = \{\mathbf{x}_1, \cdots, \mathbf{x}_J\}$ and associated labels $\mathcal{Y}_{tr} = \{y_1, \cdots, y_J\}$

obtained from the labeling experiments. To make the presentation in this chapter easier, we only consider the binary case, where the labels $y \in \{0, 1\}$. For M-ary classification, we use the "one-versus-all" rule [196, 158] to design M classifiers, each for a distinct class and all the M classifiers sharing the same basis functions ϕ. As ϕ are shared across the M classifiers, the selection criteria developed in this chapter will remain unchanged. The only change is that after the ϕ and $\mathcal{X}_{\text{tr}}$ are selected and the labels are acquired, one uses the one-versus-all rule to compute M $\mathbf{w}$'s, each associated with a distinct class.

Given $\mathcal{X}_{\text{tr}}$ and $\mathcal{Y}_{\text{tr}}$, the empirical loss is given as

$$\widehat{\psi}(\phi, \mathbf{w}) = \sum_{i=1}^{J} [f(\mathbf{x}_i; \mathbf{w}) - y_i]^2 = \sum_{i=1}^{J} [\mathbf{w}^\mathsf{T} \phi_i - y_i]^2 \quad (9.2)$$

where $\phi_i \equiv \phi(\mathbf{x}_i)$ for notational simplicity. For convenience of later discussions, we define the relative empirical loss

$$\widehat{\psi}_{\text{rel}}(\phi, \mathbf{w}) = \frac{\sum_{i=1}^{J} [\mathbf{w}^\mathsf{T} \phi_i - y_i]^2}{\sum_{i=1}^{J} y_i^2} \quad (9.3)$$

It is clear that for given $\mathcal{X}_{\text{tr}}$ and $\mathcal{Y}_{\text{tr}}$, $\widehat{\psi}_{\text{rel}}(\phi, \mathbf{w})$ is equivalent to $\widehat{\psi}(\phi, \mathbf{w})$ up to a multiplicative constant. To prevent the $\mathbf{A}$ matrix defined in (9.6) from being singular, a regularization term is added to $\widehat{\psi}(\phi, \mathbf{w})$ to yield the regularized empirical loss

$$\widehat{\psi}_{\text{reg}}(\phi, \mathbf{w}) = \widehat{\psi}(\phi, \mathbf{w}) + \lambda \|\mathbf{w}\|^2 \quad (9.4)$$

where $\|\cdot\|$ denotes the L_2 norm and λ is typically set to a small positive number. Given ϕ, $\mathbf{w}$ is found by minimizing the $\widehat{\psi}_{\text{reg}}(\phi, \mathbf{w})$ with respect to $\mathbf{w}$, yielding

$$\mathbf{w} = \mathbf{A}^{-1} \sum_{i=1}^{J} y_i \phi_i \quad (9.5)$$

and

$$\mathbf{A} = \sum_{i=1}^{J} \phi_i \phi_i^\mathsf{T} + \lambda \mathbf{I}_{n+1} \quad (9.6)$$

where $\mathbf{I}_{n+1}$ denotes the $(n+1) \times (n+1)$ identity matrix.

3. Pre-labeling Selection of Basis Functions ϕ

The analytical form of the classifier in (9.1) is not completed until the basis functions $\phi(\mathbf{x}) = [1, \phi^1(\mathbf{x}), \cdots, \phi^n(\mathbf{x})]^\mathsf{T}$ have been specified. The first basis function is already given as a constant, so our discussion in this section refers to selection of the remaining basis functions in ϕ. We are interested in selecting the basis functions from a given set of candidate basis functions $\mathcal{B} = \{b(\mathbf{x}; \alpha) : \alpha \in \Lambda\}$. The selection is based on minimizing $\widehat{\psi}_{\text{rel}}(\phi, \mathbf{w})$ with $\mathbf{w}$ produced by (9.5). Since $\widehat{\psi}_{\text{rel}}(\phi, \mathbf{w})$ is equal to $\widehat{\psi}(\phi, \mathbf{w})$ up to a multiplicative constant and $\widehat{\psi}_{\text{reg}}(\phi, \mathbf{w})$ can be made as close as possible to $\widehat{\psi}(\phi, \mathbf{w})$ by setting λ sufficiently small, $\widehat{\psi}_{\text{rel}}(\phi, \mathbf{w})$ and $\widehat{\psi}_{\text{reg}}(\phi, \mathbf{w})$ are minimized by the same $\mathbf{w}$ and ϕ to an arbitrary precision.

Define $\mathbf{y} = [y_1, \cdots, y_J]^\mathsf{T}$ and $\boldsymbol{\Phi} = [\phi_1, \cdots, \phi_J]$. Substituting the solution of $\mathbf{w}$ from (9.5) into (9.2) and making use of the notation of $\mathbf{y}$ and $\boldsymbol{\Phi}$, we obtain the empirical loss that depends only on ϕ,

$$\begin{aligned}
\widehat{\psi}_{\text{rel}}(\phi) &= \frac{\left\|\mathbf{y} - \boldsymbol{\Phi}^\mathsf{T}(\boldsymbol{\Phi}\boldsymbol{\Phi}^\mathsf{T} + \lambda \mathbf{I}_{n+1})^{-1}\boldsymbol{\Phi}\mathbf{y}\right\|^2}{\mathbf{y}^\mathsf{T}\mathbf{y}} \\
&= \frac{\left\|\left[\mathbf{I}_J - \boldsymbol{\Phi}^\mathsf{T}(\boldsymbol{\Phi}\boldsymbol{\Phi}^\mathsf{T} + \lambda \mathbf{I}_{n+1})^{-1}\boldsymbol{\Phi}\right]\mathbf{y}\right\|^2}{\mathbf{y}^\mathsf{T}\mathbf{y}} \\
&\stackrel{3}{=} \frac{\left\|\left(\mathbf{I}_J + \lambda^{-1}\boldsymbol{\Phi}^\mathsf{T}\boldsymbol{\Phi}\right)^{-1}\mathbf{y}\right\|^2}{\mathbf{y}^\mathsf{T}\mathbf{y}} \\
&= \frac{\mathbf{y}^\mathsf{T}\left(\mathbf{I}_J + \lambda^{-1}\boldsymbol{\Phi}^\mathsf{T}\boldsymbol{\Phi}\right)^{-2}\mathbf{y}}{\mathbf{y}^\mathsf{T}\mathbf{y}} \\
&= \frac{\mathbf{y}^\mathsf{T}\mathbf{C}^{-2}\mathbf{y}}{\mathbf{y}^\mathsf{T}\mathbf{y}}
\end{aligned} \quad (9.7)$$

where $\mathbf{C} = \mathbf{I}_J + \lambda^{-1}\boldsymbol{\Phi}^\mathsf{T}\boldsymbol{\Phi}$ and equality 3 is due to the Sherman-Morrison-Woodbury formula. By construction the matrix $\mathbf{C}$ is positive definite, and therefore $\mathbf{C}^{-2}$ is positive definite, too. Consequently the rightmost side of (9.7) is the Rayleigh quotient, which has the well-known property:

$$\lambda_{\min}(\mathbf{C}^{-2}) \leq \mathbf{y}^\mathsf{T}\mathbf{C}^{-2}\mathbf{y}/(\mathbf{y}^\mathsf{T}\mathbf{y}) \leq \lambda_{\max}(\mathbf{C}^{-2}),$$

where $\lambda_{\max}(\cdot)$ and $\lambda_{\min}(\cdot)$ denote the maximum and minimum eigenvalues of a matrix, respectively. Since $\lambda_{\min}(\mathbf{C}^{-2}) = \lambda_{\max}^{-2}(\mathbf{C})$ and $\lambda_{\max}(\mathbf{C}^{-2}) = \lambda_{\min}^{-2}(\mathbf{C})$, these inequalities are equivalent to $\lambda_{\max}^{-2}(\mathbf{C}) \leq \mathbf{y}^\mathsf{T}\mathbf{C}^{-2}\mathbf{y}/(\mathbf{y}^\mathsf{T}\mathbf{y}) \leq \lambda_{\min}^{-2}(\mathbf{C})$, which are applied to (9.3) to yield

$$\lambda_{\max}^{-2}(\mathbf{C}) \leq \widehat{\psi}_{\text{rel}}(\phi) \leq \lambda_{\min}^{-2}(\mathbf{C}) \leq 1 \quad (9.8)$$

where the last inequality holds because the eigenvalues of $\mathbf{C}$ are greater than 1, which is evident from definition of $\mathbf{C}$.

Equation (9.8) implies that $\widehat{\psi}_{\text{rel}}(\phi, \mathbf{w})$ is bounded between the reciprocal squares of the maximum and minimum eigenvalues of $\mathbf{C}$. Since we want to minimize $\widehat{\psi}_{\text{rel}}(\phi)$, the optimal basis functions are $\phi_{\text{opt}} = \arg\min_\phi \widehat{\psi}_{\text{rel}}(\phi)$. However, we are interested in pre-labeling selection. It is clear that before seeing the labels $\mathbf{y}$, we cannot solve this minimization problem because $\widehat{\psi}_{\text{rel}}(\phi)$ depends on $\mathbf{y}$. Therefore we turn to minimizing the bounds given in (9.8), which are independent of $\mathbf{y}$. An approximate way of doing this is to maximize the determinant of $\mathbf{C}$, which is equal to the product of all eigenvalues of $\mathbf{C}$, with respect to ϕ. Since taking the logarithm does not affect the maximization, we will maximize the logarithmic determinant of $\mathbf{C}$ instead; i.e.,

$$\phi_{\text{approx.opt}} = \arg\max_\phi \{\log\det\mathbf{C}(\phi)\} \tag{9.9}$$

where we have written $\mathbf{C}$ as $\mathbf{C}(\phi)$ to make the dependence of $\mathbf{C}$ on ϕ explicit. The logarithmic determinant of $\mathbf{C}(\phi)$ is found to be

$$\begin{aligned}
\log\det\mathbf{C}(\phi) &= \log\det(\mathbf{I}_J + \lambda^{-1}\mathbf{\Phi}^\mathsf{T}\mathbf{\Phi}) \\
&\stackrel{2}{=} \log\frac{\det(\lambda\mathbf{I}_{n+1} + \mathbf{\Phi}\mathbf{\Phi}^\mathsf{T})}{\det(\lambda\mathbf{I}_{n+1})} \\
&\stackrel{3}{=} \log\det(\lambda\mathbf{I}_{n+1} + \sum_{i=1}^{J}\phi_i\phi_i^\mathsf{T}) - \log\lambda^{n+1} \\
&\stackrel{4}{=} \log\det\mathbf{A} - \log\lambda^{n+1} \tag{9.10}
\end{aligned}$$

where equality 2 results from application of a property of matrix determinants, equality 3 from the definition of $\mathbf{\Phi}$, and equality 4 is a result of back-substitution of (9.6). It follows from (9.6) that $\mathbf{A}$ is positive definite, as a result $\det\mathbf{A} > 0$ and $\log\det\mathbf{A}$ is a valid expression.

It is clear that maximizing $\log\det\mathbf{C}(\phi)$ with respect to all basis functions in ϕ results in a search space that grows exponentially with n. To make the problem tractable we turn to an approximate method, in which we sequentially find a single basis function, say ϕ^{n+1}, as

$$\phi^{n+1} = \arg\max_\phi \{\log\det\mathbf{C}([1,\phi^1,\cdots,\phi^n,\phi]^\mathsf{T})\} \tag{9.11}$$

conditional on the n basis functions that have been previously found. The sequential procedure starts with $n = 1$ and terminates when n reaches a pre-defined number or when $\log\det\mathbf{C}$ converges.

In the following, $\phi = [1,\phi^1,\cdots,\phi^n]^\mathsf{T}$ is understood to denote the basis functions that have been selected up to the moment when the $(n+1)^{\text{st}}$ basis function ϕ^{n+1} is being determined. The dependence of ϕ on n is dropped for national simplicity.

Define $\phi^{\text{new}} = \begin{bmatrix} \phi \\ \phi^{n+1} \end{bmatrix}$, where ϕ^{n+1} is the $(n+1)^{\text{st}}$ basis function, which is being determined. We examine the difference between $\log\det\mathbf{C}(\phi^{\text{new}})$ and

$\log \det \mathbf{C}(\boldsymbol{\phi})$. Replacing $\boldsymbol{\phi}_i$ with $\boldsymbol{\phi}_i^{\text{new}}$ and $\lambda \mathbf{I}_{n+1}$ with $\lambda \mathbf{I}_{n+2}$ in (9.10), we obtain

$$\log \det \mathbf{C}(\boldsymbol{\phi}^{\text{new}}) = \log \det \left(\lambda \mathbf{I}_{n+2} + \sum_{i=1}^{J} \begin{bmatrix} \boldsymbol{\phi}_i \\ \phi_i^{n+1} \end{bmatrix} \begin{bmatrix} \boldsymbol{\phi}_i \\ \phi_i^{n+1} \end{bmatrix}^\mathsf{T} \right) - \log \lambda^{n+1}$$

$$= \log \det \begin{bmatrix} \mathbf{A} & \mathbf{c} \\ \mathbf{c}^\mathsf{T} & d \end{bmatrix} - \log \lambda^{n+1}$$

$$= \log \det \mathbf{A} + \log(d - \mathbf{c}^\mathsf{T} \mathbf{A}^{-1} \mathbf{c}) - \log \lambda^{n+1} \quad (9.12)$$

where $\mathbf{c} = \sum_{i=1}^{J} \boldsymbol{\phi}_i \phi_i^{n+1}$, $d = \lambda + \sum_{i=1}^{J} (\phi_i^{n+1})^2$, and $\mathbf{A}$ is as in (9.6). Since $\mathbf{A}$ and

$$\begin{bmatrix} \mathbf{A} & \mathbf{c} \\ \mathbf{c}^\mathsf{T} & d \end{bmatrix} \quad (9.13)$$

are both positive definite, $(d - \mathbf{c}^\mathsf{T} \mathbf{A}^{-1} \mathbf{c}) > 0$ and its logarithm exists. Subtracting (9.10) from (9.12), we obtain

$$\begin{aligned} q(\boldsymbol{\phi}^{n+1}) &= \log \det \mathbf{C}(\boldsymbol{\phi}^{\text{new}}) - \log \det \mathbf{C}(\boldsymbol{\phi}) \\ &= \log(d - \mathbf{c}^\mathsf{T} \mathbf{A}^{-1} \mathbf{c}) - \log \lambda \\ &= \log \left\{ \lambda + \sum_{i=1}^{J} (\phi_i^{n+1})^2 - \left(\sum_{i=1}^{J} \boldsymbol{\phi}_i \phi_i^{n+1} \right)^\mathsf{T} \mathbf{A}^{-1} \right. \\ &\qquad \left. \times \left(\sum_{i=1}^{J} \boldsymbol{\phi}_i \phi_i^{n+1} \right) \right\} - \log \lambda \end{aligned} \quad (9.14)$$

where the last equality follows from the definitions of $\mathbf{c}$ and d.

It clear from (9.14) that, given $\boldsymbol{\phi}$, $\log \det \mathbf{C}(\boldsymbol{\phi}^{\text{new}})$ is maximized by

$$\phi^{n+1} = \arg \max_{\phi \in \mathcal{B} \setminus \{\phi^1, \cdots, \phi^n\}} q(\phi) \quad (9.15)$$

where the search for ϕ^{n+1} is constrained within the set of candidate basis functions with $\{\phi^1, \cdots, \phi^n\}$ excluded.

The sequential procedures for basis function selection are summarized in the following. Initially we set $\mathcal{B} = \{b(\mathbf{x}; \alpha) : \alpha \in \Lambda\}$, $\boldsymbol{\phi} = 1$, and $n = 0$. We then iterate over n. At each iteration, we select ϕ_{n+1} using (9.15) and (9.14), make the updates $\boldsymbol{\phi} := \begin{bmatrix} \boldsymbol{\phi} \\ \phi^{n+1} \end{bmatrix}$ and $n := n+1$, and move to the next iteration. Although we have not established the convergence of $q(\phi^{n+1})$ theoretically, we find empirically that typically $q(\phi^{n+1})$ decreases with n until it reaches numerical convergence. In practice, one can assume convergence or set a predefined threshold for n to stop the iteration.

We conclude this section by pointing out a geometric interpretation of the selection criterion given by (9.14). As λ can be made arbitrarily small, we can write from (9.14) and (9.6) that

$$q(\phi^{n+1}) \approx \log\left\{\sum_{i=1}^{J}(\phi_i^{n+1})^2 - \left(\sum_{i=1}^{J}\phi_i\phi_i^{n+1}\right)^{\mathsf{T}}\left(\sum_{i=1}^{J}\phi_i\phi_i^{\mathsf{T}}\right)^{-1}\right.$$
$$\left.\times\left(\sum_{i=1}^{J}\phi_i\phi_i^{n+1}\right)\right\} - \text{constant}$$
$$= \log\left\{\left\|\mathbf{P}^{\perp}[\phi_1^{n+1},\cdots,\phi_J^{n+1}]^{\mathsf{T}}\right\|^2\right\} - \text{constant} \qquad (9.16)$$

where $\|\cdot\|$ denotes the Euclidean norm, $\mathbf{P}^{\perp}$ is a $J \times J$ projection matrix representing projection into the orthogonal complement of $\text{span}([\phi_1,\cdots,\phi_J]^{\mathsf{T}})$. This projection operation is illustrated in Figure 9.1.

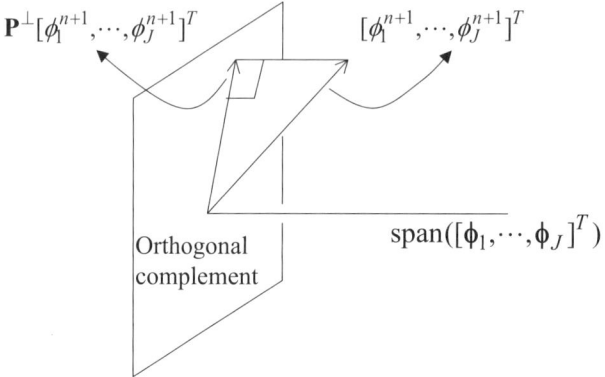

Figure 9.1. Geometric interpretation of the selection criterion given by (9.14).

It is seen that, given basis functions $\phi = [1, \phi^1, \cdots, \phi^n]^{\mathsf{T}}$, the next optimal basis function ϕ^{n+1} as determined by (9.15) is the one that has the maximum orthogonal projection on the orthogonal complement of $\text{span}([\phi_1,\cdots,\phi_J]^{\mathsf{T}})$. With this geometric interpretation, for two basis functions of equal norm, selection by (9.15) favors the one that is more nearly orthogonal to the span of the basis functions already selected.

The selection in (9.15) can therefore be understood as a sequential procedure for selecting principal components, where a principal component consists of a *single* basis function, instead of a linear combination of *all* basis function as in principal component analysis (PCA).

4. Pre-labeling Selection of Data $\mathcal{X}_{\text{tr}}$

Once the basis functions ϕ have been selected, we let them remain fixed and proceed to estimating the parameters $\mathbf{w}$ using (9.5). Though we waived the necessity of knowing any labels during selection of ϕ by working with bounds of the relative empirical loss, we need to know the labels of data while estimating $\mathbf{w}$. Since labeling incurs cost, it is important to select a set of data and label only the selected data. Selective labeling is based on the rationale that labels of different data are not independent of each other. Thus the labels of previous data will predict the label of a new datum. If the dependence is strong and hence the prediction is good enough, the necessity of labeling the new datum will be small. This notion is quantified below.

Assume we are given $J+1$ data $\{\mathbf{x}_1, \cdots, \mathbf{x}_J, \mathbf{x}_*\}$ and their respective labels $\{y_1, \cdots, y_J, y_*\}$. For given basis functions ϕ, we can obtain two classifiers $f(\mathbf{x}; \mathbf{w})$ and $f(\mathbf{x}; \mathbf{w}_*)$, with their parameters estimated in the following way. The parameter $\mathbf{w}$ is estimated using data $\{\mathbf{x}_1, \cdots, \mathbf{x}_J\}$ and their labels, while $\mathbf{w}_*$ is estimated using data $\{\mathbf{x}_1, \cdots, \mathbf{x}_J, \mathbf{x}_*\}$ and their labels. We now examine the losses that the two classifiers receive at $\mathbf{x} = \mathbf{x}_*$. Following (9.1) and (9.5) and the notational convention $\phi_i = \phi(\mathbf{x}_i)$, we have

$$f(\mathbf{x}_*; \mathbf{w}) = \phi_*^\mathsf{T} \mathbf{w} = \phi_*^\mathsf{T} \mathbf{A}^{-1} \sum_{i=1}^{J} y_i \phi_i \qquad (9.17)$$

and

$$\begin{aligned} f(\mathbf{x}_*; \mathbf{w}_*) &= \phi_*^\mathsf{T} [\mathbf{A} + \phi_* \phi_*^\mathsf{T}]^{-1} \left[\sum_{i=1}^{J} y_i \phi_i + y_* \phi_* \right] \\ &\stackrel{2}{=} \phi_*^\mathsf{T} \left[\mathbf{A}^{-1} - \frac{\mathbf{A}^{-1} \phi_* \phi_*^\mathsf{T} \mathbf{A}^{-1}}{1 + \phi_*^\mathsf{T} \mathbf{A}^{-1} \phi_*} \right] \left[\sum_{i=1}^{J} y_i \phi_i + y_* \phi_* \right] \\ &= \frac{\phi_*^\mathsf{T} \mathbf{A}^{-1}}{1 + \phi_*^\mathsf{T} \mathbf{A}^{-1} \phi_*} \left[\sum_{i=1}^{J} y_i \phi_i + y_* \phi_* \right] \\ &\stackrel{4}{=} \frac{f(\mathbf{x}_*; \mathbf{w}) - y_*}{1 + \phi_*^\mathsf{T} \mathbf{A}^{-1} \phi_*} + y_* \end{aligned} \qquad (9.18)$$

where equality 2 follows from the Sherman-Morrison-Woodbury formula and equality 4 is obtained by back-substituting (9.17). It follows from (9.18) that

$$[f(\mathbf{x}_*; \mathbf{w}_*) - y_*]^2 = \frac{[f(\mathbf{x}_*; \mathbf{w}) - y_*]^2}{[1 + \phi_*^\mathsf{T} \mathbf{A}^{-1} \phi_*]^2} \qquad (9.19)$$

An original form of this formula was given in [224]. Since $\mathbf{A}$ is positive definite by construction, we have $1 + \phi_*^\mathsf{T} \mathbf{A}^{-1} \phi_* > 1$. Hence, (9.19) implies that the loss of $f(\mathbf{x}; \mathbf{w}_*)$ received at $\mathbf{x} = \mathbf{x}_*$ is reduced by a factor $(1+\boldsymbol{\phi}_*^\mathsf{T}\mathbf{A}^{-1}\boldsymbol{\phi}_*)^2$, in reference to the corresponding loss of $f(\mathbf{x}; \mathbf{w})$. This reduction is reasonable, because we saw and used y_* during the estimation of $\mathbf{w}_*$ but did not during the estimation of $\mathbf{w}$.

The magnitude of $(1 + \boldsymbol{\phi}_*^\mathsf{T}\mathbf{A}^{-1}\boldsymbol{\phi}_*)^2$ has important implications. If $(1 + \boldsymbol{\phi}_*^\mathsf{T}\mathbf{A}^{-1}\boldsymbol{\phi}_*)^2 \approx 1$, seeing y_*, the label of $\mathbf{x}_*$, does not further reduce the loss that the classifier receives at $\mathbf{x} = \mathbf{x}_*$, implying that the labels of $\{\mathbf{x}_1, \cdots, \mathbf{x}_J\}$ already give a reasonably good prediction of y_* thanks to a strong dependence between them. On the contrary, if $(1 + \boldsymbol{\phi}_*^\mathsf{T}\mathbf{A}^{-1}\boldsymbol{\phi}_*)^2 \gg 1$, seeing y_* greatly reduces the loss of the classifier received at $\mathbf{x} = \mathbf{x}_*$, implying that y_* is weakly dependent on the labels of $\{\mathbf{x}_1, \cdots, \mathbf{x}_J\}$, and thus provides new information to improve the classifier and reduce its loss.

These rationales clearly suggest an optimality criterion for data selection: given that $\mathcal{X}_{\text{tr}} = \{\mathbf{x}_1, \cdots, \mathbf{x}_J\}$ have been selected, the next optimal datum is

$$\mathbf{x}_{J+1,\text{opt}} = \arg\max_{\mathbf{x} \in \mathcal{X} \setminus \mathcal{X}_{\text{tr}}} \left[1 + \boldsymbol{\phi}^\mathsf{T}(\mathbf{x})\mathbf{A}^{-1}\boldsymbol{\phi}(\mathbf{x})\right] \quad (9.20)$$

Data selection proceeds in a sequential manner, starting from the initial setting: $\mathcal{X}_{\text{tr}}$ is null, $\mathbf{A} = \lambda\mathbf{I}_{n+1}$, and $J = 1$. We iterate over J. At each iteration, we select $\mathbf{x}_J$ using (9.20) and move to the next iteration after completing the following updating: $\mathbf{A} := \mathbf{A} + \boldsymbol{\phi}(\mathbf{x}_1)\boldsymbol{\phi}^\mathsf{T}(\mathbf{x}_1)$, $\mathcal{X}_{\text{tr}} := \mathcal{X}_{\text{tr}} \cup \{\mathbf{x}_J\}$, and $J := J + 1$. As the iteration advances $1 + \boldsymbol{\phi}^\mathsf{T}(\mathbf{x}_J)\mathbf{A}^{-1}\boldsymbol{\phi}(\mathbf{x}_J)$ will decrease until it reaches convergence; data selection stops when convergence occurs.

It is important to note that $1+\boldsymbol{\phi}^\mathsf{T}(\mathbf{x}_J)\mathbf{A}^{-1}\boldsymbol{\phi}(\mathbf{x}_J)$ is independent of the labels y. Therefore the data selection is a pre-labeling operation. Upon completing selection of data $\mathcal{X}_{\text{tr}}$, the pre-labeling stage is ended. We then proceed to the labeling stage, in which we go out to acquire labels $\mathcal{Y}_{\text{tr}}$ for $\mathcal{X}_{\text{tr}}$, and employ (9.5) to compute $\mathbf{w}$ based on $\mathcal{X}_{\text{tr}}$ and $\mathcal{Y}_{\text{tr}}$.

5. Connection to Theory of Optimal Experiments

The plan-in-advance active learning methods presented in Sections 3 and 4 have a strong connection to the theory of optimal experiments [81] developed in statistics. From the experiment-theoretic viewpoint, the classifier in (9.1) can be regarded as a model for conditional expectation of class label y given the datum $\mathbf{x}$, i.e.,

$$\mathbb{E}(y|\mathbf{x}, \mathcal{H}_\phi) = \mathbf{w}^\mathsf{T}\boldsymbol{\phi}(\mathbf{x}) \quad (9.21)$$

where $\mathcal{H}_\phi$ denotes the hypothesis that "ϕ is the true model". We further assume

$$\text{var}(y|\mathbf{x}, \mathcal{H}_\phi) = \sigma_\phi^2 \qquad (9.22)$$

Other than (9.21) and (9.22), we make no assumptions on the form of $p(y|\mathbf{x}, \mathcal{H}_\phi)$.

D-optimal Basis Functions.

For given basis $\phi = [1, \phi^1, \cdots, \phi^n]^\mathsf{T}$, training data $\mathcal{X}_\text{tr} = \{\mathbf{x}_1, \cdots, \mathbf{x}_J\}$, and associated labels $\mathcal{Y}_\text{tr} = \{y_1, \cdots, y_J\}$, we consider the linear estimator

$$\mathbf{w} = \mathbf{J}^{-1} \sum_{i=1}^{J} y_i \phi_i \qquad (9.23)$$

where

$$\mathbf{J} = \sum_{i=1}^{J} \phi_i \phi_i^\mathsf{T}, \qquad (9.24)$$

is the Fisher information matrix [64].

Let $\mathbf{w}_0$ be the true value of $\mathbf{w}$ in (9.21), $\mathbf{y} = [y_1, \cdots, y_J]^\mathsf{T}$, and $\mathbf{T} \in \mathbb{R}^{(n+1) \times J}$. An estimator of the form $\mathbf{w} = \mathbf{T}\mathbf{y}$ satisfying $\mathbb{E}_{\mathcal{Y}_\text{tr}|\mathcal{X}_\text{tr}}(\mathbf{T}\mathbf{y}) = \mathbf{w}_0$ is a linear unbiased estimator of $\mathbf{w}$ in (9.21). It was proven in [81] that, among all linear unbiased estimators of $\mathbf{w}$ in (9.21), the one specified by (9.23) and (9.24) has the smallest covariance (dispersion) matrix and this covariance matrix is given by $\mathbf{D} = \mathbf{J}^{-1}$. Here, by "smallest" we mean $\widetilde{\mathbf{D}} - \mathbf{D}$ is nonnegative definite, with $\widetilde{\mathbf{D}}$ the covariance matrix of any linear unbiased estimator of $\mathbf{w}$ in (9.21).

Under hypothesis $\mathcal{H}_\phi$, $\mathbf{J}^{-1}$ achieves the Cramér-Rao bound (CRB) for estimation of $\mathbf{w}$. The matrix $\mathbf{D}$ is widely used in the theory of optimal experiments to generate various D-optimal criteria, among which det$\mathbf{D}$ is popular. Using det$\mathbf{D}$, we have the D-optimal selection criterion

$$\phi_\text{D-opt} = \arg\min_\phi \{\det \mathbf{D}(\phi)\} \qquad (9.25)$$

which identifies basis functions whose parameters $\mathbf{w}$ achieve the minimum Cramér-Rao bound for the given training data $\mathcal{X}_\text{tr}$.

Comparing (9.24) to (9.6), we have $\mathbf{J} \approx \mathbf{A}$, with the approximation arbitrarily accurate by setting λ small. Therefore (9.10) can be rewritten as $\log \det \mathbf{C}(\phi) \approx \log \det \mathbf{J} - \log \lambda^{n+1} = -\log \det \mathbf{D} - \log \lambda^{n+1}$, and as a result, the ϕ that minimizes det$\mathbf{D}$ approximately maximizes $\log \det \mathbf{C}(\phi)$. Thus, $\phi_\text{D-opt}$ is approximately equal to $\phi_\text{approx.opt}$, which is given in (9.9).

D-optimal Data for Label Acquisition. Assume that $\mathcal{X}_{\text{tr}} = \{\mathbf{x}_1, \cdots, \mathbf{x}_J\}$ have been selected at the time step indexed by J, with the corresponding Fisher information matrix given in (9.24). For a new datum $\mathbf{x}_*$, the Fisher information matrix corresponding to $\mathcal{X}_{\text{tr}} \cup \{\mathbf{x}_*\}$ is found to be

$$\mathbf{J}_{\text{new}} = \mathbf{J} + \phi(\mathbf{x}_*)\phi^{\mathsf{T}}(\mathbf{x}_*) \qquad (9.26)$$

and, by using properties of a matrix's determinant,

$$\log \det \mathbf{J}_{\text{new}} = \log \det \mathbf{J} + \log\left[1 + \phi^{\mathsf{T}}(\mathbf{x}_*)\mathbf{J}^{-1}\phi(\mathbf{x}_*)\right] \qquad (9.27)$$

Therefore $\log\left[1 + \phi^{\mathsf{T}}(\mathbf{x}_*)\mathbf{J}^{-1}\phi(\mathbf{x}_*)\right]$ measures the increase in $\log \det \mathbf{J}$, or equivalently, the decrease in $\log \det \mathbf{D}$, brought about by the new datum $\mathbf{x}_*$. It is clear we should maximize this increase, since we want to minimize the Cramér-Rao bound. Thus, given that $\mathcal{X}_{\text{tr}}$ have been selected D-optimally, the next D-optimal datum is

$$\mathbf{x}_{J+1,\text{D-opt}} = \arg\max_{\mathbf{x}} \left[1 + \phi^{\mathsf{T}}(\mathbf{x})\mathbf{J}^{-1}\phi(\mathbf{x})\right] \qquad (9.28)$$

Comparing (9.28) to (9.20), we have $\mathbf{x}_{J+1,\text{D-opt}} \approx \mathbf{x}_{J+1,\text{opt}}$, as $\mathbf{J} \approx \mathbf{A}$.

6. Application to UXO Detection

The active learning methodology presented in the foregoing sections may be applied to any detection problem for which the data labels are expensive to acquire, and for which there is no distinct training data. In particular, we consider the detection of buried unexploded ordnance (UXO). For UXO remediation, the label of a potential target is acquired by excavation, a dangerous and time consuming task. The overwhelming majority of UXO cleanup costs come from excavation of non-UXO items.

The results presented here were originally reported in [260]. The data used were collected at an actual UXO site: Jefferson Proving Ground in the United States. The approach developed in Sections 3 and 4 is compared with existing procedures. Specifically, the principal challenge in UXO sensing is development of a training set for design of the detection algorithm. At an actual UXO site there is often a significant quantity of UXO, UXO fragments and man-made clutter on the surface. It has been recognized that the characteristics of the surface UXO and clutter is a good indicator of what will be found in the subsurface. Consequently, in practice, a subset of the surface UXO and clutter are buried, and magnetometer and induction data are collected for these items, for which the labels are obviously known. The measured data and associated labels (UXO/non-UXO) are then used for training purposes. Of course, the process of burying, collecting data, and then excavating these emplaced items is time consuming and dangerous (for the UXO items), with this procedure eliminated by the active learning techniques developed in Sections 3 and 4.

6.1 Magnetometer and electromagnetic induction sensors

Magnetometer and electromagnetic induction (EMI) sensors are widely applied in sensing buried conducting/ferrous targets, such as landmines and UXO. The magnetometer is a passive sensor that measures the change of the Earth's background magnetic field due to the presence of a ferrous target. Magnetometers measure static magnetic fields. An EMI sensor actively transmits a time-varying electromagnetic field, and consequently senses the dynamic induced secondary field from the target. To enhance soil penetration, EMI sensors typically operate at kilohertz frequencies. We here employ a frequency-domain EMI sensor that transmits and senses at several discrete frequencies [255]. Magnetometers only sense ferrous targets, while EMI sensors detect general conducting and ferrous items.

Parametric models have been developed for both magnetometer and EMI sensors [86, 44, 259]. The target features $\mathbf{x}$ are extracted by fitting the EMI and magnetometer models to measured sensor data. The vector $\mathbf{x}$ has parameters from both the magnetometer and EMI data, and therefore in this sense the data from these two sensors are "fused." The one place where these two models have overlapping parameters is in specification of the target position. The magnetometer data often yields a very good estimation of the target position, and therefore such are used in $\mathbf{x}$. In fact, the target position specified by the magnetometer data is explicitly utilized as prior information when fitting EMI data to the EMI parametric model. Details on the magnetometer and EMI models, and on the model-fitting procedure, may be found in [259].

The features employed are as in [259], and the features are centered and normalized. Specifically, using the training data, we compute the mean feature vector $\mathbf{x}_{\text{mean}}$ and the standard deviation of each feature component (let σ_i represent the standard deviation of the ith feature). Before classification, a given feature vector $\mathbf{x}$ is shifted by implementing $\mathbf{x}_{\text{shift}} = \mathbf{x} - \mathbf{x}_{\text{mean}}$, and then the ith feature component of $\mathbf{x}_{\text{shift}}$ is divided by σ_i to effect the normalization.

6.2 Measured sensor data from the Jefferson Proving Ground

Jefferson Proving Ground (JPG) is a former military range that has been utilized for UXO technology demonstrations since 1994. We consider data collected by Geophex, Ltd. in Phase V of the JPG demonstration. The goal of the JPG V was to evaluate the UXO detection and discrimination abilities under realistic scenarios, where man-made and natural clutter coexist with UXO items. Our results are presented with the GEM-3 and magnetometer data from two adjoining areas, constituting a total of approximately five acres. There are 433 potential targets detected from sensor anomalies, 40 of which are proven to be UXO and the others are clutter. The excavated UXO items include 4.2 inch, 60 mm, and 81mm mortars; 5 inch, 57 mm, 76 mm, 105 mm, 152 mm, and 155 mm projectiles; and 2.75 inch rockets.

This test was performed with U.S. Army oversight. One of the two JPG areas was assigned as the training area, for which the ground truth (UXO/non-UXO) was given. The trained detection algorithms were then tested on the other area, and the associated ground truth was revealed later to evaluate performance. It was subsequently recognized that several UXO types were found in equal numbers in each of the two areas. This indicates an effort to match the training data to the detection data, in the manner discussed above, involving burial of known UXO and non-UXO collected on the surface.

Each sensor anomaly is processed by fitting the associated magnetometer and EMI data to the parametric models [259], and the estimated parameters define x. In addition, the model-fitting procedure functions as a prescreening tool. Any sensor anomaly failing to fit well to the model is regarded as having been generated by a clutter item. Therefore, a total of 300 potential targets remain after this prescreening stage, 40 of which are UXO. The 300 potential targets define $\mathcal{X}$, a set of 300 feature vectors, each associating one of the potential targets. In the training area, there are 128 buried items, 16 of which are UXO.

6.3 Detection results

Before presenting classification results, we examine the characteristics of the basis functions selected in the first phase of the algorithm, prior to adaptively selecting training data. In Figure 9.2 we consider the first three basis functions ϕ^1, ϕ^2 and ϕ^3 selected by the first stage of the algorithm. For each feature vector x (from all UXO and non-UXO), we compute a vector $[\phi^1(\mathbf{x}), \phi^2(\mathbf{x}), \phi^3(\mathbf{x})]^\mathsf{T}$. By examining this three-dimensional vector for all x, we may observe the degree to which UXO and non-UXO features are distinguished via the features and kernel. A radial basis function kernel is employed here, corresponding to the kernel used to select the basis functions (see discussion below concerning the selected kernel). By examining Figure 9.2 we observe that the UXO and non-UXO features are relatively separated, although there is significant overlap, undermining classification performance.

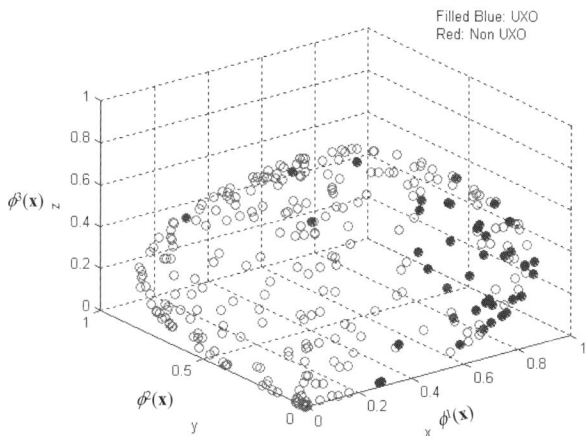

Figure 9.2. For the first three basis functions selected, ϕ^1, ϕ^2, ϕ^3, the 3-D vector $[\phi^1(\mathbf{x}), \phi^2(\mathbf{x}), \phi^3(\mathbf{x})]^\mathsf{T}$ is plotted. Considered are feature vectors $\mathbf{x}$ for all UXO and non-UXO targets considered in this study (Figure originally appeared in [260] - ©2005 IEEE used with permission).

The detection results are presented in the form of the receiver operating characteristic (ROC), quantifying the probability of detection (Pd) as a function of the false alarm count. We present ROC curves using the adaptive-training approach discussed in Sections 3 and 4, with performance compared to results realized by training on the distinct training region discussed above (the latter approach reflects current practice). With regard to conventional training, the classifier takes the same form as in (9.1), with w determined iteratively using kernel matching pursuits (KMP). Details on the KMP algorithm may be found in [242, 158]. To make the comparison appropriate, the active learning approach and the KMP approach employ the same set of candidate basis functions $\mathcal{B}$. In particular, $\mathcal{B}$ is constructed as a set of radial basis functions (RBF) [65]

$$\mathcal{B} = \left\{ \phi(\mathbf{x}) : \phi(\mathbf{x}) = \exp\left(-\frac{\|\mathbf{x} - \widetilde{\mathbf{x}}\|^2}{2\mu^2}\right), \widetilde{\mathbf{x}} \in \mathcal{X} \right\} \quad (9.29)$$

where μ is adaptively adjusted by applying gradient search to maximize the respective criteria of the active approach and the KMP with respect to μ, given that $\widetilde{\mathbf{x}}^{n+1}$ has been selected.

As indicated above, the designated training area has 128 labeled items, and conventionally classifiers are tested on the remaining signatures, in this case constituting 172 items. As a first comparison, the active learning technique developed in Sections 3 and 4 is employed to select $J = 128$ items from the original 300, with these "excavated" to learn the associated labels. This therefore defines the set $\mathcal{X}_{tr}$ and associated labels $\mathcal{Y}_{tr}$. The basis functions $\{\phi^1, \cdots, \phi^n\}$ are also determined adaptively using the original 300 signatures, and here $n = 10$. The performance of the adaptive learning algorithm is then tested on the remaining 172 $\mathbf{x} \notin \mathcal{X}_{tr}$, although these are generally not the same testing examples used by the KMP algorithm (the training sets do not overlap completely). For comparison, we also show training and testing results implemented via KMP, in which the 128 training examples are selected randomly from the original 300 signatures in $\mathcal{X}$. Performance comparisons are shown in Figure 9.3, wherein we present results for active data selection (methods in Sections 3 and 4), KMP results using the assigned 128 training examples (the training examples were carefully designed *a priori*), and average results for randomly choosing the 128 examples for KMP training (100 random selections were performed). In addition, for the latter case we also place error bars on the results; the length of the error bar is twice the standard derivation of the Pd for the associated false-alarm count. Therefore, if the result is Gaussian distributed, 95% of the values lie within the error bar.

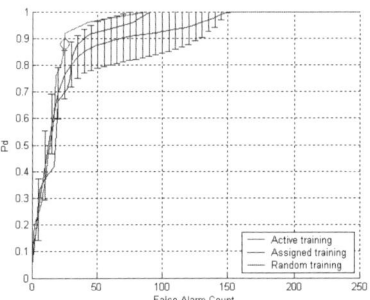

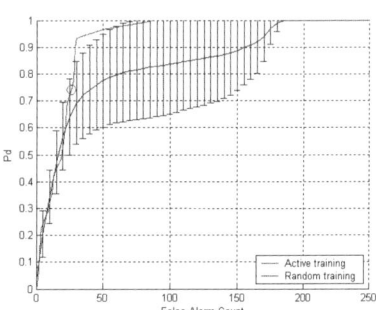

Figure 9.3. ROC curves based on 128 training examples, for which the target labels were known. Being compared are KMP with carefully or randomly designed training set (error bars shown in the latter case) and the methods in Sections 3 and 4 (Figure originally appeared in [260] - ©2005 IEEE used with permission).

Figure 9.4. As in Figure 9.3, but now results are only shown for adaptive training-data selection (Sections 3 and 4) and for random selection. Results are shown for $J = 90$ training examples (Figure originally appeared in [260] - ©2005 IEEE used with permission).

Before proceeding, we note that the ROC curves are generated by varying the threshold t, as applied to the estimated label y. For the binary UXO-classification problem considered here, by design we choose the label $y = 1$ for UXO and $y = 0$ for non-UXO. In practice one must choose one point on the ROC at which to operate. A naïve choice of the operating point would be 0.5 (i.e., if the classifier maps a testing feature vector **x** to a label $y > 0.5$ the item is declared UXO, and otherwise it is declared non-UXO). However, we must account for the fact that in practice the number of non-UXO items is often much larger than the number of UXO. We have therefore invoked the following procedure.

We assume that the error (noise) between the true label ($y = 1$ or $y = 0$) and the estimated label is i.i.d. Gaussian with variance of σ^2. Let N_0 and N_1 represent respectively the number of non-UXO and UXO items in the training set. Considering the UXO ($y = 1$) data, an unbiased estimator of the label y will yield a mean of one and a minimum variance of σ^2/N_1. Similarly, considering the non-UXO data ($y = 0$), an unbiased estimator of the label y will have zero mean and minimum variance σ^2/N_0. Let $\mathcal{H}_1$ and $\mathcal{H}_0$ correspond to the UXO and non-UXO hypotheses. Based upon the above discussion, we model the probability density function of y for the $\mathcal{H}_1$ and $\mathcal{H}_0$ hypotheses as $p(y|\mathcal{H}_1) = \mathcal{N}(1, \sigma^2/N_1)$ and $p(y|\mathcal{H}_0) = \mathcal{N}(0, \sigma^2/N_0)$. Rather than setting the threshold at $t = 0.5$, we set the threshold at that value of y for which $p(y|\mathcal{H}_1) = p(y|\mathcal{H}_0)$, yielding

$$t = \frac{N_1 - \sqrt{N_1^2 - (N_1 - N_0)(N_1 + \sigma^2 \log \frac{N_0}{N_1})}}{N_1 - N_0} \quad (9.30)$$

Assuming σ^2 is a small, we omit $\sigma^2 \log \frac{N_0}{N_1}$, obtaining

$$t = \frac{\sqrt{N_1}}{\sqrt{N_1} + \sqrt{N_0}} \quad (9.31)$$

From (9.31), the appropriate threshold is $t = 0.5$ only if $N_0 = N_1$.

For example, in Figure 9.3, only 15 of the 128 actively selected training data are UXO, and therefore $N_1 = 15$, $N_0 = 113$. If we set the threshold to be $t = 0.5$, we detect 16% of the UXO with two false alarms. By contrast, using the procedure discussed above (for which $t = 0.27$), we detect 88% of the UXO with 25 false alarms. The operating point corresponding to $t = 0.27$ is indicated in Figure 9.3. We similarly plot this point in all subsequent ROCs presented below.

We observe from the results in Figure 9.3 that the active data selection procedure produces the best ROC results (for Pd > 0.7, which is of most interest in practice), with the KMP results from the specified training area almost as good. It is observed that the average performance based on choosing the training set randomly is substantially below that of the two former approaches, with significant variability reflected in the error bars. These results demonstrate the power of the active-data-selection algorithm introduced in Sections 3 and 4, and also that the training data defined for JPG V is well matched to the testing data.

In the first example we set $J = 128$ to be consistent with the size of the training area specified in the JPG V test. The algorithm in Sections 3 and 4 can be implemented for smaller values of J, reflecting less excavation required in the training phase (for determination of target labels). It is of interest to examine algorithm performance as J is decreased from 128. In this case training is performed using signatures and labels from the J "excavated" items, and testing is performed on the remaining $300 - J$ examples. Results are presented for the active training procedure and for randomly choosing J training examples (100 random instantiations), as in Figure 9.3. In Figures 9.4, 9.5, 9.6 results are presented for $J = 90, 60$ and 40. Using $J = 90$ rather than $J = 128$ results in very little degradation in ROC performance (comparing 9.3 and 9.4), with a slight performance drop for $J = 60$, and a more substantial drop for $J = 40$. It is interesting to note that with decreasing J, the number of test items $300 - J$ increases, therefore increasing the number of false-alarm opportunities. This further highlights the quality of the results in Figures 9.4, 9.5, 9.6, vis-à-vis Figure 9.3. In all of these and subsequent examples, the number of selected basis functions is $n = 10$.

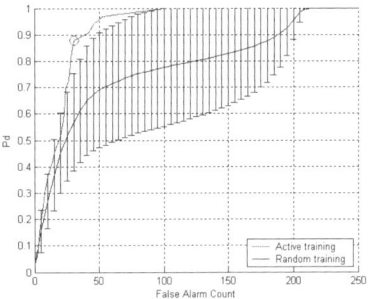
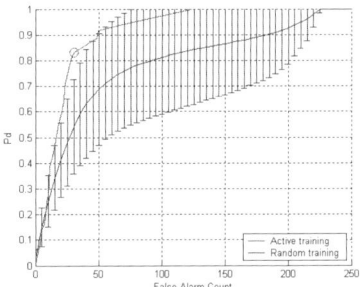

Figure 9.5. As in Figure 9.4, with $J = 60$ (Figure originally appeared in [260] - ©2005 IEEE used with permission).

Figure 9.6. As in Figure 9.4, with $J = 40$ (Figure originally appeared in [260] - ©2005 IEEE used with permission).

In the above examples J was specified to be matched to the size of a specified training set, or it was varied for comparison to such. However, the procedure in Sections 3 and 4 may be employed to adaptively determine the size of the desired training set $\mathcal{X}_{tr}$, based on the information gain as J is increased.

Specifically, we track the information gain $1 + \phi^\mathsf{T}(\mathbf{x}_J)\mathbf{A}^{-1}\phi(\mathbf{x}_J)$ for increasing J, and terminate the algorithm when the information gain converges. At this point, adding a new datum to the training dataset does not provide significant additional information to the classifier design.

For the JPG V data, the information gain $1 + \phi^\mathsf{T}(\mathbf{x}_J)\mathbf{A}^{-1}\phi(\mathbf{x}_J)$ is plotted in Figure 9.8 as a function of J. Based on Figure 9.8 the size of the training set is set to $J = 65$. In Figure 9.7 results are shown for $J = 65$, with comparison as before to KMP results in which the $J = 65$ training examples are selected randomly. Examining the results in Figure 9.7, we observe that the active selection of training data yields a detection probability of approximately 0.95 with approximately 35 false alarms; *on average* one encounters about five times this number of false alarms to achieve the same detection probability (when selecting the training data randomly).

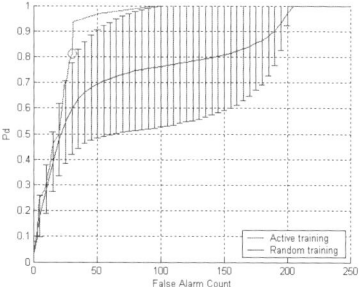
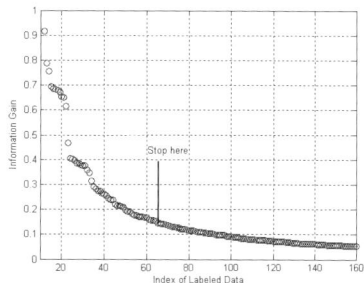

Figure 9.7. ROC curves based on $J = 65$ training examples, comparing the adaptive procedure (Sections 3 and 4) to random training data selection. Number of training examples chosen based on Figure 9.8 (Figure originally appeared in [260] - ©2005 IEEE used with permission).

Figure 9.8. Information gain, $1 + \phi^\mathsf{T}(\mathbf{x}_J)\mathbf{A}^{-1}\phi(\mathbf{x}_J)$, of adding a new datum, as a function of the number of the training examples J, selected adaptively (Figure originally appeared in [260] - ©2005 IEEE used with permission).

7. Chapter Summary

The generalization of a classifier is highly dependent on the training data set $\mathcal{X}_\mathrm{tr}$, given the underlying label generator $p(y|\mathbf{x})$. From the perspective of experimental design, $\mathcal{X}_\mathrm{tr}$ control the locations at which the labeling experiments are performed to learn the underlying label generator $p(y|\mathbf{x})$. A $\mathcal{X}_\mathrm{tr}$ drawn i.i.d.

from $p(\mathbf{x})$, the underlying data generator, is not optimal because it ignores the dependence between labels at different $\mathbf{x}$ locations.

The active learning approach presented in this chapter accounts for this between-label dependence by drawing $\mathcal{X}_{tr}$ according to some selection criteria. The selection criteria we have developed have the appealing property that they allow one to *plan in advance* the visits to the data in $\mathcal{X}_{tr}$ in performing the labeling experiments.

The *plan-in-advance* approach has particular implications in remote-sensing problems, in which one collects data from a given site and aims to specify the identity of the object responsible for the measured field. Due to the variability and site-dependent characteristic of many target signatures, it is often difficult to have reliable training data *a priori*. Using the plan-in-advance approach developed in this chapter, one can determine a set of training data, without having to see the labels. The data are selected based on criteria that are directly related to loss minimization and yet are independent of labels. It is after the data selection is completed that one actually sets out to acquire the labels. Thus the labeling experiments can be performed by following an optimal route of visits to the selected data. In the context of landmine and UXO sensing, labeling experiments correspond to excavating the respective buried items. This is a reasonable procedure, since landmines and UXO need be excavated ultimately anyway, and therefore active learning essentially prioritizes the order in which items are excavated, with the goal of ultimately excavating fewer non-targets (false alarms) via proper data selection. The plan-in-advance approach has been demonstrated successfully on measured magnetometer and EMI data from an actual former bombing range, addressing the sensing of UXO.

Chapter 10

APPLICATION OF SENSOR SCHEDULING CONCEPTS TO RADAR

William Moran

University of Melbourne, Melbourne, Australia

Sofia Suvorova

University of Melbourne, Melbourne, Australia

Stephen Howard

Defence Science and Technology Organisation, Edinburgh, Australia

1. Introduction

In this chapter, we spend some time illustrating the ideas on sensor scheduling in a specific context: that of a radar system. A typical pulse radar system operates by illuminating a scene with a short pulse of electromagnetic energy at radio frequency (RF) and collecting the energy reflected from the scene. Newer radar systems have the capability to modify the method of illumination in several ways. Parameters such as time between pulses or pulse repetition interval (PRI), carrier frequency, the shape of the transmitted waveform, the shape and direction of the beam, etc., which were typically hard-wired in previous generations of radars, may now be regarded as controllable during system operation. All of these parameters admit the possibility of modification based on knowledge of the scene, so as best to extract the required information. We remark that scheduling "on the transmit side" is quite different from the possibility of changing processing after receipt of the return. If different methods of processing are contemplated for the radar return, then in principle these can be done in parallel, so that in some sense no choice between the two approaches is required. On the transmit side, in particular, parameter selections may have to be made in an especially timely way to keep pace with the system's operating tempo.

Ideally we would like, on the basis of what we now know about the scene, including "targets of interest" and uninteresting reflections, to schedule the transmission parameters over the long-term to maximize the operationally important information returned to the operator on time scales required for operational purposes. Such a goal would demand scheduling algorithms capable of undertaking non-myopic scheduling on a pulse-by-pulse basis. As we have seen in earlier chapters, computational considerations make such a goal infeasible at the moment. Simplifications that permit suboptimal solutions are required. Ideas associated with these are discussed in the following sections; but first we describe the basic mathematics of radar.

2. Basic Radar

For our purposes, a radar system is comprised of a transmitter and a receiver of electromagnetic radiation at a common frequency and with a common clock. The clock maintains accurate times for transmission and reception of signals, and importantly it keeps a record of phase of the carrier.

2.1 Range-Doppler Ambiguity

The transmitter is assumed to emit a signal $\mathbf{w}_{Tx}(t)$. This signal is a rapidly oscillating sinusoid (the "carrier") on which is imposed a relatively slowly varying modulation $\mathbf{s}_{Tx}(t)$ (the "waveform"):

$$\mathbf{w}_{Tx}(t) = \mathbf{s}_{Tx}(t) \sin(2\pi f_c t), \qquad (10.1)$$

where f_c is the carrier frequency.

For the moment we are not concerned with the direction of transmission or polarization of the transmit beam. While the total amount of energy depends on a number of factors, some of it hits each reflecting element in the scene and a small proportion of this energy is reflected back to the antenna of the receiver. To put energy issues in perspective, the total energy returning from an element of the scene falls off as the inverse fourth power of the distance of the element from the radar. This means that the amount of energy entering a radar from a small target at distance is often tiny even relative to the thermal noise in the receiver. Significant processing is needed to extract the signal from the noise in such circumstances.

Sensor Scheduling in Radar

The return to the receiver is generally assumed to be a linear superposition of reflections from the different elements of the scene. As a result, in the current analysis of the response of the radar to the scene, it is enough to consider the response to a single reflector, which we shall refer to as the *target*.

We assume that the receiver is collocated with the transmitter, so that the waveform returns to the receiver a time delay of $\tau = 2R/c$ later, where R is the distance to the target and c is the speed of light. The return signal is $\mathbf{w}_{Tx}(t-\tau)$ multiplied by an attenuation and phase shift factor due to the distance involved, the nature of the target, and atmospheric conditions. At this stage, we ignore this factor. On return to the receiver, the carrier is stripped off (demodulated).

There are two channels in the receiver; the in-phase (I) channel corresponding to demodulation by $\sin(2\pi f_c t)$, and the quadrature (Q) channel corresponding to demodulation by $\cos(2\pi f_c t)$. This duplication of effort is required to sense the phase (or at least the phase-differences) in the return. After demodulation, the signal in the I channel of the receiver is $\mathbf{s}_{Tx}(t-\tau)\cos(2\pi f_c \tau)$ and in the Q channel is $-\mathbf{s}_{Tx}(t-\tau)\sin(2\pi f_c \tau)$. We can now regard this as a complex waveform

$$\mathbf{s}_{Tx}(t-\tau)e^{-2\pi i f f_c \tau}. \tag{10.2}$$

In this way, we can think even of the transmitted signal $\mathbf{w}_{Tx}(t)$ as having values in the complex domain. The real and complex parts of the signal are imposed on the I and Q channels respectively to form the transmit waveform, so that again the complex return is

$$\mathbf{s}_{Tx}(t-\tau)e^{-2\pi i f f_c \tau}. \tag{10.3}$$

At this point, the signal is filtered against another signal $\mathbf{s}_{Rx}(t)$. From a mathematical perspective, this produces the filtered response[1]

$$\int_{\mathbb{R}} \mathbf{s}_{Tx}(t-\tau)\overline{\mathbf{s}_{Rx}(t-x)}\, dt. \tag{10.4}$$

So far we have assumed that the target is stationary. When the target moves the return is modified by the Doppler effect. If this is done correctly, it results in a "time dilation" of the return. For our purposes, where the signals have small bandwidth relative to the carrier frequency, and the pulses are relatively short, the narrowband approximation allows this time dilation to be closely approximated by a frequency shift, so that if the target has a radial velocity v,

---

[1] In this chapter, overbar is used to denote complex conjugation.

the return becomes

$$s_{\text{Txc}}(t-\tau)e^{2\pi i f_c(1-2v/c)(t-\tau)}.$$

This is demodulated and filtered as in equation (10.4) to obtain

$$A_{s_{\text{Txc}},s_{\text{Rx}}}(\tau, f_d) = \int_{\mathbb{R}} e^{-2\pi i f_d(t-\tau)} s_{\text{Txc}}(t-\tau)\overline{s_{\text{Rx}}(t-x)}\, dt$$

$$= \int_{\mathbb{R}} e^{-2\pi i f_d t} s_{\text{Txc}}(t)\overline{s_{\text{Rx}}(t-(x-\tau))}\, dt. \quad (10.5)$$

This function of time delay τ and Doppler f_d is called the *ambiguity function* of the transmit and receive signals. To be more precise, this is referred to as the *cross ambiguity function* since the filtering kernel used by the receiver is, in principle, different from the transmitted signal. When the filtering kernel and transmit signal are the same, this filtering process is called *matched filtering* and its result the *auto-ambiguity function*. Propagation through the transmission medium and electronic processing in the receiver add noise to the signal and the effect of the matched filter is to maximize the signal power relative to the noise power in the processed signal.

No discussion of the ambiguity function is complete without a mention of Moyal's Identity. This states that if f and g are finite energy signals (i.e., in $L^2(\mathbb{R})$) then $A_{f,g}$ is in $L^2(\mathbb{R}^2)$ and for f_1, g_1 also in $L^2(\mathbb{R})$,

$$\langle A_{f_1,g_1}, A_{f,g}\rangle = \langle f_1, f\rangle\overline{\langle g_1, g\rangle}. \quad (10.6)$$

In particular this says that if consider the auto-ambiguities of f and f_1, we obtain

$$\langle A_{f_1,f_1}, A_{f,f}\rangle = |\langle f_1, f\rangle|^2. \quad (10.7)$$

Finally, letting $f = f_1$ we find

$$||A_{f,f}||^2 = ||f||^4. \quad (10.8)$$

It is clear also that $A_{f,f}(0,0) = ||f||^2$ and that this is the maximum value of the ambiguity (by the Cauchy-Schwarz Inequality). This implies that the ambiguity function must spread out; it cannot be concentrated close to the origin. Thus, there is always a non-trivial trade-off between Doppler and time delay (range) measurements. The choice of waveforms determines where this trade-off is pitched. At one extreme, for example, long sinusoidal waveforms will give good Doppler measurements but bad range measurements.

We return to a general range-Doppler scene; again for the moment no azimuth and, in this treatment will not discuss polarization. Such a scene is a

Sensor Scheduling in Radar   225

linear combination $\sum_k \sigma(\tau_k, \phi_k)\delta(t - \tau_k, f - \phi_k)$ of reflectors (scatterers) at delays $\tau_k = 2r_k/c$ and Doppler shifts $\phi_k = 2f_c v_k/c$, where the range and radial velocity of the kth target are r_k and v_k respectively. The processed return is

$$\sum_k \alpha_k A_{s_{\text{Txc}}, s_{\text{Rx}}}(t - \tau_k, f - \phi_k). \tag{10.9}$$

We end this section with a few remarks on the use of waveforms in practice. First, as we have already remarked, there are good reasons to choose $s_{\text{Rx}} = s_{\text{Txc}}$. While this is by no means always the case, there have to be other good reasons (in terms of achievable side-lobes in the ambiguity, for example) not to do this. For the remainder of this section we make that assumption and restrict our attention to matched filtering and the auto-ambiguity function.

In view of the nature of the ambiguity function, it may seem that an optimal choice of waveform to achieve very accurate range resolution is a very short pulse; as sharp as the available bandwidth will allow. While this solution is sometimes adopted, it has disadvantages, the most important of which is that such a pulse carries very little energy. In view of the $1/r^4$ decay in return energy from a scatterer at distance r from the radar, it is important to transmit as much energy as other considerations will allow. Accordingly a "pulse compression" approach is adopted. Waveforms s_{Txc} are chosen that are relatively long pulses (though, as we shall see in section 2.3, not too long) in time but are such that their auto-correlations (i.e., their ambiguity at zero Doppler) has a sharp peak at zero time delay and has small side-lobes away from this peak. In other words, the ambiguity function at zero Doppler is close to a "thumbtack" in the range direction.

We remark that for the kind of radar system we are interested in here (so-called "pulse-Doppler radars"), the pulse is short enough that Doppler has an insignificant effect on it. Effectively then the ambiguity is constant in the Doppler direction. One might ask why we have spent so much time on the ambiguity if it is so simple in this situation. The first reason is that this is totally general to any kind of radar and exhibits the limits of radar as a sensor. The second reason is that, while the effects of Doppler are relatively trivial in this context, the study of the ambiguity function quickly shows why this is the case. The final reason is that when we do Doppler processing in Section 2.3 it will be important to see that theory in the context of radar ambiguity.

Probably the most commonly used (complex) waveform in radar is a linear frequency modulated (LFM) or "chirp" pulse:

$$s_{\text{Txc}}(t) = \exp 2\pi i \gamma t^2. \tag{10.10}$$

The "chirp rate" γ determines the effectiveness of the pulse. We illustrate the ambiguity with this example. The key feature is that the ambiguity has a ridge

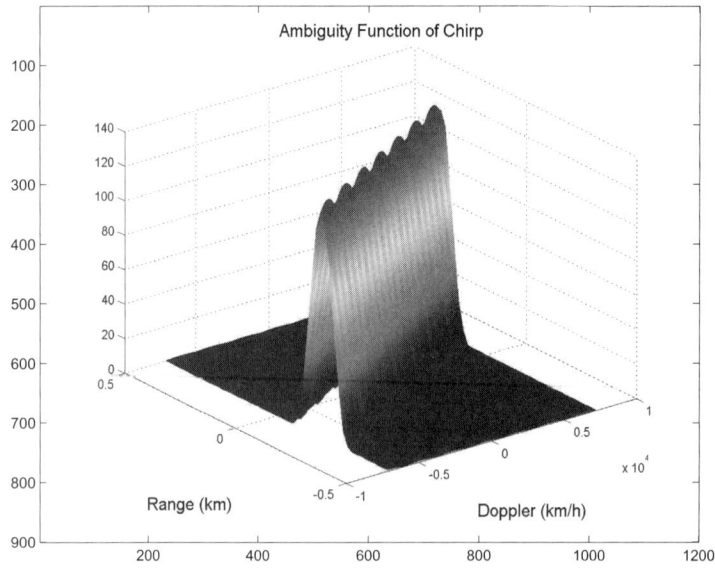

Figure 10.1. Absolute value of ambiguity of a linear frequency modulated pulse.

at an angle to the range axis and decays rapidly away from this ridge. Along the ridge is a slight ripple. The angle made by this ridge is a function of γ; in fact, in the correct units, the tangent of the angle is γ.

2.2 Elevation and Azimuth

In addition to the measurements of range and, potentially, Doppler discussed in the previous section, it is usually important to have a good estimate of the azimuth of the target. In some applications the elevation is of interest too. The focus of this discussion will be on azimuth, while we remark that the elevation can be treated in essentially the same way. Many radars have a dish antenna of some kind that has a fixed beam-pattern, but it is in the philosophy of this book to consider systems in which the radar has some flexibility in the shape of its beam, as well as the ability to rapidly switch the direction of the beam. This is the case for systems using phased-array antennas and we discuss ambiguity for these systems.

Typically we are interested in scenes consisting of complex (that is, range and phase) information at each azimuth angle, range bin and Doppler bin.[2] Thus we shall regard a *scene* as a complex function $\sigma(r, v, \alpha)$ of range r, radial velocity v and azimuth α. More often, it will be described by a collection of scatterers at points (r_k, v_k, α_k) where the complex magnitude of a scatterer is $\sigma(r_k, v_k, \alpha_k)$. It will also be convenient, with some abuse of notation, to think of the scene in terms of delay $\tau_k = 2r_k/c$ and Doppler shift $\phi_k = 2f_c v_k/c$. The phase of the complex number represents the phase shift on the return induced by the carrier and the scatterer.

The array is assumed to have M antenna elements forming a horizontal linear array, which are separated by a uniform distance δ. The transmitter and receiver share the same antenna elements. Minor variants of this analysis will accommodate alternative schemes. The mth antenna element transmits a relatively slowly varying signal $s_{\text{Tx}}(t)$ modulated onto a rapidly varying carrier to produce the waveform $s_{\text{Tx}}(t)e^{2\pi i f_c t}$. For this analysis, it is important that a time delay corresponding to the distance across the array face makes an insignificant difference to the signal $s_{\text{Tx}}(t)$; this is the role of the "slowly varying" assumption. With this approximation, the kth scatterer receives the waveform

$$s_{\text{Tx}}(t - r_k)e^{2\pi i f_c m \delta \sin \alpha_k} e^{2\pi i f_c(t - r_k)}, \tag{10.11}$$

from the mth antenna element.

We may assign a weight a_m (which can be complex) to each antenna element. This weight includes both a phase-shift and attenuation. Assuming omni-directional antenna elements, the kth scatterer sees the total transmission from all channels (omitting the terms associated with carrier frequency) as

$$\sum_{m=1}^{M} a_m s_{\text{Tx}}(t - r_k) e^{2\pi i f_c m \delta \sin \alpha_k} = s_{\text{Tx}}(t - r_k) \sum_{m=1}^{M} a_m e^{2\pi i f_c m \delta \sin \alpha_k}$$
$$= s_{\text{Tx}}(t - 2r_k)\phi_{\text{tr}}(\alpha_k), \tag{10.12}$$

where $\phi_{\text{tr}}(\alpha) = \sum_{m=1}^{M} a_m e^{2\pi i f_c m \delta \sin \alpha}$ denotes the transmit pattern of the antenna array as a function of azimuth. If the elements are not omni-directional, the transmit pattern will need to be modified to accommodate the element pattern. We remark that the phase and attenuation serve different purposes in the design of the transmit pattern. The phase is used to point the beam in a

---

[2]Processed returns are typically quantized into discrete azimuth, range and Doppler values called *bins*.

given direction. Thus, to produce a beam with a main lobe at an angle α from boresight, a_m should equal $|a_m| = \exp 2\pi i \theta_m$ where $\theta_m = -m \sin \alpha$. The attenuation is used to shape the beam. Typically, this is manifested as a trade-off between sharpness of the main beam on the one hand, and height and spread of side-lobes on the other.

The return in channel m' due to the kth scatterer will be, assuming that the scatterers are far enough away from the radar that the return can be regarded as a plane wave,[3]

$$\sigma(r_k, v_k, \alpha_k) s_{\text{Txc}}(t - 2r_k) e^{4\pi i t f_c v_k/c} \phi_{\text{tr}}(\alpha_k) e^{-2\pi i f_c m' \delta \sin \alpha_k}, \quad (10.13)$$

where the $e^{4\pi i t f_c v_k/c}$ term in the product represents the effect of Doppler in shifting the frequency of the carrier. The data received then will be the sum over all scatterers in the scene

$$\sum_k \sigma(r_k, v_k, \alpha_k) s_{\text{Txc}}(t - 2r_k) \phi_{\text{tr}}(\alpha_k) e^{4\pi i t f_c v_k/c} e^{-2\pi i f_c m' \delta \sin \alpha_k}. \quad (10.14)$$

In order to extract azimuthal information, the signal is Fourier transformed in the m' variable. It may also be "tapered" (or "windowed") by additional weights $w_{m'}$. Thus we obtain

$$\sum_m w_m e^{2\pi i f_c m \delta \sin \alpha} \sum_k \sigma(r_k, v_k, \alpha_k) s_{\text{Txc}}(t - 2r_k) \phi_{\text{tr}}(\alpha_k) e^{4\pi i t f_c v_k/c} e^{-2\pi i f_c m \delta \sin \alpha_k}$$

$$= \sum_k \sigma(r_k, v_k, \alpha_k) s_{\text{Txc}}(t - 2r_k) \phi_{\text{tr}}(\alpha_k) e^{4\pi i t f_c v_k/c} \sum_m w_m e^{2\pi i f_c m \delta (\sin \alpha - \sin \alpha_k)}$$

$$(10.15)$$

If we write $\theta(\alpha, \alpha_k) = \sum_m w_m e^{2\pi i f_c m \delta (\sin \alpha - \sin \alpha_k)}$, we obtain the processed return

$$\sigma(r_k, v_k, \alpha_k) s_{\text{Txc}}(t - 2r_k) \phi_{\text{tr}}(\alpha_k) e^{4\pi i t f_c v_k/c} \theta(\alpha, \alpha_k). \quad (10.16)$$

Finally this is filtered as in Section 2.1 to obtain

$$\sum_k \sigma(r_k, v_k, \alpha_k) A_{s_{\text{Txc}}, s_{\text{Rx}}}(u - \tau_k, f - \phi_k) \phi_{\text{tr}}(\alpha_k) \theta(\alpha, \alpha_k). \quad (10.17)$$

The important conclusion to draw from this analysis is that the range-Doppler-azimuth ambiguity splits into a product of terms corresponding to azimuthal ambiguity $\theta(\alpha, \alpha_k) \phi_{\text{tr}}(\alpha_k)$ on the one hand and range-Doppler ambiguity $A(u - \tau_k, f - \phi_k)$ on the other.

---

[3]This is the so-called "far-field" assumption.

Range-Doppler ambiguity has been discussed at length in Section 2.1. The azimuthal ambiguity is

$$\theta(\alpha,\alpha_k)\phi_{\text{tr}}(\alpha_k)=\sum_m w_m e^{2\pi i f_c m\delta(\sin\alpha-\sin\alpha_k)}\sum_{m'=1}^{M} a_{m'}e^{2\pi i f_c m'\delta\sin\alpha_k}$$

$$=\sum_{m,m'=1}^{M} a_{m'}w_m \exp\Big(2\pi i f_c\delta\big(m\sin\alpha + (m'-m)\sin\alpha_k\big)\Big). \qquad (10.18)$$

Typically if we are looking in the direction α, we would want to point the transmit beam in that direction. This corresponds to a choice $a_{m'} = r_{m'}e^{-2\pi i f_c m'\sin\alpha}$ where $r_{m'} > 0$ for all m'. This gives an azimuthal ambiguity of

$$\sum_{m,m'=1}^{M} r_{m'}w_m \exp\Big(2\pi i f_c\delta\big((m-m')(\sin\alpha-\sin\alpha_k)\big)\Big) = \Gamma(\sin\alpha - \sin\alpha_k). \qquad (10.19)$$

A choice of $w_m = r_m$ so that the transmit and receive beam-patterns are the same will give

$$\Gamma(s) = |\sum_{m=1}^{M} w_m \exp 2\pi i\delta f_c m s|^2. \qquad (10.20)$$

The choice $w_m = 1$ for all m produces

$$\Gamma(s) = |\sum_{m=1}^{M} \exp 2\pi i\delta f_c m s|^2 = \Big(\frac{\sin\pi\delta f_c(M+1)s}{\sin\pi\delta f_c}\Big), \qquad (10.21)$$

which is a "periodic sinc" response. Alternative choices of attenuation produce reduced sidelobes but at the expense of wider central beam. Roughly speaking, the flatter the coefficients w_m the more pronounced the sidelobes are. If the coefficients are tapered toward the ends of the antenna, then typically the sidelobes will be lower but the main lobe is broader. Thus a more sharply focused beam results in greater sidelobes. There is a considerable literature on the choice of these "spatial filters" to shape the beam. We refer the reader to standard references, such as [180, 217], for more details.

2.3 Doppler Processing

As we have seen in considering the ambiguity function, an individual waveform is capable of detecting the movement of a target toward or away from the radar. However, in practice this is rarely the way the detection of range-rate is achieved. If only range-rate (and not range) is required, as is the case, for example, with radars for the detection of highway speeding infringements, then a standard solution is to transmit a continuous sinusoid. The shift in frequency is detected by a filter that eliminates the zero-Doppler component usually dominating the return and resulting from the direct return of the transmitted signal to the receiver from the transmitter and fixed elements in the scene.

When knowledge of range is required along with range-rate, the choice of waveforms represents a compromise to accommodate various conflicting requirements. On the one hand they should be long to produce "energy on target" and to allow Doppler to have a measurable effect. On the other hand the longer the pulse (in a *monostatic* mode; i.e., when the transmitter and receiver are collocated) the longer the "blind range." Since in this scheme the receiver can only receive the return when the transmitter is not illuminating the scene, there is a distance of half the length of the pulse (in meters) that cannot be viewed by the radar. In any case, within a pulse there should be some modulation to produce good pulse compression.

Since an acceptable blind range conflicts with the pulse length required to produce measurable intra-pulse Doppler, radar engineers have had to resort to a trick that overcomes both the Doppler and blind range constraints. The method of "Doppler processing" involves the transmission of a short waveform over multiple pulses, with a significant "listening period" between the pulses. While on each pulse the Doppler has negligible effect, it does have a measurable effect over the sequence of pulses. It is important for this process that the oscillator used to time the operation of the system maintain coherence over the extent of these multiple pulses.

We assume that N_{dp} pulses $\mathbf{s}_{\mathrm{Txc}}(t - n\Delta)$ are transmitted at a uniform time separation of Δ in such a way as to retain coherence; i.e., the phase information from the carrier is retained.

We have learned from the preceding two sections that

1. We can consider a single target;

2. We can separate the effects of range-Doppler on the one hand and azimuth (and elevation if required) on the other.

Accordingly we treat the case of a single scatterer at delay τ_1 and Doppler frequency ϕ_1 and ignore the azimuthal dependence. As before, this exposition can take place "at baseband;" i.e., without inclusion of the carrier term $\exp 2\pi i f_c t$. The return from the nth pulse is then

$$\mathbf{s}_{\mathrm{Txc}}(t - \tau_1 - n\Delta)e^{2\pi i \phi_1 t}, \qquad (10.22)$$

with an appropriate multiplicative factor to account for the decay due to range and the reflectivity of the scatterer. The intra-pulse time variable t is known as "fast time" and the pulse count n as "slow time."

It is assumed that the inter-pulse time Δ is long enough that most of the return from a given transmit pulse arrives back within the subsequent listening period. Of course this cannot be guaranteed, especially if there are large distant scatterers. The problem of "range-aliasing" is a significant one in radar. We shall return to it briefly later.

We collect the N_{dp} returns and stack them alongside each other so that ranges match. The effect is to appear to shift the pulses in time by an appropriate multiple of Δ. Thus we obtain

$$s'_{\text{rec},n}(t, m) = \mathsf{s}_{\text{Txc}}(t - \tau_1) e^{2\pi i \phi_1 (t + n\Delta)}. \tag{10.23}$$

Now we take a Fourier transform in the n (slow time) direction to obtain

$$\begin{aligned} S_{\text{rec}}(t, \nu) &= \sum_{n=0}^{N_{\text{dp}}-1} e^{-2\pi i n \frac{\nu}{N_{\text{dp}}-1}} e^{2\pi i \phi_1 (t - n\Delta)} \mathsf{s}_{\text{Txc}}(t - \tau_1) \theta \\ &= \mathsf{s}_{\text{Txc}}(t - \tau_1) e^{2\pi i \phi_1 t} \sum_{n=0}^{N_{\text{dp}}-1} e^{2\pi i n (\phi_1 \Delta - \frac{\nu}{N_{\text{dp}}})} \\ &= \mathsf{s}_{\text{Txc}}(t - \tau) e^{t\pi i \phi_1 t} \psi(\nu, \phi_1), \end{aligned} \tag{10.24}$$

where $\psi(\nu, \phi) = \sum_{n=0}^{N_{\text{dp}}-1} e^{2\pi i n (\phi_1 \Delta - \frac{\nu}{N_{\text{dp}}})}$. Note that this is (up to a phase factor) just a "periodic sinc" function with a peak when $\nu = 2 N f_c \frac{v_k}{c} \Delta$ and ν's which differ from this by multiples of N_{dp}.

Finally we perform a matched filter against the original waveform $\mathsf{s}_{\text{Rx}}(t)$ to obtain

$$\int_{\mathbb{R}} \mathsf{s}_{\text{Txc}}(t - \tau_1) e^{2\pi i \phi_1 t} \mathsf{s}_{\text{Rx}}(t - u) \, dt \, \psi(\nu, \phi_1)$$

$$= e^{2\pi i u \phi_1} A(u - \tau_1, \phi_1) \psi(\nu, \phi_1), \tag{10.25}$$

where

$$A_{\mathsf{s}_{\text{Txc}}, \mathsf{s}_{\text{Rx}}}(u, d) = \int_{\mathbb{R}} \mathsf{s}_{\text{Rx}}(t - u) e^{2\pi i d t} \mathsf{s}_{\text{Txc}}(t) \, dt, \tag{10.26}$$

is the ambiguity function as described in Section 2.1.

232 *FOUNDATIONS AND APPLICATIONS OF SENSOR MANAGEMENT*

Observe now that with the assumption of small Dopplers and short pulses, the ambiguity $A(u - \tau_1, \phi_1)$ is approximately $A(u - \tau_1, 0)$. This results in a separation of the Doppler and range ambiguities in the same way as we split off the azimuthal ambiguity. The Doppler ambiguity happens in slow time, the range ambiguity in fast time.

It might be asked now what happened to Moyal's identity and the constraints it imposes on the range and Doppler ambiguities. Moyal's identity still applies, of course. We observe that the Doppler has been sampled at the rate given by Δ. This means that there is again an aliasing problem for targets of which the range rate gives a frequency in excess of the Nyquist frequency for this sampling rate. The smaller Δ is, the less the effect of Doppler aliasing, but the greater the effect of range aliasing. The total ambiguity of the pulse train in this case has peaks at points of the form $(n\Delta, \frac{m}{\Delta})$ in the range-Doppler plane.

There is a trade-off then between Doppler aliasing and range aliasing. This multi-pulse concept can be fitted into the ambiguity picture given in Section 2.1 and the trade-off between range and Doppler aliasing is just a consequence of Moyal's Identity. From the scheduling viewpoint, the time Δ and the number of pulses N_{dp} are both available for modification in real time.

2.4 Effects of Ambiguity

Given a scene, the radar return is produced by "convolving" it with the *generalized ambiguity* function.

$$G(t; \nu, d; \alpha, \beta) = A(t, 2f_c\frac{v}{c})\phi_{\text{tr}}(\beta)\theta(\alpha, \beta)\psi(\nu, v_k), \quad (10.27)$$

where, by convolution, we mean here that the processed return is just

$$S_{\text{pr}}(u, \nu, \alpha) = \sum_k \sigma(r_k, v_k, \alpha_k) G(u - 2r_k; \nu, v_k; \alpha, \alpha_k). \quad (10.28)$$

In assessing the effect of a given radar mode, it is enough then to store for each waveform, taper and transmit pattern the generalized ambiguity, including (where appropriate) Doppler processing. Moreover, when we wish to test a waveform against a predicted scene, so as to choose one that is optimal in some quantifiable sense, it is this object that we use.

In fact the stored information can be somewhat reduced by first taking out the transmit pattern $\phi_{\text{tr}}(\beta)$, since it is scene independent. Then we may regard θ as a function of $\sin(\alpha)$; i.e., we can define $\theta_r(y) = \sum_m w_m e^{2\pi i m \delta y}$, so that

$$\theta(\alpha, \beta) = \theta_r(\sin(\alpha) - \sin(\beta)). \quad (10.29)$$

Thus we may make the generalized ambiguity

$$G_r(u, \nu, v, y) = A(u, 2f_c\frac{v}{c})\theta_r(y)\psi(\nu, v), \qquad (10.30)$$

and then the convolution equation (10.28) becomes

$$S_{\text{pr}}(t, \nu, \alpha) = \phi_{\text{tr}}(\beta) \sum_k \sigma(r_k, v_k, \alpha_k) G_r(t - r_k, \nu, v_k, \sin(\alpha) - \sin(\alpha_k)). \qquad (10.31)$$

3. Measurement in Radar

Here we discuss the issue of measurement. As we have seen in Section 2.4, the radar measurement process amounts to a smearing of the scene by the ambiguity function. In order to incorporate these ideas into tracking and sensor scheduling problems, we need to find a simplification of the measurement process. We restrict attention for the moment to a radar that is undertaking only range and Doppler measurements, as in Section 2.1.

Our treatment here follows very closely that of van Trees in [237, Ch. 10]. As shown there, the log-likelihood associated with the measurement of range and Doppler of a single point target at a delay τ_a and Doppler ω_a resulting from the use of a waveform $\mathbf{w}(t)$ and a matched-filter is (leaving aside constants associated with the radar parameters that are unchanged throughout our calculations),

$$\begin{aligned}\Lambda(\tau, \omega) = &|\mathbf{b}|^2 |A_\mathbf{w}(\tau - \tau_a, \omega - \omega_a)|^2 \\ &+ 2\Re\left\{\mathbf{b} A_\mathbf{w}(\tau - \tau_a, \omega - \omega_a) \mathbf{n}^*(\tau, \omega)\right\} + |\mathbf{n}(\tau, \omega)|^2,\end{aligned} \qquad (10.32)$$

where $\mathbf{b}$ is a complex Gaussian random variable representing the reflectivity of the target, and $\mathbf{n}(\tau, \omega)$ is the integral of the receiver noise $\mathbf{N}(t)$ emanating from the filtering and Doppler processing; i.e.,

$$\mathbf{n}(\tau, \omega) = \int_{-\infty}^{\infty} \mathbf{N}(t) \mathbf{w}^*(t - \tau) e^{-i\omega t} \, dt. \qquad (10.33)$$

The Fisher information matrix $\mathbf{J}$ associated with this measurement is

$$\mathbf{J} = \begin{pmatrix} J_{\tau\tau} & J_{\tau\omega} \\ J_{\omega\tau} & J_{\omega\omega} \end{pmatrix} = -\mathbb{E}\left(\begin{bmatrix} \frac{\partial^2 \Lambda(\tau, \omega)}{\partial \tau^2} \end{bmatrix} \begin{bmatrix} \frac{\partial^2 \Lambda(\tau, \omega)}{\partial \tau \partial \omega} \end{bmatrix} \\ \begin{bmatrix} \frac{\partial^2 \Lambda(\tau, \omega)}{\partial \omega \partial \tau} \end{bmatrix} \begin{bmatrix} \frac{\partial^2 \Lambda(\tau, \omega)}{\partial \omega^2} \end{bmatrix}\right), \qquad (10.34)$$

where the expectation is taken over the independent random processes, $\mathbf{b}$ and $\mathbf{N}(t)$. In our situation, it can be assumed that $J_{\tau\omega} = J_{\omega\tau}$. Computations yield, again within a constant factor,

$$J_{\tau\tau} = \int_{-\infty}^{\infty} \omega^2 |W(\omega)|^2 \frac{d\omega}{2\pi} - \left(\int_{-\infty}^{\infty} \omega |W(\omega)|^2 \frac{d\omega}{2\pi} \right)^2$$

$$J_{\tau\omega} = \int_{-\infty}^{\infty} t\mathbf{w}(t)\mathbf{w}'(t)\, dt - \int_{-\infty}^{\infty} \omega |W(\omega)|^2 \frac{d\omega}{2\pi} \int_{-\infty}^{\infty} t|\mathbf{w}(t)|^2\, dt \quad (10.35)$$

$$J_{\omega\omega} = \int_{-\infty}^{\infty} t^2 |\mathbf{w}(t)|\, dt - \left(\int_{-\infty}^{\infty} t|\mathbf{w}(t)|^2\, dt \right)^2.$$

For any given waveform this can be calculated. Of course the Fisher information matrix determines the Cramér-Rao lower bound for an estimator of the variables τ, ω. In [236], van Trees argues that in many circumstances it provides a reasonable approximation to the inverse of the covariance matrix of the measurement process. As is easily seen, $\mathbf{J}^{-1}$ is the inverse of the covariance matrix of a Gaussian that is the best approximation to $|A(\tau, \omega)|^2$ at its peak at $(0, 0)$.

Accordingly $\mathbf{J}$ has been used as the basis for assessing the performance of a waveform in measuring the time delay and Doppler of a target for the purposes of scheduling. The sensor is characterized by a measurement noise covariance matrix

$$\mathbf{R} = \mathbf{T}\mathbf{J}^{-1}\mathbf{T}, \quad (10.36)$$

where $\mathbf{T}$ is the transformation matrix between the time delay and Doppler measured by the receiver and the target range and velocity.

In particular it forms the basis of the seminal work of Kershaw and Evans on this topic. We describe this work next.

4. Basic Scheduling of Waveforms in Target Tracking

The aim of this section is to describe the work of Kershaw and Evans [131, 130] for scheduling of waveforms. This work has been at the basis of much subsequent work in the area of waveform scheduling.

4.1 Measurement Validation

Our assumption is that the radar produces measurements of the form

$$\mathbf{y}_r^k = \mathbf{H}\mathbf{x}_k + \omega_r^k, \tag{10.37}$$

where $r = 1, \ldots m_k$ are the detections at each time instance k, $\mathbf{H}$ is measurement matrix and the noise ω_r^k is Gaussian with zero mean and covariance $\mathbf{R}_k$ as discussed in Section 3, identically distributed for all $r = 1, \ldots m_k$.

At time k the system will have produced, based on the previous measurements, an estimated position $\mathbf{x}_{k|k-1}$ of the target and an innovation covariance matrix $\mathbf{S}_k$. The details about how this is done are given in the next section. Based on these estimates we establish a *validation gate*. This will be the ellipsoid

$$\{\mathbf{y} : (\mathbf{y} - \mathbf{H}\mathbf{x}_{k|k-1})^\mathsf{T}(\mathbf{S}_k)^{-1}(\mathbf{y} - \mathbf{H}\mathbf{x}_{k|k-1}) < g^2\}, \tag{10.38}$$

centered on the estimated measurement $\mathbf{H}\mathbf{x}_{k|k-1}$ of the target at time k, where g is a threshold specified in advance. We shall need the volume of this ellipsoid for later calculations. In the range-Doppler case, it is

$$V_k = \pi g^2 \det(\mathbf{S}_k)^{1/2}. \tag{10.39}$$

Only measurements $\mathbf{y}_r^k$ lying in this ellipsoid will be considered; the rest are discarded. This will facilitate computation and in any case eliminates outliers that might corrupt the tracking process.

The measurements that result then are:

1. A finite number of points $\mathbf{y}_r^k$.

2. For each of these points a covariance matrix $\mathbf{R}_k$.

4.2 IPDA Tracker

For convenience, we first describe the PDA tracker. Then we add the features specific for the IPDA tracker. We refer to the paper of Musicki, Evans and Stankovic [177] for a more detailed description of this.

For the moment we fix a waveform $\mathbf{w}$ used to make the measurements. The target is assumed to follow a standard Gauss-Markov model. Target position $\mathbf{x}_k = (x_k, \dot{x}_k, \ddot{x}_k)$ in range-Doppler-acceleration space at time k moves with essentially zero acceleration in range according to the dynamics

$$\mathbf{x}_{k+1} = \mathbf{F}\mathbf{x}_k + \nu_k \tag{10.40}$$

where ν_k is Gaussian with zero mean and covariance matrix $\mathbf{Q}$ (both $\mathbf{Q}$ and $\mathbf{F}$ are independent of k). The matrix $\mathbf{F}$ is of the form

$$\mathbf{F} = \begin{pmatrix} 1 & \Delta & \frac{\Delta^2}{2} \\ 0 & 1 & \Delta \\ 0 & 0 & 1 \end{pmatrix}, \tag{10.41}$$

where Δ is the time step between the epochs.

Each measurement has associated with it an *innovation*

$$\mathbf{v}_r^k = \mathbf{y}_r^k - \mathbf{H}\mathbf{x}_{k|k-1}, \tag{10.42}$$

where $\mathbf{x}_{k|k-1}$ is the estimate of $\mathbf{x}_k$ given the previous measurements, $\mathbf{y}_r^j$ for $j < k$ and all r.

The covariance matrix of this statistic is the *innovation covariance matrix*

$$\mathbf{S}_k = \mathbf{H}\mathbf{P}_{k|k-1}\mathbf{H}^\mathsf{T} + \mathbf{R}_k. \tag{10.43}$$

Note that this is independent of the particular measurement $\mathbf{y}_r^k$ and only depends on the waveform used at time k through the error covariance matrix $\mathbf{R}_k$ and prior data. Here $\mathbf{P}_{k|k-1}$ is the error covariance associated with the state estimator $\mathbf{x}_{k|k-1}$ and is calculated using the standard Riccati equations

$$\mathbf{x}_{k|k-1} = \mathbf{F}\mathbf{x}_{k-1|k-1}, \qquad \mathbf{P}_{k|k-1} = \mathbf{F}\mathbf{P}_{k-1|k-1}\mathbf{F}^\mathsf{T} + \mathbf{Q}. \tag{10.44}$$

We write P_D for the target detection probability and P_G for the probability that the target, if detected, is in the validation gate. Note that P_G is just the total Gaussian probability inside the validation gate and can be assumed to be unity in practice, since P_G is greater than 0.99 when $g > \sqrt{2} + 2$.

We assume that clutter is uniformly distributed with density ρ and write

$$b_k = 2\pi\rho\sqrt{\det(\mathbf{S}_k)}\frac{(1 - P_D P_G)}{P_D} = 2\frac{\rho V_k}{g^2}\frac{(1 - P_D P_G)}{P_D} \tag{10.45}$$

and

$$e_r^k = \exp\left(-\mathbf{v}_r^k(\mathbf{S}_k)^{-1}\mathbf{v}_r^k\right). \tag{10.46}$$

Then the probabilities that none of the measurements resulted from the target b_0^k and that the rth measurement is the correct one are, respectively,

$$b_0^k = \frac{b_k}{b_k + \sum_r e_r^k}, \qquad b_r^k = \frac{e_r^k}{b_k + \sum_r e_r^k}. \tag{10.47}$$

•

Note that (of course) $\sum_{r=0}^{m_k} b_r^k = 1$.

We now (using the standard PDA methodology — see [83] for the most lucid version of this) replace the innovation $\mathbf{v}_r^k$ by the combined innovation

$$\mathbf{v}_k = \sum_{r=1}^{m_k} b_r^k \mathbf{v}_r^k. \tag{10.48}$$

Next we use the Kalman update to obtain an estimate of the state vector $\mathbf{x}_{k|k}$:

$$\mathbf{x}_{k|k} = \mathbf{x}_{k|k-1} + \sum_{r=1}^{m_k} b_r^k \mathbf{K}_k \mathbf{v}_r^k, \text{ where } \mathbf{K}_k = \mathbf{P}_{k|k-1}\mathbf{H}^\mathsf{T}(\mathbf{S}_k)^{-1}. \tag{10.49}$$

The update of the error covariance matrix is

$$\mathbf{P}_{k|k} = \mathbf{P}_{k|k-1} - \sum_r b_r^k \mathbf{K}_k \mathbf{S}_k (\mathbf{K}_k)^\mathsf{T} + \mathbf{P}_k, \tag{10.50}$$

where

$$\mathbf{P}_k = \sum_{r=1}^{m_k} b_r^k \left(\mathbf{K}_k \mathbf{v}_r^k (\mathbf{K}_k \mathbf{v}_r^k)^\mathsf{T} - \mathbf{K}_k \mathbf{v}_k (\mathbf{K}_k \mathbf{v}_k)^\mathsf{T} \right). \tag{10.51}$$

A standard approximation to it is used to obtain an estimate for the error covariance. This technique of Fortmann et al. in [83], replaces the random elements in (10.50) by their expectations. In fact, we replace $\mathbf{P}_k$ and b_0^k by

$$\overline{\mathbf{P}}_k = \mathbb{E}[\mathbf{P}_k | Y^{k-1}], \qquad \overline{b}_0^k = \mathbb{E}[b_0^k | Y^{k-1}], \tag{10.52}$$

where Y^{k-1} represents the measurement history at time $k-1$.

We refrain from repeating the calculations of [83] here. The results are that, if we write

$$q_1 = \frac{P_D}{2} \int_0^g r^3 e^{-r^2/2} \, dr$$

$$I_2(m,b,g) = \frac{P_D}{2} \sum_{m=1}^{\infty} \frac{e^{-\rho V_k}(\rho V_k)^{m-1}}{(m-1)!} \left(\frac{2}{g^2}\right)^{m-1} \tag{10.53}$$

$$\times \int_0^g \cdots \int_0^g \frac{e^{-u_1^2} u_1^2}{b + \sum_{r=1}^m e^{-u_r^2/2}} \, du_1 \ldots du_r$$

where b_k is defined in equation (10.45), V_k in equation (10.39) and g is the "radius" of the validation gate, then

$$\overline{\mathbf{P}}_{k|k}(\theta_k) \approx \mathbf{P}_{k|k-1} - (P_D P_G - q_1 + q_2) \sum_r b_r^k \mathbf{K}_k \mathbf{S}_k (\mathbf{K}_k)^\mathsf{T}. \quad (10.54)$$

Note that Fortmann et al. introduce further approximations in [83]. Specifically, they note that for typical values of g, $P_G \approx 1$ and $q_1 \approx P_D$. Thus equation (10.54) simplifies to

$$\overline{\mathbf{P}}_{k|k} \approx \mathbf{P}_{k|k-1} - q_2 \sum_r b_r^k \mathbf{K}_k \mathbf{S}_k (\mathbf{K}_k)^\mathsf{T}. \quad (10.55)$$

Calculation of $I_2(m, b, g)$ in equation (10.53) requires a numerical integration scheme for a range of values of m, b and g. Kershaw and Evans [131] use an approximation to q_2 in the two-dimensional case:

$$q_2 \approx \frac{0.997 P_D}{1 + 0.37 P_D^{-1.57} \rho V_k}. \quad (10.56)$$

The IPDA tracker introduces the notion of *track existence* into the PDA tracker. We write χ_k to be the event of track existence at time k. This is assumed to behave as a Markov chain in the sense that

$$\mathrm{P}(\chi_k | Y^{k-1}) = p_1 \, \mathrm{P}(\chi_{k-1} | Y^{k-1}) + p_2 (1 - \mathrm{P}(\chi_{k-1} | Y^{k-1})), \quad (10.57)$$

where p_1 and p_2 are between 0 and 1. The choice of these is to some extent arbitrary, though simulations (R. Evans, private communication) indicate that the performance of the system is not very sensitive to the choice. On the other hand, if we assume that p_1 and p_2 are equal, then $\mathrm{P}(\chi_{k+1}|Y^k) = p_1$ and is independent of the previous history. A more appropriate choice then would have $p_1 > p_2$. There is clearly room for more work here. We write $f(\mathbf{y}|Y^{k-1})$ for the measurement Gaussian density corresponding to the mean $\mathbf{x}_{k|k-1}$ and covariance matrix $\mathbf{P}_{k|k-1}$, so that this represents the density of the predicted next measurement. Define

$$P_G = \int_{V_k} f(\mathbf{y}|Y^{k-1}) \, d\mathbf{y}. \quad (10.58)$$

This is the probability that the next measurement will be inside the validation gate. Let

$$p(\mathbf{y}|Y^{k-1}) = \frac{1}{P_G} f(\mathbf{y}|Y^{k-1}) \quad (\mathbf{y} \in V_k). \quad (10.59)$$

This is the conditional probability density conditioned on the measurement falling in the gate.

Now let

$$\delta_k = \begin{cases} P_G P_D & \text{if } m_k = 0; \\ P_G P_D \left(1 - \sum_{r=1}^{m_k} \frac{p(\mathbf{y}_r^k|Y^{k-1})}{\rho_k(\mathbf{y}_r^k)}\right) & \text{if } m_k > 0; \end{cases} \quad (10.60)$$

where P_D is the *a priori* probability of detection of a target.

We update the probability of track existence by

$$P(\chi_k|Y^k) = \frac{1 - \delta_k}{1 - \delta_k P(\chi_k|Y^{k-1})} P(\chi_k|Y^{k-1}). \quad (10.61)$$

The conditional probabilities that measurement r at time k originated from the potential target are now given by

$$b_r^k = \begin{cases} \frac{1 - P_D P_G}{1 - \delta_k} & \text{if } r = 0; \\ \frac{P_D P_G \frac{p(\mathbf{y}_r^k|Y^{k-1})}{\rho_k(\mathbf{y}_r^k)}}{1 - \delta_k} & \text{if } r > 0. \end{cases} \quad (10.62)$$

These are then used, as explained in the preceding part of this section, to calculate the state estimate $\mathbf{x}_{k|k}$ and the error covariance $\mathbf{P}_{k|k}$ (see equations (10.49–10.52)). The Markov chain property (10.57) is used to update the track existence probability. It is this value which is thresholded to provide a true or false track detection test.

5. Measures of Effectiveness for Waveforms

To optimize the choice of waveform at each epoch it is necessary to calculate a cost function for all available waveforms in the library. This cost function should be a function of:

1 the predicted clutter distribution at that epoch based on a clutter mapper as described in, for example, [177];

2 the estimated position of the potential target at that epoch, based on the IPDA tracker.

This section discusses two potential waveform evaluators: the single noise covariance (SNC) matrix and the integrated clutter measure (ICM).

5.1 Single Noise Covariance Model

The approximate equation (10.54) for the error covariance gives a measure of the effectiveness of each waveform. The parameter θ is used to range over a collection of waveforms, and so the error covariances $\mathbf{R}_k = \mathbf{R}_k(\theta)$ depend on this parameter θ, as do all objects calculated in terms of them. As already indicated in Section 4.2, the covariance of the innovation $\mathbf{v}_r^k$ equation (10.43) and the gain matrix $\mathbf{K}_k$ (see equation (10.49)) are, respectively:

$$\mathbf{S}_k(\theta) = \mathbf{H}\mathbf{P}_{k|k-1}\mathbf{H}^\mathsf{T} + \mathbf{R}_k(\theta), \qquad \mathbf{K}_k(\theta) = \mathbf{P}_{k|k-1}\mathbf{H}^\mathsf{T}\mathbf{S}_k(\theta)^{-1}. \quad (10.63)$$

Substituting these in the approximate equation for the gain matrix, we obtain

$$\overline{\mathbf{P}}_{k|k}(\theta) = \mathbf{P}_{k|k-1} - q_2 \mathbf{P}_{k|k-1}\mathbf{H}^\mathsf{T}\left(\mathbf{H}\mathbf{P}_{k|k-1}\mathbf{H}^\mathsf{T} + \sum_r b_r^k \mathbf{R}_k(\theta)\right)^{-1} \mathbf{H}\mathbf{P}_{k|k-1}^\mathsf{T}. \quad (10.64)$$

A suitable choice of waveform will be made to minimize the estimated track error $\overline{\mathbf{P}}_{k|k}(\theta)$, or rather its determinant (or trace as it was originally used in [131, 130]), over all choices of θ:

$$\theta_k^* = \underset{\theta}{\operatorname{argmin}} \det(\overline{\mathbf{P}}_{k|k}(\theta)). \quad (10.65)$$

This choice of waveform is used for the next measurement.

5.2 Integrated Clutter Measure

In this section, we assume that an estimate is available of the clutter distribution in the range-Doppler plane. A simple method for doing this is given in [177]. Write $\gamma_k(t, f)$ for the power estimate of clutter at range (time) t and Doppler (frequency) f at discrete time k. We assume too, as in the work of Kershaw and Evans, that an estimate of the target state is available.

Our aim, for any potential waveform w with ambiguity A_w, is to calculate

$$F_k(w) = \left| \iint_{V_k} \left(\iint_{\mathbb{R}^2} A_w(t-t', f-f')\gamma_k(t', f') dt' df' \right) dt\, df \right|. \quad (10.66)$$

This is the integrated "spill-over" into the validation window of the sidelobes of the waveform due to the clutter. It makes sense to minimize this quantity over all waveforms. That is, the waveform optimizer we envisage will choose the optimal waveform at the kth time instance by

$$w_k^* = \underset{w}{\operatorname{argmin}} F_k(w). \quad (10.67)$$

We remark that the implementation of this measure requires significant approximation that we shall discuss further.

Before going into the details of these approximations, we note that, because phase information from complex scatterers is usually uninformative, this waveform evaluator is replaced by

$$F_k(w) = \iint_{V_k} \left(\iint_{\mathbb{R}^2} |A_w(t - t', f - f')| |\gamma_k(t', f')| dt' df' \right) dt\, df. \quad (10.68)$$

This means that no account is taken of cancellation effects over multiple sidelobes, but these are so phase dependent as to be unlikely to exist robustly. By making this change from equation (10.66) to equation (10.68), we reduce the possibility of large waveform sidelobes (with the center of the ambiguity on the target) being in areas of high clutter.

5.3 Approximation of ICM

Assume that the clutter and target dynamics and that the measurements are modeled by linear equation, perturbed by Gaussian noise

$$\begin{aligned} \mathbf{x}_k &= \mathbf{F}\mathbf{x}_{k-1} + \nu_k; \\ \mathbf{y}_k &= \mathbf{H}\mathbf{x}_k + \omega_k. \end{aligned} \quad (10.69)$$

The noise ν_k in the dynamical model and the measurement noise ω_k are assumed to be stationary, and mutually independent, zero-mean with covariance, in the case of ν_k, $\mathbf{Q}$. The measurements $\mathbf{y}_k$ are used to estimate the target and clutter probability densities recursively. At time k these densities are described by the mean $\mathbf{x}_{k|k-1}$ and covariance $\mathbf{P}_{k|k-1}$. This, in particular allows us to calculate the validation window V_k and the clutter power $|\gamma_k(t', f')|$ for Equation (10.68).

To make Equation (10.68) quickly calculable, we use the following approximations. First, the tracking is assumed to be performed by a PDA filter based tracker. Thus the estimate of target state is given by a single Gaussian distribution or by the mixture of Gaussians, all described by individual means and covariances. The validation window V_k, usually thought of as an ellipsoid, will also be approximated by a single Gaussian. Also the clutter power $|\gamma_k(t', r')|$ is given by its estimate at time k by discrete values for $\gamma = \{\gamma_1, \gamma_2, \ldots, \gamma_m\}$, $t' = \{t'_1, t'_2, \ldots, t'_m\}$ and $f' = \{f'_1, f'_2, \ldots, f'_m\}$, where each γ_i is the power of the clutter in the ith range-Doppler cell given by (t_i, f_i).

The second approximation is the approximation of the absolute value of the ambiguity function A_w as a Gaussian mixture

$$|A_w(\mathbf{y})| = \sum_{i=-n}^{n} \alpha_i \mathcal{N}(\mathbf{y}; \delta_j, \mathbf{R}_j). \qquad (10.70)$$

This can be obtained by a least squares fitting algorithm and is performed off-line once for each waveform. The finer the resolution of the fit, the more realistic the algorithm is. In the context of tracking this approximation is interpreted as an approximation of the measurement noise w distribution by any of the Gaussians $\mathcal{N}(\mathbf{y}; \delta_j, \mathbf{R}_j)$ with mean δ_j and covariance $\mathbf{R}_j$. Note, that δ_0 is centered in the main lobe of the ambiguity function and is equal to zero and $\mathbf{R}_0$ is given by covariance $\mathbf{R}$ as discussed in Section 3. Each δ_j for $j \in \{-n, \ldots, -1, 1, \ldots, n\}$ is centered in the jth sidelobe of the ambiguity function. Note, that $2n+1$ is the number of "significant" sidelobes and is determined by a fitting algorithm. We assume that the measurements originated by the sidelobes can be detected with probability relative to α_j and to the power of the clutter scatterer γ, from which the radar measurement is originated. Thus, we postulate that at time k there could be up to $2n+1$ measurements originated from a single scatterer.

The target measurement probability density, used for calculating the validation window, is given by the mean $\mathbf{Hx}_{k|k-1}$ and covariance $\mathbf{S} = \mathbf{HP}_{k|k-1}\mathbf{H}^\mathsf{T} + \mathbf{R}_0$. This too is an approximation, justified since in a typical situation target power is much smaller than that of clutter and its detection in the sidelobes can be neglected.

The validation gate is given by an ellipsoid, for which probability that the measurement is in the gate is equal some P_G, usually greater than 0.99.

The probability of target detection in the main lobe is affected by the amount of "spill-over" from the clutter into the validation window; the less spill-over, the easier it is to detect the target.

With these approximations, Equation (10.68) is simplified as follows:

$$F_k(w) = \iint_{\mathbb{R}^2} \chi(V_k) \left(\iint_{\mathbb{R}^2} |A_w(t-t', f-f')| \|\gamma_k(t', f')| dt' df' \right) dt\, df$$

$$\approx \iint_{\mathbb{R}^2} \mathcal{N}(\mathbf{y}; \hat{\mathbf{y}}, \mathbf{S}) \sum_{j=1}^{m} \gamma_j \sum_{i=-n}^{n} \alpha_i \mathcal{N}(\mathbf{y}; (t_j, f_j)^\mathsf{T} + \delta_i, \mathbf{R}_i) d\mathbf{y}$$

$$= \sum_{j=1}^{m} \sum_{i=-n}^{n} \gamma_j \alpha_i \mathcal{N}(\hat{\mathbf{y}}; (t_j, f_j)^\mathsf{T} + \delta_i, \mathbf{S} + \mathbf{R}_i).$$

$$(10.71)$$

In the second line, the characteristic function $\chi(V_k)$ is replaced by a Gaussian function. In the last line the formula for the Gaussian product is used. Without a loss of generality we can assume that the target power is 1, then γ_j in Equation (10.71) can be seen as the clutter-to-target power ratio and $F_k(w)^{-1}$ represents SNR in the validation gate. While this is not rigorously justified, it appears to be a reasonable assumption that works well in simulations. Assuming that the detection was obtained by a likelihood ratio test realized via, for example, a Neyman-Pearson type of detector, we write well-known formula for probability of target detection P_d as

$$P_d = 1 - \Phi(\Phi^{-1}(1 - P_f) - F_k(w)^{-1}), \qquad (10.72)$$

where Φ is a standard normal probability distribution function, and P_f is desired probability of false alarm.

5.4 Simulation Results

We illustrate the ideas described in Section 5.3 with a simple example. Given a scenario with a constant velocity small target trajectory and random, large, slow moving clutter, two waveforms are scheduled using ICM and compared to the results with those for the same waveforms used in a non-scheduled way. The clutter-to-target power ratio γ is similar for all clutter and about 60 dB. The two waveforms are up-sweep and down-sweep linear frequency modulated (LFM) waveforms, their ambiguity function is approximated by a Gaussian mixture with five Gaussian terms as shown in Figure 10.2. The tracking is performed using an IPDA filter [178], as described earlier. Target measurement using i^{th} waveform is present with probability of detection given in Equation (10.72) with $P_f = 10^{-5}$.

To generate clutter detections, we proceeded as follows. For a given (clutter) scatterer, if it is of sufficient amplitude it is detected at the main lobe of the ambiguity function with probability $\alpha_0 = 1$ and in the ith sidelobe it is detected with probability α_i, $i = -n, -(n-1), ldots, -1, 1, \ldots, (n-1), n$. For a given scenario, 1000 runs were performed and the results averaged over them. The cost as calculated in Equation (10.71). Position RMS error and probability of target existence are represented in Figure 10.3. We observe that, with scheduling, probability of target existence is maintained close to 1, in contrast to the two non-scheduled cases. A similar conclusion holds for RMS error: with scheduling the track error is smaller than without scheduling.

244 *FOUNDATIONS AND APPLICATIONS OF SENSOR MANAGEMENT*

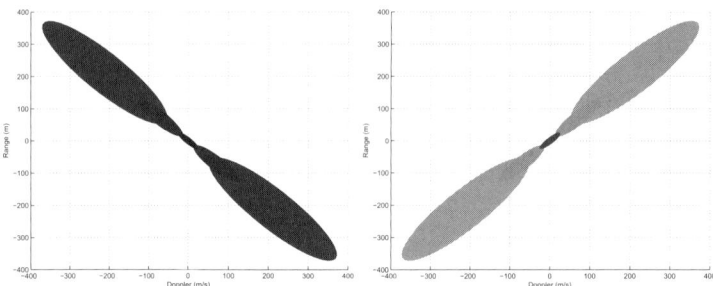

Figure 10.2. Approximation of ambiguity function by Gaussian mixture for up and down sweep waveforms.

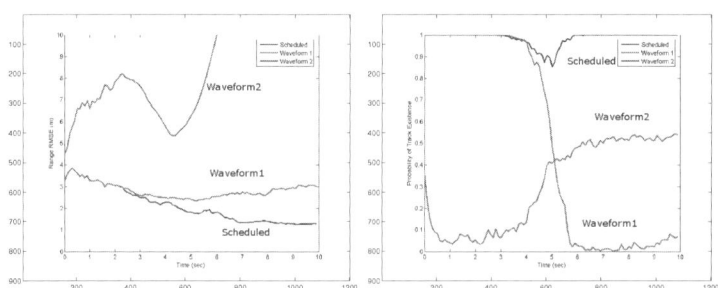

Figure 10.3. Position root-mean-square error and probability of track existence.

6. Scheduling of Beam Steering and Waveforms

In this section a scene consists of a collection of scatterers varying over range, Doppler and azimuth, that have been spread (in range-Doppler) by the waveform and (in azimuth) by the beam-shape. These ideas are described earlier. We investigate a radar system capable of rapid beam steering and of waveform switching. The transmit waveform is chosen from a small library of such. The operational requirement of the radar is to track a number of maneuvering targets while performing surveillance for new potential targets. Tracking is accomplished by means of an Linear Multi-target Integrated Probabilistic Data Association (LMIPDA) tracker as described in [176], which is a variant of the IPDA tracker described in Section 4 that is designed to handle many targets efficiently. Interacting multiple models (IMM) is used to model maneuvering targets in the tracker. LMIPDA provides a probability of track existence, permitting a "track-before-detect" technique to be adopted. "False alarm" tracks are maintained until the probability of track existence falls below a threshold. In derivations of formulae in this section, the SNC case (described in Section 5.1) is assumed. LMIPDA-IMM is a recursive algorithm combining multi-target data association (LMIPDA) with maneuvering target state estimation implemented using IMM.

The aim is to maintain the tracks of the existing targets to within a specified accuracy as determined by the absolute value of the determinant of the track error covariance matrix. However, this has to be done within the time available given that a full scan has to be performed within a prescribed interval. In this section we give an algorithm for scheduling revisits to measure the targets while maintaining surveillance.

6.1 Tracking of Multiple Maneuvering Targets

We assume that a radar system tracks T targets where T is a random variable $0 \leq T \leq T_0$ and the tth target is in state $\mathbf{x}_k^t$ at epoch k. In addition, the radar undertakes surveillance to discover new targets. This surveillance is assumed to require a certain length of time, say τ_{scan} within every interval of length τ_{total}. The remainder of the time is spent measuring targets being tracked. We aim to schedule revisit times to targets within these constraints.

For IMM-based tracking algorithms, for example, the target trajectory is approximated by an average over a finite number of given dynamic models. In this context, we assume that the dynamical models are independent of the target and associated to each is a corresponding state propagation matrix $\mathbf{F}_m$ ($m =$

$1, 2, \ldots, M$). The recursion for state transition is

$$\mathbf{x}^t(k) = \mathbf{F}_m(k)\mathbf{x}^t(k-1) + \nu_m^t(k), \qquad (10.73)$$

where the index m is a possible value of a random variable $M(k)$, the *dynamical model* which takes any discrete value $[1, 2, \cdots, M]$. Process noise $\nu_1^t(k), \cdots, \nu_M^t(k)$ is Gaussian, depends on both target and dynamical model and is independent between different values of each of these indices. The covariance matrix of $\nu_m^t(k)$ is denoted by $\mathbf{Q}_m^t(k)$.

In the tracker, the dynamical model of the tth target $M^t(k)$ is assumed to evolve as a Markov chain with given transition probabilities, denoted by

$$\pi_{m,\ell}^t = P\{M^t(k) = m | M^t(k-1) = \ell\} \quad j, \ell \in [1, \cdots, M]. \qquad (10.74)$$

It is assumed that N different *measurement modes* are available for each target, each given by a measurement matrix $\mathbf{H}_n^t$ $n = 1, 2 \ldots, N$:

$$\mathbf{y}^t(k) = \mathbf{H}_n^t(k)\mathbf{x}^t(k) + \omega_n^t(k) \qquad (10.75)$$

where here $\mathbf{y}^t(k)$ is the measurement to be obtained from the tth target at time k, $\omega_n^t(k)$ is the measurement noise, and $n = n(k)$ is a control variable for the measurement mode. The variable $\tilde{t} = \tilde{t}(k)$ represents the choice of target to which the beam is steered at the kth epoch. The measurement noise $\omega_1^t(k), \cdots, \omega_N^t(k)$ consists of zero-mean white and uncorrelated Gaussian noise sequences with the covariance matrix of $\omega_n^t(k)$ denoted by $\mathbf{R}_n^t(k)$.

The waveforms impinge on the measurement process through the covariance matrix of the noise $\omega_n^t(k)$, as described in Section 3.

The choice of measurement is made using the control variable $n(k)$. In this example, two choices are made at each epoch: the target to be measured and the waveform used. More than one target may be in the beam and then measurements of each target will be updated using the LMIPDA-IMM algorithm.

6.2 Scheduling

As already stated, at each epoch a target track and a beam direction have to be selected. The scheduler has a list $\Delta = \{\delta_1, \delta_2, \ldots, \delta_K\}$ of "revisit intervals." Each of the numbers δ_k is a number of epochs representing the possible times between measurements of any of the existing targets. It is assumed for the purposes of scheduling and tracking that during any of these revisit intervals the target dynamics do not change.

In order to determine which target to measure and which waveform to use, for each existing target and each waveform the track error covariance $P^t_{k-1|k-1}$ is propagated forward using the Kalman update equations and assuming each of the different potential revisit intervals in the list Δ in the dynamics. In the absence of measurements the best we can do is to use the current knowledge to predict forward and update the covariance matrix, dynamic model probability density and probability of track existence. The algorithms is now modified as follows. *Forward prediction* is performed separately for each dynamical model. Because the dynamics of the target depend on the revisit time $\delta \in \Delta$, these calculations are performed for each revisit time. *Covariance update* is normally done with the data, but since we are interested in choosing the best sensor mode at this stage, the following calculations are required. If the target does not exist there will be no measurements originating from the target and the error covariance matrix is equal to the prior covariance matrix. If the target exists, is detected, and the measurement is received, then the error covariance matrix is updated using the Kalman update equations. The expected covariance update is calculated using Bayes' rule:

$$\mathbf{P}_{k|k}(j,\delta) = (I - \psi_{k|k-1} P_D P_G \mathbf{K}(\phi,\delta)\mathbf{H})\mathbf{P}_{k|k-1}(j,\delta), \qquad (10.76)$$

where $\psi_{k|k-1}$ is the *a priori* probability of track existence, and $P_D P_G$ is the probability that target is detected and its measurement is validated. $\mathbf{K}(\phi,\delta)$ is a Kalman gain calculated for each sensor mode; i.e., for the waveform ϕ and revisit time δ. Both ϕ and δ take discrete values from the waveform library and the revisit time set Δ.

$$\mathbf{K}(\phi,\delta) = \mathbf{P}_{k|k-1}(j,\delta)\mathbf{H}\mathbf{S}^{-1}(\phi), \qquad (10.77)$$

where $\mathbf{S}$ is the innovation covariance matrix, calculated as usual:

$$\mathbf{S}(\phi) = \mathbf{H}\mathbf{P}_{k|k-1}\mathbf{H}^\mathsf{T} + \mathbf{R}_\phi.$$

The above calculations are made for all combinations of revisit times in Δ and waveforms in the library. In considering a non-myopic approach, the number of combinations grows exponentially in the number of steps ahead, and soon becomes impractical for implementation. Having obtained the error covariance matrix for all possible combinations of sensor modes, the optimal sensor mode (waveform) is then chosen for each target to be the one which gives the longest re-visit time, while constraining the absolute value of the determinant of the error covariance matrix to be smaller than the prescribed upper limit K. When no feasible solution exists satisfying the constraint, we take the solution yielding the smallest absolute value for the determinant of

$\mathbf{P}_{k|k}$; i.e., the objective is to obtain

$$\phi, \delta = \arg\max \Delta, \text{ subject to } |\mathbf{P}_{k|k}(\phi, \delta)| \leq \max\{K, \min_{\phi,\delta} |\mathbf{P}_{k|k}(\phi, \delta)|\}. \tag{10.78}$$

Scheduling is then done to permit a full scan over the prescribed scan period while also satisfying the constraints imposed by the revisit times obtained by the sensor scheduler. Once a target is measured, its revisit time is re-calculated.

The solution to the above equation is not necessary unique for ϕ. In cases when there is more than one ϕ, the waveform that gives the smallest determinant of the error covariance matrix is chosen.

For the case of N-step-ahead scheduling, the revisit times and waveforms are calculated while the target states are propagated forward over N measurements. The cost function is the absolute value of the determinant of the track error covariance after the Nth measurement. Only the first of these measurements is done before the revisit calculation is done again for that target, so the second may never be implemented.

6.3 Simulation results

Here we demonstrate the effects, in simulation, of scheduling as described in Sections 4, 5 and 6. Specifically, we compare random choice of waveform with a scanning beam against one-step-ahead and two-step-ahead beam and waveform scheduling. All three simulations were performed 100 times on the same scenario. In the first case, measurements were taken at each scan with no further measurements beyond the scan measurements permitted. In these experiments we used a small waveform library consisting of three waveforms: an up-sweep chirp, a down-sweep chirp and an unmodulated pulse. In the unscheduled case, waveforms were chosen randomly from this library. The simulated scene corresponded to a surveillance area of 15 km by 15 km. The scene contained two maneuvering land targets in stationary land clutter, which had small random Doppler to simulate movement of vegetation in wind. While the level of fidelity of the clutter is low, it is sufficient to demonstrate the principles of scheduling. The number of clutter measurements at each epoch was generated by samples from a Poisson distribution with mean ~ 5 per scan per square kilometer. Target measurements were produced with probability of detection 0.9. The target state x^t consisted of target range, target range rate and target azimuth. The targets were performing the following maneuvers: constant velocity, constant acceleration, constant deceleration and coordinated turns with constant angular velocity. In the scheduling cases, surveillance time used ap-

Sensor Scheduling in Radar

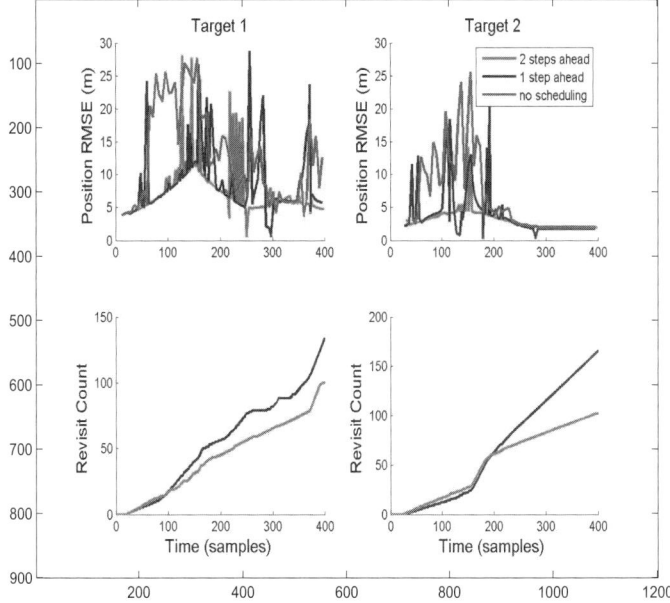

Figure 10.4. Root-mean-square error (RMSE) and revisit count for one-step-ahead versus two-step-ahead beam and waveform scheduling.

proximately 80% of each scan period; the remaining 20% was allocated, as described above, to maintaining tracks of existing targets.

The outcome of these experiments suggests that in the presence of clutter tracking performance can be improved with scheduling and even more with multiple-step-ahead scheduling as opposed to one-step-ahead. The results are represented in Figure 10.4.

It should be observed that in Figure 10.4 the RMS error was considerably worse, especially during the early part of the simulation, for the unscheduled case. In fact the RMS error in the unscheduled case is larger immediately after significant maneuvers as can be expected. Of course, in this case the revisit time is fixed and is not plotted in the second subplot. One observes that, for the two-step-ahead case, tracking accuracy is improved (top plots) slightly over the one-step-ahead case, but with a significant reduction in revisits necessary to maintain those tracks.

7. Waveform Libraries

We have discussed how to estimate the effectiveness of individual waveforms in Section 5. We envisage that the choice of waveforms will be from a relatively small library. Others have considered the design of waveforms by the choice of parameters in a generic waveform type. For example, see [216], one could choose waveforms of the form

$$w_{\alpha,\beta}(t) = e^{2\pi i \alpha t^\beta}. \tag{10.79}$$

This has the form of a *power frequency modulated* waveform. Choice of different values of α and β provides waveforms with different range and Doppler properties. This is effectively an infinite library. Since our interest is mainly in understanding the concepts of sensor management, we shall restrict attention to the situation where the number of waveforms is finite. In the context of radar, where the time between pulses is very small, computational issues are crucial. Scheduling problems are notoriously computationally intensive, and so it makes sense to reduce that problem by minimizing the number of waveforms from which to make the choice. Here we discuss how to choose libraries, bearing in mind computational constraints.

In order to choose between different radar modes, or more specifically waveforms, at a given epoch, a method is needed to measure the effectiveness of such a mode in a given context. There are three issues associated with this *cost function* assignment that need to be addressed. The first is the apparently obvious statement that optimization of the cost should, in principle, produce a desired operational outcome. However, it is often the case that, for reasons of feasibility, cost functions represent only part of the total operational requirement. For example, in the context of tracking a target such as a missile, it is fairly normal to measure the cost in terms of an estimate of the track accuracy such as the estimated mean-square error of the estimated target position. On the other hand, from an operational perspective, at least for some of the engagement between the missile and its target, it may be much more important that the track be maintained; i.e., that the missile be not lost by the radar. While it would appear that track error is in some sense related to track maintenance, the optimal choice of radar mode to reduce track error may be different from that required to maintain track. For real radar systems, we propose that cost functions be based on the precise operational needs at the time with the objective of maximizing the probability of achieving the operational goals.

The second issue is that the cost function is a function both of the library of radar modes and of the environment. The information available about the environment is available in two forms: information that is acquired through exogenous sources (perhaps prior to the current collection) and information that has been acquired in the current sequence of measurements. The former are extremely important, especially in the context of radars observing an urban situation where the effects on the return due to buildings, fences, and other aspects of the built environment can be significant. We shall, nevertheless, ignore them in the current discussion. We focus only on the information that is acquired during the deployment of the radar system. Of course if computational constraints were not at issue, we would base cost functions on all the data that has been acquired so far. But this is infeasible, except in rare cases. Some processing has to be done, itself limited by computational considerations, to extract key features that can be fed to the cost function. In much of the literature on scheduling of waveforms for tracking, this "extracted information" is just the track error covariance.

The third and final issue we wish to address, would seem to contradict the first. There we have espoused the idea that cost functions should be operationally significant. However, this means that theoretical developments will be fragmented according to the cost function used. We propose therefore, for the purposes of developing a theory of waveform and other mode libraries, a generic cost function that has some relevance to operational costs.

One approach to this, at least in the context of tracking, is to calculate the *expected information* to be obtained from a measurement with a given waveform, given the current state of knowledge of the scene. In this context, it is expressed by knowledge of the target parameters – typically velocity and position. This is defined as the mutual information between the target variable (range and Doppler) and the processed (e.g., matched filtered) radar return resulting from the use of the waveform. This approach is along the lines discussed in Chapter 3.

The *utility* of a waveform library is defined by averaging over a distribution on the possible state covariance matrices, the maximum of this expected information over all waveforms. A subset of a library is *irrelevant* if it does not contribute to the utility in the sense that the library has the same utility with or without that subset. It is possible to use this concept to obtain parsimonious libraries involving LFM waveforms and more general "chirped" collections of waveforms.

We point out that it is usually difficult to find an explicit and usable expression for this distribution of state covariance matrices in practice. Nonetheless the very nature of this formalism permits us to make some general statements about waveform libraries.

We use the model described in Section 3. In the context of our discussion in this section, we represent the measurement obtained using the waveform ϕ as a

Gaussian measurement with covariance $\mathbf{R}_\phi$. The current state of the system is represented by the state covariance matrix $\mathbf{P}$. Of course, the estimated position and velocity of the target is also important for the tracking function of the radar. But in this context they play no role in the choice of waveforms. In a clutter-rich (and varying) scenario, the estimate of the target parameters will clearly play a more important role. The *expected information* obtained from a measurement with such a waveform, given the current state of knowledge of the target, is

$$I(X;Y) = \log \det(\mathbf{I} + \mathbf{R}_\phi^{-1}\mathbf{P}). \qquad (10.80)$$

This is the mutual information between the target variables (range and Doppler) X and the processed radar return Y resulting from the use of the waveform ϕ. $\mathbf{I}$ is the identity matrix. We use this expected information as the *measure of effectiveness* of the waveform ϕ in this context. The more information we extract from the situation the better.

We assume a knowledge of the possible state covariances $\mathbf{P}$ generated by the tracking system. This knowledge is statistical and is represented by a probability distribution $F(\mathbf{P})$ over the space of all positive definite matrices. This distribution will be a function of the previous choices of waveform since their Fisher matrices play a part in the calculation of this distribution, so that there is an inherent circularity in the definition. Indeed every update of the tracker combines the distribution $F_n(\mathbf{P})$ of the covariances at time n propagated forward by the dynamics with the system and measurement noises, as in, for example the Kalman equations. Thus the distribution of state covariances at time $n+1$ is a function of the measurement policy. We make an assumption that the distribution $F(P)$ is unaffected by the choice of policy.

We define the *utility* of a waveform library $\mathcal{L} \subset L^2(\mathbb{R})$, with respect to a distribution F, to be

$$G_F(\mathcal{L}) = \int_{\mathbf{P}>0} \max_{\phi \in \mathcal{L}} \log \det(\mathbf{I} + \mathbf{R}_\phi^{-1}\mathbf{P}) \, dF(\mathbf{P}). \qquad (10.81)$$

Thus we have assumed that the optimal waveform is chosen in accordance with the measure of effectiveness defined in equation (10.80) and have averaged this over all possible current states, as represented by the covariance matrices $\mathbf{P}$ and in accordance with their distribution $F(\mathbf{P})$.

We consider two libraries $\mathcal{L}$ and $\mathcal{L}'$ to be *weakly equivalent*, with respect to the distribution F, if $G_F(\mathcal{L}) = G_F(\mathcal{L}')$, and *strongly equivalent* if $G_F(\mathcal{L}) = G_F(\mathcal{L}')$ for all F.

We call the subset $\mathcal{S} \subset \mathcal{L}$ *weakly irrelevant* with respect to the distribution F if $G_F(\mathcal{L}\backslash\mathcal{S}) = G_F(\mathcal{L})$, and *strongly irrelevant* if $G_F(\mathcal{L}\backslash\mathcal{S}) = G_F(\mathcal{L})$ for all F. Here, $\setminus$ denotes the usual set difference. In what follows we will work in receiver coordinates, i.e., treat $\mathbf{T}$ above as $\mathbf{I}$. This amounts to a change in parameterization of the positive definite matrices in the integral in (10.81). Again, an invariance assumption (under the dynamics of the Gauss-Markov system) on the distribution $F(\mathbf{P})$ will make this choice of coordinates irrelevant.

While this theory continues to be explored, we discuss here just one result. As we said in section 2.1, LFM "chirps" are important waveforms in radar applications.

7.1 LFM Waveform Library

We investigate an LFM ("chirp") waveform library. In this case the library consists of

$$\mathcal{L}_{\text{chirp}} = \{\phi_0(t) \exp(i\lambda t^2/2) \mid \lambda_{\min} \leq \lambda \leq \lambda_{\max}\} \tag{10.82}$$

where $\phi_0(t)$ is an unmodulated pulse, $\lambda_{\min}$ and λ_{max} are the minimum and maximum chirp rates supported by the radar. For this library the corresponding measurement covariance matrices are of the form [182, 131]

$$\mathbf{R}_\phi = S(\lambda)\mathbf{R}_{\phi_0}S(\lambda)^\mathsf{T}, \tag{10.83}$$

where

$$S(\lambda) = \begin{pmatrix} 1 & 0 \\ \lambda & 1 \end{pmatrix}. \tag{10.84}$$

We have been able to prove (see [228]) that this library is strongly equivalent to

$$\mathcal{L}'_{\text{chirp}} = \{\phi_0(t) \exp(i\lambda_{\min} t^2/2), \phi_0(t) \exp(i\lambda_{\max} t^2/2)\}. \tag{10.85}$$

That is, we do just as well if we keep only the LFMs with the minimum and maximum rates. In the case of the LFM waveform library, the error covariance matrices (despite a popular assumption [185]) are not rotations of each other, but the results of shearing transformations. For $\mathbf{R}_{\phi_0}$ a diagonal matrix with ρ_1, ρ_2 on the diagonal, direct computation yields the following expression for the mutual information $I(X;Y)$:

$$I(X;Y) = \frac{P_{11}}{\rho_2}\lambda^2 - 2\frac{P_{12}}{\rho_2}\lambda + \frac{|\mathbf{P}|}{|\mathbf{R}_0|} + 1 + \frac{P_{11}}{\rho_1} + \frac{P_{22}}{\rho_2}. \tag{10.86}$$

This is a quadratic in λ with positive second derivative since $\mathbf{P}$ and $\mathbf{R}_{\phi_0}$ are both positive definite, and therefore achieves its maximum at the end points;

i.e., at maximum or minimum allowed sweep rate. The optimal sweep rate is chosen to be

$$\lambda_\phi = \begin{cases} \lambda_{\max}, & \text{if } \lambda_{\min} + \lambda_{\max} > \frac{P_{12}}{P_{11}} \\ \lambda_{\min}, & \text{otherwise} . \end{cases} \qquad (10.87)$$

7.2 LFM-Rotation Library

For the library under consideration here, we start with an unmodulated waveform ϕ_0 and allow both the "chirping" transformations (10.82) and the fractional Fourier transformations, i.e.,

$$\mathcal{L}_{\text{FrFT}} = \{\exp(i\theta(\mathbf{t}^2 + \mathbf{f}^2)/2)\phi_0 \mid \theta \in \Theta\}, \qquad (10.88)$$

That is, we consider all transformations of the following form.

$$\mathcal{L}_{\text{FrFT}} = \{\exp(i\theta(\mathbf{t}^2 + \mathbf{f}^2)/2)\exp(i\lambda\mathbf{t}^2/2)\phi_0 \mid \lambda_{\min} \le \lambda \le \lambda_{\max}, \theta \in \Theta\} \qquad (10.89)$$

where the set Θ is chosen so as not to violate the bandwidth constraints of the radar, and $\mathbf{f}$ is the operator on $L^2(\mathbb{R})$ defined by

$$\mathbf{f}\phi(t) = i\frac{d}{dt}\phi(t), \qquad (10.90)$$

Note that $\mathbf{f}$ and $\mathbf{t}$ commute up to an extra additive term (the "canonical commutation relations"). To be precise,

$$[\mathbf{t}, \mathbf{f}] = \mathbf{tf} - \mathbf{ft} = -i\mathbf{I}. \qquad (10.91)$$

For this library the corresponding measurement covariance matrices are given by (10.83) with

$$S(\theta, \lambda) = \begin{pmatrix} \cos\theta & -\sin\theta \\ \sin\theta & \cos\theta \end{pmatrix} \begin{pmatrix} 1 & 0 \\ \lambda & 1 \end{pmatrix}. \qquad (10.92)$$

In the case of a finite number of waveforms in the library, we observe that the utility of the rotation library improves with the number of waveforms in the library. There exists a unique θ that maximizes the mutual information $I(X;Y)$ and, similar to the pure chirp library case, maximum and minimum allowed chirp rate λ

$$\mathcal{L}'_{\text{FrFT-chirp}} = \{\exp(i\theta(\mathbf{t}^2+\mathbf{f}^2)/2)\exp(i\lambda\mathbf{t}^2/2)\phi_0 \mid \lambda \in \{\lambda_{\min}, \lambda_{\max}\}, \theta \in \Theta\} \qquad (10.93)$$

is strongly equivalent to $\mathcal{L}_{\text{FrFT-chirp}}$.

8. Conclusion

Our aim in this chapter has been to illustrate some of the ideas of sensor scheduling in the context of radar. This has been a highly subjective view of the issues and ideas, based to a large extent on our own ideas and work on this subject. To understand the sensor scheduling problem for radar, and elsewhere, it is important to be aware of the various functionalities of the sensors and their capabilities. To do this for radar, we have described the way in which a pulse-Doppler radar works, at least roughly. This has given some indication of what level of flexibility is available to the designers of such a system, and what aspects might be dynamically and adaptively tunable. We remark that, in reality, radar engineers have to work at a much finer level of detail. Issues associated with the limitations of the electronics play a significant role in the design of a radar, and robustness is a key objective.

We have discussed some relatively simple techniques in radar scheduling. Our focus has been to show, in a few areas, that scheduling is useful. It will improve detection and tracking performance and reduce revisit time to tracked targets while maximizing surveillance time. Some recent work by Sira et al. [215], has shown, for example, that scheduling will significantly improve detection of small targets in sea clutter. Nehorai [244] has also shown similar results.

An important issue that has been largely neglected in the sensor scheduling community is that of library design. In whatever sensor scheduling context, the library of available sensor modes will be a key design component in the system. What principles should guide the system designer in this context? We have touched on this subject albeit briefly and only in the context of radar waveforms in Section 7. A closely related aspect of the design of a scheduling system, which has also not received much attention in the literature, is that of the design of measures of effectiveness for the various sensor modes, and in the radar context for waveforms in particular. Our belief is that ultimately these measures will be based on operational criteria, but that these criteria will drive "local" more easily computable cost functions that are more closely tied to the system. We have initiated a discussion of measures of effectiveness within the context of waveforms for radar in Section 5.

This is an embryonic subject and much more needs to be done. In a sense, all of the work done so far really addresses idealizations of the true problems. More work is needed, in particular, on the problems associated with implementation. The short time between pulses in a radar system make it very difficult to schedule on a pulse-to-pulse basis. The calculations associated with, in particular, clutter mapping which involve large amounts of data, are probably too computationally intensive to be done on this time scale. Compressive sensing ideas [71] may play a role in reducing this complexity.

Work is needed too on the potential gains of non-myopic scheduling in this context. A real system will be required to work across multiple time scales, integrating multiple tracking functions, and scheduling over many pulses. Scheduling will be used both to optimize allocation of the resource between different operational functions, as we have discussed in Section 6, and to improve operational performance by, for example, waveform and beam-shape scheduling. For military functions, issues such as jamming are important, and scheduling is destined to play a role here too. Ultimately, we anticipate that, against a sophisticated adversary, game-theoretic techniques will be used to drive radar scheduling to mitigate jamming.

Radar scheduling is replete with problems, some of which we have touched on, and many of which we have not, and very few solutions. It represents an exciting and challenging area of research that will drive much of the research activity in radar signal processing over the next few years.

Chapter 11

DEFENSE APPLICATIONS

Stanton H. Musick

Sensors Directorate, Air Force Research Laboratory, Wright-Patterson Air Force Base, OH, USA

1. Introduction

Over the last several decades, US defense and intelligence agencies have developed and deployed a wide variety of sensor systems for monitoring enemy activity. Two well-known examples are the Keyhole family of reconnaissance satellites that first came on-line in 1960 to obtain photo images of ground targets using high-resolution cameras, and JSTARS, a radar-equipped reconnaissance and surveillance aircraft first fielded for the war with Iraq in 1991. Today such sensors provide intelligence data to support nearly every type of combat mission. To illustrate, sensors are critical in the search/track mission that finds and maintains awareness of targets in a specified region, in the trip-wire/warning mission that monitors and classifies traffic crossing a designated boundary, and in the attack phase of most missions where dispersed sensors can act in concert to reduce target tracking error and improve visibility in zones masked by terrain.

Intelligence, reconnaissance and surveillance (ISR) sensors gather data to detect, localize and identify targets, to assess threat levels, and to deduce enemy intent. Such sensors help with these functions by gathering the enemy's emitted signals or by measuring physical properties that characterize the objects of interest, including reflective, radiative, magnetic, acoustic, seismic, and motion phenomena. For example, a radar system can measure the range and range rate of a moving vehicle in order to localize it in a geospatial coordinate system, and then switch waveforms to measure the vehicle signature (e.g., reflective intensity vs. range) in order to assess vehicle class, as discussed in Chapter 10.

ISR sensors include radars, sonars, electro-optical and infrared devices, warning receivers, and electronic signal monitors. Newer sensors with a potential for tactical deployment include laser radars to provide shape information at night, foliage-penetrating (FOPEN) radars to detect vehicles under vegetation, and hyperspectral imagers for characterizing surface materials to aid in target identification (ID). Any of these newer sensors could conceivably be deployed on an unmanned aerial vehicle (UAV). In addition, unattended ground sensors (UGSs) are ideal devices for passive target detection, location and/or ID in remote areas.

ISR sensors are often deployed onboard platforms that move, e.g., the camera in the Keyhole spacecraft or the radar onboard the JSTARS aircraft. Since what a platform sensor sees is constrained by where its platform goes, platform route planning is a critical part of the puzzle. Thus, the sensor deployment problem generally consists of both platform route control and sensor scheduling, and these functions must be considered concurrently to achieve effective performance.

Generally those sensors deployed on satellites in deep space are centrally controlled by national organizations (national assets). For example, the National Reconnaissance Organization fields and controls US reconnaissance satellites like those in the Keyhole family to warn of potential trouble and help plan military activities. Those sensors deployed on airborne platforms that remain at safe distances from the battle (stand-off ranges) are controlled by theater commanders (theater assets). And those sensors deployed close-in for tactical purposes in both ground and airborne configurations are organic assets controlled by local commanders at brigade, battalion and lower levels (tactical assets). Although there are many assets that a local commander does not control directly, she can nonetheless access both national and theater data and make requests of those assets for new collections; to her, those other assets are partially controllable.

Because ISR sensors are so integral to modern warfare and also because they are so often scarce assets, there is an urgent need to manage them effectively. This need is widely recognized, and concerted efforts to address the technical problems associated with planning and coordinating sensor deployments have been underway for at least two decades. These efforts continue today in the US, especially in foundational research and development, as illustrated by the state-of-the-art techniques and example applications presented in each previous chapter of this book.

This chapter provides a concise picture of the sensor management problem in a realistic military setting. In order to limit the problem scope while retaining a useful essence, we focus on ISR applications against ground vehicles, a crucial function that serves many different military operations. We begin with a brief summary of the history and state of sensor management for vehicle ISR, and conclude by discussing a recent defense program in which sensor management played a vital role.

2. Background

In the oldest form of ISR sensor control, an operator on a particular platform employs her onboard sensors to uncover and engage targets of opportunity. The operator is pre-briefed on where to search, on what sorts of targets and backgrounds to expect, and on how to react to what is found, including how to verify target identity, position for weapon release, and employ the sensors needed by the weapon to produce a kill. The effectiveness of this manual approach varies widely from one situation to the next according to factors such as the chance nature of enemy actions, operator skills and workloads, and the utility of the sensor in performing its functions. To illustrate, in the mission called combat air patrol (CAP), a fighter aircraft is assigned to guard a designated object (e.g., a fleet at sea or an aerial tanker) and the pilot uses her radar to find, identify and then engage incoming enemy aircraft. CAP is an example of this highly free form, human intensive, and goal driven approach to sensor control.

Now contrast that free-form employment philosophy with its polar opposite, an approach in which the sensors on an ISR platform gather data on a preplanned route and schedule, data that will require further off-line analysis before it is applied. Historically, most national and some theater assets have been used in this manner. Under this philosophy, each sensor on each platform is given a schedule of sensing mode and pointing direction indexed to platform location along a planned route, and both the route and the sensor schedule remain fixed during that mission. Although such fixed plans are not able to respond to spontaneous events such as popup threats that could force route changes or revamp sensor use schedules, fixed plans are used today, mostly in situations where the transport platform is out of harm's way. To illustrate with U-2, a reconnaissance aircraft controlled as a national asset, missions are planned to satisfy theater-wide needs for photographic, electro-optic, infrared and radar imagery, as well as signals intelligence. Once a mission is launched, U-2's route does not usually change, even though sensor taskings can and do.

Traditionally, many ISR sensor systems have been *vertically integrated*, meaning that raw sensor data is received, processed and analyzed by a trained operator who in turn determines how best to use the sensor next. Vertically integrated systems are common in space-based intelligence gathering, and in situations requiring operator analysis of imagery or signals. Generally, operator intervention takes time, which complicates the sensor data exploitation process, but produces smaller location errors and better identifications. Vertically integrated systems are quite prevalent today, the JSTARS radar system being a good example.

3. The Contemporary Situation

In military operations today, a commander allocates both her platforms and sensor assets to detect, track, identify, fix, and engage targets of high value. To complicate matters, high-value targets often appear among a multitude of low-value targets and civilian components. A commander's challenge is to employ her sensing resources effectively to eventually engage (strike) the high-value targets before they can meld into the background and disappear[1]. The available time to prosecute this challenge can be quite short because an intelligent enemy usually employs tactics designed to minimize both his chances of being detected and his exposure to strike.

Although a large and growing inventory of national, theater, and tactical sensors exists, no single system suffices by itself. For example, the ground moving target indicator (GMTI) mode of JSTARS, which measures Doppler-shifted frequencies relative to the stationary ground, can defeat evasion of moving vehicles by continually monitoring their motion over a large area of interest. As good as GMTI is at following ground vehicle motion, it does not provide the unambiguous vehicle ID that terminal engagement requires. Although a video sensor can supply confident ID, its need to be inside the area of interest and its limited field-of-view lead to a kind of disorienting myopia. However, when radar and video are combined by cross-cueing the video sensor to suspicious targets in the radar returns, high-value targets can be exposed and identified, and a synergism of sensor types is realized. As the examples in Chapters 3 and 5 have demonstrated, objective approaches recognize and exploit such cross-cueing opportunities by properly balancing sensor modes.

As just illustrated, the benefits of dynamic sensor cross-cueing are manifest. Even so, currently it is not operationally possible to harness several collection systems together to search, track and identify large numbers of targets in a realistic environment that includes a complex mix of confuser objects. To be sure, many of the required elements exist, e.g., sensors of diverse type,

---

[1]ISR problems intensify when personnel, not vehicles, comprise the target set because personnel are more mobile and more able to hide and thereby deceive sensors in complex environments. Fortunately, analysis approaches for personnel and vehicles are quite similar, but be aware that what we describe here does not pertain directly to personnel as targets.

communication links, a network-centric architecture, track maintenance algorithms, capable registration algorithms, adaptive methods for hypothesis management, methods for relieving the computational intractability of planning under uncertainty, and so forth. Nevertheless, when diverse platform sensor elements need to be combined to produce a unified picture, the current intelligence structure utterly fails because platform sensor collection systems cannot cooperate sufficiently.

Success in a dynamic, complex and time-sensitive ISR situation requires a collections management system that continually updates its allocations for platforms and sensors in response to changing battlefield conditions. The base reasons for the current deficit in this arena range from profound technical problems to potentially large costs for engineering a systems-level solution. Some of the foremost operational and technical problems include: latency in almost every sensor data feed; incompatible and incomplete data interfaces arising from independently developed systems that were not designed to work together; data misaligned in the common grid (gridlock); the combinatoric explosion of potential resource assignments that overwhelms any human and challenges even the best optimization algorithms; data improperly correlated with like data from other sources; inaccurate predictions of future target motion; insufficient scalability in core algorithms; insufficient recognition of operational realities such as failed sensors or platform airspace conflicts; lack of effective means for commanders to query the automated software system to oversee its operation and gain confidence in its consistency; and sensor tasking that fails to balance mission needs, prioritize demands, and/or predict the value of sensing tasks. Although many of these problems do not relate directly to sensor management in isolation, most affect the all-source fused solution upon which the sensor manager must reason, and are therefore important in the larger sense.

Thus, a capability is needed to dynamically control military ISR systems to produce situation estimates that are timely, complete, and accurate. Such capability must overcome individual system weaknesses, be responsive to commander's needs, and operate automatically in harsh environments. The US defense and intelligence communities have recognized this challenge and are today pursuing a variety of investigations and developments designed to improve ISR sensor exploitation. An exemplar is the program named Dynamic Tactical Targeting (DTT), an effort sponsored by the US Defense Advanced Research Agency (DARPA) and completed in 2007. The material that follows synopsizes the problem, the solution approach, and the results of DTT [99].

4. Dynamic Tactical Targeting (DTT)

The mission of DTT was to develop sensor control and data fusion technologies that would make possible a tactically responsive targeting process that could be managed by warfighters. Built as an experimental system, DTT processed ISR data from national, theater and tactical assets to detect, identify and track large numbers of ground vehicles in a large surveillance region. To do this, DTT fused data from all ISR sources, located and identified regions/targets of interest, and dynamically tasked the tactical ISR assets to fill coverage gaps and provide relevant sensor observations. Treating tactical assets as resources and mission needs as prioritized demands, DTT assigned resources to demands using a model-based closed-loop approach that adapted to a dynamic world while accommodating a variety of problem constraints, including the shooter's need to engage time-sensitive targets.

Here is a motivating scenario for DTT that highlights several types of cross-cueing opportunities. An electronic intelligence aircraft (e.g., Rivet Joint) detects the radar of a mobile missile battery (e.g., SA-6) and cross-cues a theater asset (e.g., U-2) to collect a synthetic aperture radar (SAR) image of the detected area. An image analyst examines the SAR image and precisely locates the potential missile battery. This location information then cues a video-equipped UAV (e.g., Predator) to approach the target to obtain high-resolution imagery, from which an analyst can establish a positive visual ID. If a JSTARS aircraft were simultaneously tasked to detect target motion using occasional GMTI dwells (while the UAV moves on to investigate other SAR detections), the onset of motion could be discovered, GMTI dwell rate increased, and a track initiated on the missile battery vehicle. Kinematic tracking would continue with GMTI measurements until the vehicle approached other vehicles, at which time the UAV could be reassigned to maintain track through a town containing dense civilian traffic. When the target vehicle eventually reached a location where collateral damage was expected to be low, it could be attacked. This scenario exemplifies effective birth-to-death tracking against a high-value target.

Note that this motivating scenario integrates complementary capabilities of existing platform sensors to achieve a force-level capability that exceeds the individual platform-level capabilities. Thus, DTT was a system that sought to employ existing ISR assets to achieve its ends through smart sensing, effective data fusion, and on-line re-planning. Although new software and doctrine were required, new hardware was not.

The synergy derived from sensor cross-cueing exists for many of the situations experienced in typical combat. The problem for DTT was being nimble

Defense Applications 263

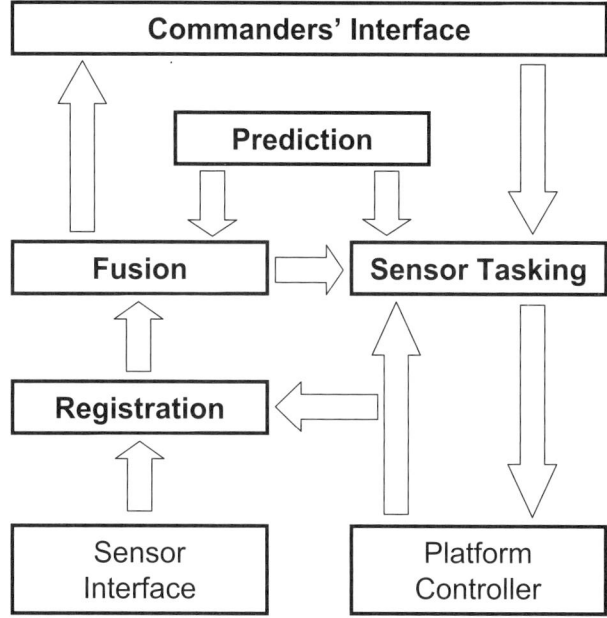

Figure 11.1. The major functional components of the DTT system

enough to recognize and seize these cross-cueing opportunities before they disappeared. Solving this problem involved distributing platforms to favorable locations and tasking sensors in such a way that the match of sensors to targets responded well to the evolving combat needs. Not one of the cross-cueing examples presented above could have been entirely preplanned because each arose spontaneously and depended on target actions that were both non-deterministic and non-cooperative. However, deceptive enemy actions can often be thwarted by deploying complementary sensors that concurrently view targets at multiple aspects and in many spectral bands.

Figure 11.1, which depicts the organization of the DTT system, shows it comprised of five principal components: Commanders' Interface, Registration; Fusion; Prediction; and Sensor Tasking. The DTT user can enter guidance and make requests to DTT through the Commanders' Interface, a GUI-based component that serves as the commanders' portal. Sensor input enters through Registration, which reduces errors in sensor reports by estimating and removing the systematic biases in each collection asset. Fusion is an "all-source" function that incorporates all available data, initiates new tracks, maintains existing tracks, and identifies targets when ID data become available. In order to anticipate future gaps in coverage where tracks could be lost, Prediction projects target motion based on fused track estimates and target mobility models. Sensor Tasking generates and maintains synchronized collection plans for all organic sensor assets in response to evolving information needs, focusing coverage to favor high-valued targets, and re-planning as the situation dictates. The DTT ISR Testbed (not shown) uses capability models of the sensors and platforms to emulate the external ISR environment and generate the synthetic input needed for conducting Monte Carlo closed-loop testing.

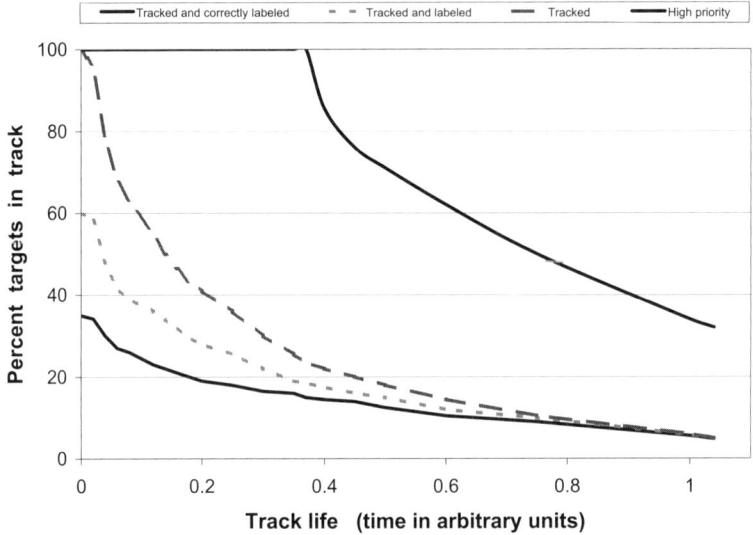

Figure 11.2. Preliminary results from simulation testing of DTT. Note that the rightmost curve is the one labeled "high priority."

Figure 11.2 shows DTT performance in terms of track life, a key metric for DTT. A variety of scenarios and operating conditions were employed in forming these preliminary results, some benign and some stressful. Three clarifying remarks about Figure 11.2 are in order. First, tracks are initially formed from kinematics data and must wait to be labeled until sufficient identity data have accumulated. Of course, labeling may be correct or incorrect. Thus, the correctly labeled tracks are a subset of the labeled tracks, which themselves are a subset of the tracked set. Second, target mislabeling occurs more frequently and lasts longer when targets are in close proximity to one another. Sometimes mislabeling adversely affects two tracks when labels swap as targets pass one another. In order to reduce the likelihood of mislabeling, DTT gathers identity data more often on closely spaced targets than on isolated targets. Third, the Fusion component of DTT performs track stitching, meaning that participating collection systems typically input their (labeled or unlabeled) tracks, not their raw sensor reports. When many input tracks are erroneous, i.e. composed from only clutter and false alarms, Fusion's estimate of the world state can deteriorate rapidly and significantly. Since erroneous tracks are generally quite short, discounting all short tracks provides a simple means to weed out erroneous tracks, at the expense of also eliminating good tracks that happen to be short. The rightmost curve in Figure 11.2 shows how track life performance can improve for high-value tracks in the favorable conditions created when Fusion discounts short tracks, target identification performance is nearly perfect, and Sensor Tasking strongly favors the highest priority objects.

Recognizing the need to reduce the response timeline against high-value targets and also to relieve the human workload of assigning platform sensor resources to targets, DTT built an automated system that focused on closing the loop between the fusion engine and the sensor manager. A key element of DTT's solution is the generation of "information needs" as the means of communicating sensing requirements from the other four system components into the Sensor Tasking component. In this manner, DTT took an important step towards objective collections management.

The second phase of work on DTT was called DTT:TEST (DTT: Tactical Exercises and System Testing). DTT:TEST built on DTT with the goal of demonstrating the operational utility of the system to the warfighter. Where DTT validated a network-centric collection strategy against simulated data, DTT:TEST evaluated DTT system performance in live exercises that employed actual data that contained many of the artifacts and errors present in the real world.

In addition, where DTT confined its attention to fairly traditional force-on-force scenarios, DTT:TEST extended DTT technology to deal with newer ISR challenges. Examples included: newer sensors in the ISR collection force, e.g., UGSs that can intercept remote enemy traffic to extend the coverage region; more diverse target types, e.g., civilian vehicles being used for military purposes; more evasive targets, e.g., adversaries that hide in the terrain (caves, forests, valleys) to avoid exposure to sensing; mixed environments, e.g., coalition vehicles unavoidably intermingled with civilian and threat vehicles; and smaller staffs leading to a greater need for automation. Thus, in addition to its efforts to prove DTT's robustness using real data, DTT:TEST asked how DTT could be extended to tackle these newer challenges.

Here are quantified statements of the technical objectives for the Fusion, Control, and Command components in DTT:TEST.

- Fusion - Maintain location and identity estimates on 25 military targets moving evasively among 1,000 civilian vehicles in an area of 1,000 km^2.

- Control - Generate tasking for 25 controlled airborne assets and 10 uncontrolled overhead assets to maintain 95% coverage and median track life of 1 hour for military vehicles operating in rough terrain during inclement weather.

- Command - Elicit prioritized mission needs spanning the full range of Army Priority Information Requests, and present them to three collaborating commanders.

Both phases of DTT are now complete. Although quantitative results have not been released, the following general qualitative statements may be made. Over a range of more than a dozen scenarios that varied in difficulty from easy to stressful, median track life for the entire collection of targets was over an hour. This result should be contrasted with the standard established in AMSTE (Affordable Moving Surface Target Engagement, [241]) of roughly twelve minutes, although the AMSTE and DTT scenarios were not closely matched. In addition, in side-by-side comparisons with and without sensor management, there were significant improvements in all measures of performance when sensor management was turned on. The benefits of using sensor management were greatest in the most difficult scenarios, i.e. those with the greatest vehicular congestion and the most topographic relief.

5. Conclusion

This chapter discussed the role of sensor management in defense applications. In a subject so vast, one can only hope to convey a representative slice that suggests some of the challenges and approaches. We have done this by focusing on vehicle detection, tracking and identification using ISR sensors, and by providing information on history, contemporary approaches, and problems that exist in building a responsive system. We illustrated the current state of system-level research in this field with a discussion of DTT, a prototype system for fusing data and controlling sensors that was evaluated in both simulation and live exercises. General results from DTT were presented.

As with any emerging discipline, technical issues persist in defense sensor management and deserve further research attention. The following list briefly outlines six of the most difficult issues.

1 Tracking personnel in urban areas - In this system-level challenge, the sensor manager attempts to achieve tracking and identification in complex urban environments populated by personnel who can easily flee and/or hide.

2. Timely tracking of hidden vehicles - Here the system must provide sufficient sensor coverage to expose partially hidden and/or time-sensitive vehicular targets before they elude prosecution by virtue of their camouflage or agility.

3. Sensor diversity for improved effectiveness - This refers to managing sensors with diverse spatial and spectral capabilities to produce multiple looks on a difficult target to allow its early detection and successful discrimination. Geometry, geography, weather and environment all enter into the reasoning needed to select the right sequence of sensing actions. Some believe that target recognition will profit most from such ability.

4. Layered sensing to achieve tailored effects - This concept imagines many sensors widely dispersed in both location and phenomena (e.g., passive and active sensors) that can be used when needed to rapidly achieve a particular operational outcome, e.g., identification of a popup threat or geolocation of a potential target.

5. Cooperation across regions - As a sensor control region grows in size, its coverage can begin to overlap with neighboring regions and interactions inevitably occur, or should occur. This leads to questions regarding process scaling for both fusion and control, such as the information-sharing paradigm and the control model, e.g., distributed, decentralized, or something else. Although there is a significant body of work in large-scale system control, it has yet to be applied in significant measure in the arena of managing sensors, where practical problems like data delays abound.

6. Anytime/anywhere algorithms - Many computations in any sensor allocation system use optimized algorithms at various stages. When these algorithms run slowly, there is a risk that they won't complete before their output is needed to make a time-sensitive decision. Algorithms that can deliver a best available answer at all times are greatly preferred.

We hope this chapter has helped the reader recognize that practical applications in sensor management often come with extenuating circumstances that compound the original problem and undo the simple solution. Still, we trust that the rest of this book has instilled an appreciation for the power that modern scientific methods can bring to solving problems of great complexity, including those that initially appear intractable or beyond objective formulation. In the words of Bertrand Russell (1872-1970), British author, mathematician and philosopher, "The greatest challenge to any thinker is stating the problem in a way that will allow a solution."

Acknowledgement: The author acknowledges his colleague Raj Malhotra of the Sensors Directorate whose insights enhanced the conclusions of this chapter.

Chapter 12

APPENDICES

Alfred O. Hero, Aditya Mahajan, Demosthenis Teneketzis
University of Michigan, Ann Arbor, MI, USA

Edwin Chong
Colorado State University, Fort Collins, CO, USA

1. Information Theory

Given the option of making one of several different types of measurements, e.g., $Y_1, \ldots, Y_m$, one would generally prefer making the measurement that leads to maximal uncertainty reduction or, equivalently, maximal information gain about a signal of interest S. This is one of the prime motivations behind information theoretic sensor management since information theory provides a way to systematically quantify uncertainty and information. In this appendix we present those elements of information theory pertinent to sensor management. We will limit our coverage to Shannon's definitions of entropy and conditional entropy, the data processing theorem and mutual information, and information divergence.

1.1 Entropy and Conditional Entropy

Let Y be a measurement and S be a quantity of interest, e.g. the position of a target or the target id. We assume that Y and S are random variables or random vectors with joint distribution $p_{Y,S}(y,s)$ and marginal distributions p_Y and p_S, respectively. The entropy of S, denoted $\mathcal{H}(S)$, quantifies uncertainty in the value of S before any measurement is made, called the prior uncertainty in S. High values of $\mathcal{H}(S)$ imply high uncertainty about the value of S. The of a continuous random variable with density p_S is defined as

$$\mathcal{H}(S) = -\int p_S(s) \log p_S(s) ds, \qquad (12.1)$$

where p_S denotes the probability density of S. If $S = \mathbf{S}$ is a continuous random vector the definition of is similar except that the expression on the right involves a multidimensional integral over each component of the vector valued s. For a discrete random variable the Shannon entropy is

$$\mathcal{H}(S) = -\sum_{s \in \mathcal{S}} p_S(s) \log p_S(s),$$

where p_S is now a probability mass function and $\mathcal{S}$ is its support set, i.e., the discrete set of values s for which $p_S(s) > 0$.

Oftentimes one is interested in the entropy of a random variable S conditioned on another random variable Y. For example, the amount by which an observation of Y reduces the entropy of S indicates the value of this observation in predicting S. There are two possible ways of defining such an entropy quantity: the entropy of the conditional distribution $p_{S|Y}$ of S given Y, which is a function of Y, and the of S given Y.

The Shannon entropy of the conditional distribution of S given Y, also called the point conditioned Shannon entropy, is denoted $\mathcal{H}(S|Y = y)$ and is defined as follows. We assume for simplicity that, given $Y = y$, S is a conditionally continuous random variable with conditional (posterior) density $p_{S|Y}(s|y)$ and define the entropy of this conditional density as

$$\mathcal{H}(S|Y=y) \stackrel{\text{def}}{=} -\int p_{S|Y}(s|y) \log p_{S|Y}(s|y) ds.$$

The point conditioned entropy is a function of y. It becomes a random variable when y is replaced by the random variable Y.

The conditional Shannon entropy $\mathcal{H}(S|Y)$ of S given Y is defined as the Shannon entropy of the conditional distribution $p_{S|Y}$. This conditional entropy can be interpreted as the uncertainty in S after the measurement Y is made, called the posterior uncertainty. When S and Y are continuous random variables with joint density $p_{S,Y}$ and conditional (posterior) density $p_{S|Y}$

$$\mathcal{H}(S|Y) = -\int dy\, p_Y(y) \int ds\, p_{S|Y}(s|y) \log p_{S|Y}(s|y).$$

The conditional entropies $\mathcal{H}(S|Y)$ and $\mathcal{H}(S|Y = y)$ are related

$$\mathcal{H}(S|Y) = \int \mathcal{H}(S|Y=y) p_Y(y) dy.$$

Appendices 271

When S and Y are discrete random variables an analogous expression holds for $\mathcal{H}(S|Y)$ with conditional and marginal densities replaced by conditional and marginal probability mass functions and integrals replaced by summations. A special "mixed discrete-continuous" case that frequently arises in target tracking problems is: S is a continuously evolving random vector, e.g., a target state vector, while Y is a discrete random vector, e.g., the binary output of the signal detector of a radar receiver. In this case the conditional entropy is

$$\mathcal{H}(S|Y) = -\sum_{y \in \mathcal{Y}} p_Y(y) \int p_{S|Y}(s|y) \log p_{S|Y}(s|y) ds ,$$

where $\mathcal{Y}$ is a discrete set containing all possible measurement values, p_Y is the probability mass function for Y, and $p_{S|Y}(s|y)$ is the (assumed continuous) posterior density of S given Y.

There are subtle but important differences between entropy for discrete vs continuous random variables. For discrete S the entropy is always non-negative, while for continuous S the entropy can be negative. For discrete random variables the entropy is directly related to the maximal attainable compression-rate without loss of information about S.

1.2 Information Divergence

Let p and q be two candidate probability densities of a real random variable S. The Kullback-Liebler (KL) divergence between p and q is defined as [152]

$$\mathrm{KL}(p\|q) = \int p(s) \log \frac{p(s)}{q(s)} ds.$$

The KL divergence is not symmetric in p and q and is thus is not true measure of distance between densities. However, it does behave like a similarity measure, sometimes called a pseudo-distance, in that it is concave(convex) in p (q), it is non-negative, and it is equal to zero when $p = q$.

1.3 Shannon's Data Processing Theorem

The average reduction in uncertainty about S due to observing Y can be quantified by the difference:

$$\Delta \mathcal{H}(S|Y) = \mathcal{H}(S) - \mathcal{H}(S|Y).$$

The data processing theorem asserts that this difference is always non-negative regardless of whether S is continuous or discrete. This theorem is easily proven by invoking convexity of the log function and the Jensen inequality [64] and mathematically captures the obvious: observations are never harmful in that they can never increase uncertainty about a signal.

1.4 Shannon Mutual Information

The difference $\Delta \mathcal{H}(S|Y)$ is better known as the Shannon mutual information, denoted $I(S;Y)$, between S and Y. The more reduction there is in uncertainty the higher is the mutual information. An equivalent expression for Shannon mutual information that applies to continuous random variables is

$$I(S;Y) = \int dy \int ds \, p_{S,Y}(s,y) \log \frac{p_{S,Y}(s,y)}{p_S(s)p_Y(y)}.$$

An analogous expression applies to discrete random variables. Shannon's mutual information can be recognized as the Kullback-Liebler (KL) divergence between $p_{S,Y}$ and $p_S p_Y$ and can be interpreted as a measure of closeness to independence of the joint density of S and Y.

The obvious symmetry of the mutual information expression in the random variables S and Y implies that

$$I(S;Y) = \mathcal{H}(S) - \mathcal{H}(S|Y) = \mathcal{H}(Y) - \mathcal{H}(Y|S).$$

This relation is often used in the implementation of mutual information driven strategies of sensor management since the quantities on the right hand side of the equality are usually more easily computed than those on the left hand side.

1.5 Further Reading

Information theory is a mature subject and there are many good sources for the beginner. One of the most popular textbooks used in introductory graduate courses on information theory is the textbook by Cover and Thomas [64] that is accessible to electrical engineers. The book by MacKay [164] covers the topic from the unique perspective of machine learning and contains many interesting applications. The classic book by Kullback [152] is a treatment of information theory that is firmly motivated by mathematical statistics. More mathematically advanced treatments of the subject are the books by Csiszár and Korner [67] and Yeung [257].

2. Markov Processes

Our ability to make effective sensor-management decisions is based fundamentally on our access to models. In particular, nonmyopic decision making relies on modeling the random processes that represent uncertainty in the system, such as target motion in tracking problems. In this section, we review a framework for uncertainty modeling based on Markov processes. The material discussed here provides the necessary background for understanding the methods discussed throughout this book, including Kalman filtering and partially observable Markov decision processes (POMDPs). We also provide some pointers to sources for further study.

2.1 Definition of Markov Process

A Markov process is a stochastic process satisfying a particular property called the Markov property, which we will define precisely below. Here, we consider only discrete-time Markov processes, and use $k = 0, 1, \ldots$ as the time index. (So a stochastic process here is no different from a sequence of random variables.)

We use $\mathcal{X}$ to denote the *state space* of the process, which is the set of values that the process can take at each time step. We also assume that associated with $\mathcal{X}$ is a collection of subsets $\mathcal{F}$ forming a σ-algebra. In the case where $\mathcal{X}$ is countable (discrete), we take $\mathcal{F}$ to be the power set of $\mathcal{X}$. If $\mathcal{X}$ is a Euclidean space, we take $\mathcal{F}$ to be the Borel σ-algebra. Throughout this appendix, we will refer to these two special cases simply by the terms *discrete* and *continuous*.

A stochastic process $X_0, X_1, \ldots$ is a *Markov process* (also called a *Markov chain*) if for each $k = 1, 2, \ldots$ and $E \in \mathcal{F}$,

$$P(X_{k+1} \in E | X_k, \ldots, X_0) = P(X_{k+1} \in E | X_k).$$

We call X_k the *state* of the Markov process at time k.

The condition above is called the *Markov property*, which boils down to this: the conditional distribution of X_{k+1} given the entire history up to time k depends only on X_k. In other words, the future of the process is conditionally independent of the past, given the present. To put it a different way, the "memory" in the process lasts only one time step.

The Markov property is in fact not as stringent a requirement as it may first appear to be. Indeed, suppose we are given a stochastic process that fails to satisfy the Markov property, but instead satisfies, for each $k = 1, 2, \ldots$ and

$E \in \mathcal{F}$,

$$P(X_{k+1} \in E | X_k, \ldots, X_0) = P(X_{k+1} \in E | X_k, X_{k-1});$$

in other words, the memory in the process is two instead of one. Then, it is easy to see that this process gives rise to a Markov process $\{Y_k\}$ by defining $Y_k = (X_k, X_{k-1})$. Indeed, using this construction, any stochastic process with memory lasting only a bounded time into the past gives rise to a Markov process. It turns out that many scenarios in practice can be modeled as Markov processes provided we define the state spaces appropriately.

2.2 State-transition Probability

In the discrete case, the Markov property can be expressed more simply as follows: for each $k = 1, 2, \ldots$ and $i, j, i_0, \ldots, i_{k-1} \in \mathcal{X}$,

$$P(X_{k+1} = j | X_k = i, \ldots, X_0 = i_0) = P(X_{k+1} = j | X_k = i).$$

It is clear that for a Markov process, once we specify $P(X_0 = i)$ and $P(X_{k+1} = j | X_k = i)$ for each $k = 0, 1, \ldots$ and $i, j \in \mathcal{X}$, the probability law of the process is completely specified. In many problems of interest, the conditional probabilities $P(X_{k+1} = j | X_k = i)$ do not depend on k. In this case, we say that the Markov process is *time-homogeneous* (or simply *homogeneous*).

For a homogeneous Markov process with discrete state space, write $p_{ij} = P(X_{k+1} = j | X_k = i)$, $i, j \in \mathcal{X}$. We call this set of probabilities the *state-transition law* (or simply the *transition law*) of the Markov process. Each p_{ij} is called a *state-transition probability* (or simply a *transition probability*). It is often convenient to represent the transition law using a graph, where the nodes are the states and the arcs are labeled with transition probabilities. In the case where $\mathcal{X}$ is finite (whence we can write, without loss of generality, $\mathcal{X} = \{1, 2, \ldots, N\}$), the transition law can also be represented by a square matrix $[p_{ij}]$, called the *state-transition matrix* (or *transition matrix*). Note that any transition matrix has the property that each entry is nonnegative and each row sums to one. Any matrix satisfying this property is called a *stochastic matrix*, and is in fact the transition matrix of some Markov chain.

In the continuous case, we assume that the state-transition law can be written in terms of conditional densities $p_{X_{k+1}|X_k}(x_{k+1}|x_k)$, $x_k, x_{k+1} \in \mathcal{X}$. (The reader may assume for simplicity that $\mathcal{X}$ is the real line or a subset of it; the case of multidimensional Euclidean spaces involves treating all densities as multivariable functions.) We also assume that X_0 has a density p_{X_0}. If $p_{X_{k+1}|X_k}$ does not depend on k, then we say that the Markov process is time-homogeneous. In the remainder of this discussion, we consider only time-

Appendices

homogeneous Markov processes. Also, for simplicity, we will drop the subscripts in the notation for the conditional densities.

2.3 Chapman-Kolmogorov Equation

Consider a discrete-state Markov process. Given $n = 0, 1, \ldots$, we define the n-step transition law by $p_{ij}^{(n)} = P(X_n = j | X_0 = i)$, $i, j \in \mathcal{X}$. The n-step transition law satisfies the *Chapman-Kolmogorov equation*:

$$p_{ij}^{(n+m)} = \sum_{k \in \mathcal{X}} p_{ik}^{(n)} p_{kj}^{(m)}, \qquad i, j \in \mathcal{X}.$$

In the case of a finite state space, the Chapman-Kolmogorov equation has a natural interpretation in terms of the transition matrix: the n-step transition law is given by the nth power of the transition matrix.

In the continuous case, we can similarly define the n-step transition law in terms of the conditional density $f^{(n)}(x_n|x_0)$, $x_n, x_0 \in \mathcal{X}$. The Chapman-Kolmogorov equation then takes the form

$$f^{(n+m)}(x_{n+m}|x_0) = \int_{\mathcal{X}} f^{(n)}(x_n|x_0) f^{(m)}(x_{n+m}|x_n)\, dx_n, \quad x_{n+m}, x_0 \in \mathcal{X}.$$

2.4 Markov reward processes

Given a Markov process, suppose we associate with each state $x \in \mathcal{X}$ a real number $R(x)$, called a *reward*. A Markov process so endowed with a reward function is called a *Markov reward process*. We define the *mean total reward* over a horizon H as

$$\mathbb{E}\left[\sum_{k=0}^{H-1} R(X_k)\right].$$

Many problems in practice can be modeled as Markov reward processes—in such a model, the mean total reward represents some quantity of interest (such as the value of a performance metric).

It is often the case that the horizon H is very large. In such cases, for technical reasons relevant to the analysis of Markov processes, the objective function is often expressed as a limit (i.e., with an infinite horizon). A sensible limiting objective function is the *infinite horizon* (or *long-term*) *average* reward:

$$\lim_{H \to \infty} \mathbb{E}\left[\frac{1}{H} \sum_{k=0}^{H-1} R(X_k)\right].$$

Another common limiting objective function is the *infinite horizon discounted reward*:

$$\lim_{H \to \infty} \mathbb{E}\left[\sum_{k=0}^{H-1} \beta^k R(X_k)\right],$$

where β is a number between 0 and 1 called the *discount factor*. .

2.5 Partially Observable Markov Processes

Given a Markov process and a set $\mathcal{Y}$ (with a σ-algebra $\mathcal{F}$ of subsets of $\mathcal{Y}$), suppose we associate with each state $x \in \mathcal{X}$ a conditional distribution $P(Y \in E|x)$, $E \in \mathcal{F}$. We call $\mathcal{Y}$ the *observation space*, and the conditional distribution $P(Y \in E|x)$, $E \in \mathcal{F}$, the *observation law*. A Markov process so endowed with an observation law is called a *partially observable Markov processes* or a *hidden Markov model*.

The reason we use the terms "partially observable" and "hidden" is that we think of $\mathcal{Y}$ as the set of observations we have of the "underlying" Markov process, but we cannot directly observe the underlying process. If $X_0, X_1, \ldots$ represents the underlying Markov process, then all we can observe is the random sequence $Y_0, Y_1, \ldots$, where each Y_k has conditional distribution given X_k specified by the observation law. The sequence $Y_0, Y_1, \ldots$ is assumed to be conditionally independent given $X_0, X_1, \ldots$. Many practical processes, especially in sensing applications, are well modeled by hidden Markov models. For example, X_k may represent the location of a target at time k, and Y_k may be a radar measurement of that location. The transition law in this case represents the motion of the target, and the observation law represents the relationship between the target location and the radar measurement.

Even though we cannot directly access X_k, the observations provide us with some information on X_k. In fact, at each k, we can compute the *a posteriori* (or posterior) distribution of X_k given the history of observations $\mathcal{I}_k = \{Y_0, \ldots, Y_{k-1}\}$. We call this posterior distribution the *belief state* or *information state* at time k, and here it is denoted π_k. The sequence of belief states satisfies the Markov property, and is therefore a legitimate Markov process, albeit with a rather unwieldy state space—the set of all distributions on $\mathcal{X}$. It turns out that given the belief state at time k and the observation Y_k, we can calculate the belief state at time $k+1$ using a simple update procedure, as we will show below.

For the case of a discrete Markov process with discrete observation space, suppose the transition law is given by p_{ij}, $i, j \in \mathcal{X}$. Suppose $y_0, y_1, \ldots$ are the observations. Let π_k represent the belief state at time k, which is a conditional

Appendices

probability mass function:

$$\pi_k(i) = P(X_k = i | Y_0 = y_0, \ldots, Y_{k-1} = y_{k-1}), \quad i \in \mathcal{X}.$$

Also define the "updated belief state" taking into account the observation y_k:

$$\hat{\pi}_k(i) = P(X_k = i | Y_0 = y_0, \ldots, Y_k = y_k), \quad i \in \mathcal{X}.$$

Then π_{k+1} can be derived from π_k and Y_k using the following two-step procedure:

1. Calculate the "updated belief state" $\hat{\pi}_k$ (taking into account the observation y_k) using Bayes' rule:

$$\hat{\pi}_k(j) = \frac{P(Y_k = y_k | X_k = j)\pi_k(j)}{\sum_{\ell \in \mathcal{X}} P(Y_k = y_k | X_k = \ell)\pi_k(\ell)}, \quad j \in \mathcal{X}.$$

2. Calculate the belief state π_{k+1} based on $\hat{\pi}_k$ and the transition law:

$$\pi_{k+1}(j) = \sum_{i \in \mathcal{X}} \hat{\pi}_k(i) p_{ij}, \quad j \in \mathcal{X}.$$

By reversing the order of 1. and 2. one obtains an equivalent algorithm for updating $\hat{\pi}_k$ to $\hat{\pi}_{k+1}$.

For the case of a continuous Markov process with continuous observation space (real numbers), suppose the transition law is given by the conditional density $p(x_{k+1}|x_k)$, and the observation law is given by the conditional density $q(y_k|x_k)$. The belief state at time k is then represented by a density function π_k. The two-step update procedure to calculate π_{k+1} based on π_k and y_k is given analogously as follows:

1. Calculate the "updated belief state" taking into account the observation y_k, using Bayes' rule:

$$\hat{\pi}_k(x_k) = \frac{q(y_k|x_k)\pi_k(x_k)}{\int_{\mathcal{X}} q(y_k|x)\pi_k(x)\,dx}, \quad x_k \in \mathcal{X}.$$

2. Calculate the belief state π_{k+1} based on $\hat{\pi}_k$ and the transition law:

$$\pi_{k+1}(x_{k+1}) = \int_{\mathcal{X}} \hat{\pi}_k(x_k) p(x_{k+1}|x_k)\,dx_k, \quad x_{k+1} \in \mathcal{X}.$$

If the transition and observation laws both arise from linear equations, and the initial density p_{X_0} is Gaussian, then the belief states remain Gaussian over time. In this case, the above update procedure can be reduced to a procedure to update just the mean and variance (or covariance in the multidimensional case) of the belief state. This procedure is called the *Kalman filter*.

If we augment the definition of a partially observable Markov process with control actions, then we obtain a partially observable Markov *decision* process (POMDP), as defined in Chapter 2. The two-step update procedure for belief states remains valid provided we include the action into the observation and transition laws.

2.6 Further Reading

Our discussion here assumes a basic understanding of probability and stochastic processes. An excellent recent book that provides this background and that also includes a chapter on Markov chains is by Gubner [97]. Many books focusing on Markov processes exist. A small but useful book by Ross [197] remains an accessible classic. The book by Çinlar [53] provides an excellent in-depth treatment of discrete state space Markov processes. Meyn and Tweedie [170] treat continuous (and other even more general) state space Markov processes—their treatment necessarily involves heavy mathematical machinery.

3. Stopping Times

We briefly present the concept of stopping time which plays an important role in the solution of the MAB problem discussed in Chapter 6. We proceed as follows: We first present all relevant definitions in Section 3.1. We give an example of a stopping time in Section 3.2. Finally, we characterize the stopping times that achieve the Gittins index in the classical MAB problem in Section 3.3 and suggest some further reading material for the advanced reader in Section 3.4.

3.1 Definitions

DEFINITION 12.1 (PROBABILITY SPACE) *A probability space* $(\Omega, \mathcal{F}, P)$ *consists of a sample space* Ω, *a* σ-*field* (σ-*algebra*) $\mathcal{F}$ *of the subsets of* Ω, *and a probability measure* P *on the elements of* $\mathcal{F}$.

DEFINITION 12.2 (σ-FIELD GENERATED BY A RANDOM VARIABLE) *Let* $X : (\Omega, \mathcal{F}, P) \to (\mathbb{R}, \mathcal{B}(\mathbb{R}), \hat{P})$ *be a random variable. Denote by* $\sigma(X)$ *the*

Appendices

smallest σ-field with respect to which X is measurable. Then

$$\sigma(X) = \{A \in \mathcal{F} : \exists B \in \mathcal{B}(\mathbb{R}), X^{-1}(B) = A\}.$$

The σ-field $\sigma(X)$ represents the "information" obtained about the experiment described by $(\Omega, \mathcal{F}, P)$ after observing X. This can be explained as follows. First consider a probability space $(\Omega, \mathcal{F}, P)$ which represents a random experiment. An event $E \in \mathcal{F}$ can be thought of as a "yes-no question" that can be answered after we observe the outcome of the experiment. Then $\sigma(X)$ is the collection of all "yes-no questions" that can be answered after observing X.

DEFINITION 12.3 (INCREASING FAMILY OF σ-FIELDS) *A family* $\{\mathcal{F}, \mathcal{F}_t; t = 0, 1, 2, \ldots\}$ *of σ-fields is called* increasing *if* $\mathcal{F}_t \subset \mathcal{F}_{t+1} \subset \mathcal{F}$ *for all* $t = 0, 1, 2, \ldots$.

$\mathcal{F}_t$ represents the information about the evolution of a system that is available to an observer/decision-maker at time t, $t = 0, 1, 2, \ldots$. When the observer has perfect recall, (that is, it remembers everything that it has seen and everything that it has done in the past) then $\mathcal{F}_t \subset \mathcal{F}_{t+1}, \forall t$ and $\{\mathcal{F}, \mathcal{F}_t; t = 0, 1, 2, \ldots\}$ is an *increasing family* of σ-fields.

DEFINITION 12.4 (STOPPING TIME) *Let* $\bar{N} := \{0, 1, 2, \ldots, +\infty\}$. *A random variable* $\tau : (\Omega, \mathcal{F}, P) \to (\bar{N}, 2^{\bar{N}}, \hat{P})$ *is a* stopping time *with respect to the increasing family of σ-fields* $\{\mathcal{F}, \mathcal{F}_t; t = 0, 1, 2, \ldots\}$ *if the event* $\{\tau = t\} := \{\omega : \tau(\omega) = t\} \in \mathcal{F}_t$ *for all* $t = 0, 1, 2, \ldots$.

Any constant random variable equal to a non-negative integer or $+\infty$ is a stopping time. A stopping time can be thought of as the time when a given random event E happens, with the convention that it takes the value $+\infty$ if E never happens. Alternatively, τ can be thought of as the time when a gambler playing a game decides to quit. Whether or not he quits at time t depends only on the information up to and including time t; so $\{\tau = t\} \in \mathcal{F}_t$.

3.2 Example

Let $\{X_t; t = 0, 1, 2, \ldots\}$ be a time-homogeneous finite-state Markov chain defined on $(\Omega, \mathcal{F}, P)$ with state space S, and matrix of transition probabilities $\{Q_{ij}; i, j \in S\}$. Assume that the evolution of the Markov chain is perfectly observed by an observer that has perfect recall. The observer has perfect recall, its information $\mathcal{F}_t$ at time t is given by $\sigma(X_0, X_1, \ldots, X_t)$, since $\sigma(X_0, X_1, \ldots, X_t)$ represents all the "yes-no questions" about events in $\mathcal{F}$

that the observer can answer after observing $X_0, X_1, \ldots, X_t$. Furthermore, $\{\mathcal{F}, \mathcal{F}_t; t = 0, 1, 2, \ldots\}$ is an increasing family of σ-fields. Consider a nonempty subset A of the state of space, that is, $A \subset S$, $A \neq \emptyset$. Define for all $\omega \in \Omega$,

$$\tau_A(\omega) := \min\{t : X_t(\omega) \in A\}.$$

The random variable τ_A defines the first instant of time the Markov chain enters set A, and is called the *hitting time* of A. It is a stopping time with respect to the family of σ-fields $\{\mathcal{F}, \mathcal{F}_t; t = 0, 1, 2, \ldots\}$.

3.3 Stopping Times for Multi-armed Bandit Problems

Consider the classical MAB problem of Chapter 6 with finite-state Markovian machines. The Gittins index of machine i, $i = 1, \ldots, k$ in this case is given by Eq (6.13). The stopping time that maximizes the RHS of (6.13) is the hitting time of an appropriate subset of the state space $\{1, 2, \ldots, \Delta_i\}$ of machine i. This set is the stopping set $S_i(x_i(N_i(\tau_l)))$ determined in Section 2.4.

In the case of non-Markovian machines the Gittins index of machine i, $i = 1, \ldots, k$ is given the (6.8). The stopping time that maximizes the RHS of (6.8) can be described as follows: Let $\mathbf{x}_i^{\tau_l} := (x_i(0), \ldots, x_i(N_i(\tau_l)))$. Define an appropriate family $\mathcal{S}(\tau_l)$ as $\{S_i^{N_i(\tau_l)+r}; r = 1, 2, 3, \ldots\}$ where,

$$S_i^{N_i(\tau_l+r)}(\mathbf{x}_i^{\tau_l}) \subset \mathbb{R}^{N_i(\tau_l)+r}, \quad r = 1, 2, \ldots$$

Let

$$\hat{\tau}_{l+1}(\mathcal{S}(\tau_l)) = \min\{t > \tau_l : \mathbf{x}_i^t \in S_i^{N_i(t)}; N_i(t) = N_i(\tau_l) + t - \tau_l + 1\}$$

Define $\mathcal{S}^*(\tau_l)$ by

$$\mathcal{S}^*(\tau_l) = \arg\max_{\mathcal{S}(\tau_l)} \frac{\mathbb{E}\left[\sum_{t=\tau_l(\omega)}^{\hat{\tau}_{l+1}(\mathcal{S}(\tau_l))-1} \beta^t R_i(X_i(N_i(\tau_l) + t - \tau_l(\omega)))|\mathbf{x}_i^{\tau_l}\right]}{\mathbb{E}\left[\sum_{t=\tau_l(\omega)}^{\hat{\tau}_{l+1}(\mathcal{S}(\tau_l))-1} \beta^t |\mathbf{x}_i^{\tau_l}\right]}.$$

(12.2)

Then $\tau_{l+1} = \hat{\tau}_{l+1}(\mathcal{S}^*(\tau_l))$. In general, for non-Markovian machines, maximizing RHS of (12.2) over all choices of $\mathcal{S}(\tau_l)$ is difficult, and computing the index is non-trivial.

3.4 Further Reading

Even though the notion of stopping times is intuitively simple, its formal treatment tends to be at an advanced level. Most graduate-level textbooks on probability theory contain a treatment of stopping times. See, for example, Billingsley [32], Shireyaev [210], and Jacod and Protter [117]. Stopping times is a fundamental concept in martingale theory and in optimal stopping problems. Reference books on these topics contain a more exhaustive treatment of stopping times. We refer the reader to Dellacherie and Meyer [69] for a treatment of stopping times in the context of martingales, and to Chow, Robins, and Siegmund [59], and Shireyaev [209] for a treatment of stopping times in the context of optimal stopping problems.

References

[1] R. Agrawal, M. V. Hegde, and D. Teneketzis. Asymptotically efficient adaptive allocation rules for the multiarmed bandit problem with switching cost. *IEEE Transactions on Automatic Control*, 33:899–906, 1988.

[2] R. Agrawal, M. V. Hegde, and D. Teneketzis. Multi-armed bandits with multiple plays and switching cost. *Stochastics and Stochastic Reports*, 29:437–459, 1990.

[3] R. Agrawal and D. Teneketzis. Certainty equivalence control with forcing: revisited. *Systems and Control Letters*, 13:405–412, 1989.

[4] R. Agrawal, D. Teneketzis, and V. Anantharam. Asymptotically efficient adaptive allocation schemes for controlled Markov chains: finite parameter space. *IEEE Transactions on Automatic Control*, 34:1249–1259, 1989.

[5] R. Agrawal, D. Teneketzis, and V. Anantharam. Asymptotically efficient adaptive control schemes for controlled I.I.D. processes: finite parameter space. *IEEE Transactions on Automatic Control*, 34:258–267, 1989.

[6] S.-I. Amari. *Methods of Information Geometry*. American Mathematical Society - Oxford University Press, Providence, RI, 2000.

[7] V. Anantharam, P. Varaiya, and J. Walrand. Asymptotically efficient allocation rules for the multiarmed bandit problem with multiple plays — part I: I.I.D. rewards. *IEEE Transactions on Automatic Control*, 32:968–976, 1987.

[8] V. Anantharam, P. Varaiya, and J. Walrand. Asymptotically efficient allocation rules for the multiarmed bandit problem with multiple plays — part II: Markovian rewards. *IEEE Transactions on Automatic Control*, 32:977–982, 1987.

[9] P. S. Ansell, K. D. Glazebrook, J. Niño-Mora, and M. O'Keefe. Whittle's index policy for a multi-class queueing system with convex holding costs. *Mathematical Methods of Operations Research*, 57:21–39, 2003.

[10] M. S. Arulampalam, S. Maskell, N. Gordon, and T. Clapp. A tutorial on particle filters for online nonlinear/non-Gaussian Bayesian tracking. *IEEE Transactions on Signal Processing*, 50:174–188, 2002.

[11] M. Asawa and D. Teneketzis. Multi-armed bandits with switching penalties. *IEEE Transactions on Automatic Control*, 41:328–348, 1996.

[12] J. Banks and R. Sundaram. Switching costs and the Gittins index. *Econometrica*, 62:687–694, 1994.

[13] Y. Bar-Shalom. *Multitarget Multisensor Tracking: Advanced Applications*. Artech House, Boston, MA, 1990.

[14] Y. Bar-Shalom and W. D. Blair. *Multitarget-Multisensor Tracking: Applications and Advances, Volume III*. Artech House, Boston, MA, 2000.

[15] A. R. Barron. Complexity regularization with application to artificial neural networks. In *Nonparametric Functional Estimation and Related Topics*, pages 561–576. Kluwer Academic Publishers, 1991.

[16] A. G. Barto, W. Powell, and J. Si, editors. *Learning and Approximate Dynamic Programming*. IEEE Press, New York, NY, 2004.

[17] M. Beckmann. *Dynamic Programming of Economic Decisions*. Springer-Verlag, New York, NY, 1968.

[18] R. Bellman. On the theory of dynamic programming. *Proceedings of the National Academy of Sciences*, 38:716–719, 1952.

[19] R. Bellman. A problem in the sequential design of experiments. *Sankhia*, 16:221–229, 1956.

[20] R. Bellman. *Adaptive Control Processes: a Guided Tour*. Princeton University Press, Princeton, NJ, 1961.

[21] R. Bellman and S. Dreyfus. *Applied Dynamic Programming*. Princeton University Press, Princeton, NJ, 1962.

[22] D. A. Berry and B. Fristedt. *Bandit problems: sequential allocation of experiments*. Chapman and Hall, 1985.

[23] D. P. Bertsekas. *Dynamic Programming and Optimal Control*, volume 1. Athena Scientific, 1995.

[24] D. P. Bertsekas. *Dynamic Programming and Optimal Control*, volume 2. Athena Scientific, 1995.

[25] D. P. Bertsekas. *Dynamic Programming and Optimal Control, Vols. I-II*. Athena Scientific, Belmont, MA, 3rd edition, 2005.

[26] D. P. Bertsekas and D. A. Castañón. Rollout algorithms for stochastic scheduling. *Heuristics*, 5:89–108, 1999.

[27] D. P. Bertsekas and S. E. Shreve. *Stochastic Optimal Control: The Discrete Time Case*, volume 1. Academic Press, 1978.

[28] D. P. Bertsekas and J. N. Tsitsiklis. *Neuro-Dynamic Programming*. Athena Scientific, Belmont, MA, 1996.

[29] D. Bertsimas and J. Niño-Mora. Conservation laws, extended polymatroids and multiarmed bandit problems; a polyhedral approach to indexable systems. *Mathematics of Operations Research*, 21:257–306, 1996.

[30] D. Bertsimas and J. Niño-Mora. Restless bandits, linear programming relaxations, and a primal-dual index heuristic. *Operations Research*, 48:80–90, 2000.

[31] D. Bertsimas, I. C. Paschalidis, and J. N. Tsitsiklis. Branching bandits and Klimov's problem: achievable region and side constraints. *IEEE Transactions on Automatic Control*, 40:2063–2075, 1995.

[32] P. Billingsley. *Probability and Measure*. John Wiley and Sons, New York, NY, 1995.

[33] S. S. Blackman. *Multiple-Target Tracking with Radar Applications*. Artech House, Boston, MA, 1986.

[34] D. Blackwell. Discrete dynamic programming. *Annals of Mathematical Statistics*, 33:719–726, 1962.

[35] D. Blackwell. Discounted dynamic programming. *Annals of Mathematical Statistics*, 36:226–235, 1965.

[36] W. D. Blair and M. Brandt-Pearce. Unresolved Rayleigh target detection using monopulse measurements. *IEEE Transactions on Aerospace and Electronic Systems*, 34:543–552, 1998.

[37] G. Blanchard and D. Geman. Hierarchical testing designs for pattern recognition. *Annals of Statistics*, 33(3):1155–1202, 2005.

[38] D. Blatt and A. O. Hero. From weighted classification to policy search. In *Neural Information Processing Symposium*, volume 18, pages 139–146, 2005.

[39] D. Blatt and A. O. Hero. Optimal sensor scheduling via classification reduction of policy search (CROPS). In *International Conference on Automated Planning and Scheduling*, 2006.

[40] H. A. P. Blom and E. A. Bloem. Joint IMMPDA particle filter. In *International Conference on Information Fusion*, 2003.

[41] A. G. B. S. J. Bradtke and S. P. Singh. Learning to act using real-time dynamic programming. *Artificial Intelligence*, 72:81–138, 1995.

[42] L. Breiman, J. Friedman, R. Olshen, and C. J. Stone. *Classification and Regression Trees*. Wadsworth, Belmont, CA, 1983.

[43] M. V. Burnashev and K. S. Zigangirov. An interval estimation problem for controlled observations. *Problems in Information Transmission*, 10:223–231, 1974. Translated from *Problemy Peredachi Informatsii*, 10(3):51–61, July-September, 1974.

[44] L. Carin, H. Yu, Y. Dalichaouch, A. R. Perry, P. V. Czipott, and C. E. Baum. On the wideband EMI response of a rotationally symmetric permeable and conducting target. *IEEE Transactions on Geoscience and Remote Sensing*, 39:1206–1213, June 2001.

[45] A. R. Cassandra. *Exact and Approximate Algorithms for Partially Observable Markov Decision Processes*. PhD thesis, Department of Computer Science, Brown University, 1998.

[46] A. R. Cassandra, M. L. Littman, and L. P. Kaelbling. Incremental pruning: A simple, fast, exact method for partially observable Markov decision processes. In *Uncertainty in Artificial Intelligence*, 1997.

[47] D. A. Castañón. Approximate dynamic programming for sensor management. In *IEEE Conference on Decision and Control*, pages 1202–1207. IEEE, 1997.

[48] D. A. Castañón. A lower bound on adaptive sensor management performance for classification. In *IEEE Conference on Decision and Control*. IEEE, 2005.

[49] D. A. Castañón and J. M. Wohletz. Model predictive control for dynamic unreliable resource allocation. In *IEEE Conference on Decision and Control*, volume 4, pages 3754–3759. IEEE, 2002.

[50] R. Castro, R. Willett, and R. Nowak. Coarse-to-fine manifold learning. In *IEEE International Conference on Acoustics, Speech and Signal Processing*, May, Montreal, Canada, 2004.

[51] R. Castro, R. Willett, and R. Nowak. Faster rates in regression via active learning. In *Neural Information Processing Systems*, 2005.

[52] R. Castro, R. Willett, and R. Nowak. Faster rates in regression via active learning. Technical report, University of Wisconsin, Madison, October 2005. ECE-05-3 Technical Report.

[53] E. Çinlar. *Introduction to Stochastic Processes*. Prentice-Hall, Englewood Cliffs, NJ, 1975.

[54] H. S. Chang, R. L. Givan, and E. K. P. Chong. Parallel rollout for online solution of partially observable Markov decision processes. *Discrete Event Dynamic Systems*, 14:309–341, 2004.

[55] H. Chernoff. Sequential design of experiments. *Annals of Mathematical Statistics*, 30:755–770, 1959.

[56] H. Chernoff. *Sequential Analysis and Optimal Design*. SIAM, 1972.

[57] A. Chhetri, D. Morrell, and A. Papandreou-Suppappola. Efficient search strategies for non-myopic sensor scheduling in target tracking. In *Asilomar Conference on Signals, Systems, and Computers*, 2004.

[58] E. K. P. Chong, R. L. Givan, and H. S. Chang. A framework for simulation-based network control via hindsight optimization. In *IEEE Conference on Decision and Control*, pages 1433–1438, 2000.

[59] Y. S. Chow, H. Robins, and D. Siegmund. *Great Expectations: The theory of Optimal Stopping*. Houghton Mifflin Company, Boiston, MA, 1971.

[60] D. Cochran. Waveform-agile sensing: opportunities and challenges. In *IEEE International Conference on Acoustics, Speech, and Signal Processing*, pages 877–880, Philadelphia, PA, 2005.

[61] D. Cochran, D. Sinno, and A. Clausen. Source detection and localization using a multi-mode detector: a Bayesian approach. In *IEEE International Conference on Acoustics, Speech, and Signal Processing*, pages 1173–1176, Phoenix, AZ, 1999.

[62] D. A. Cohn, Z. Ghahramani, and M. I. Jordan. Active learning with statistical models. *Advances in Neural Information Processing Systems*, 7:705–712, 1995.

[63] D. A. Cohn, Z. Ghahramani, and M. I. Jordan. Active learning with statistical models. *Journal of Artificial Intelligence Research*, pages 129–145, 1996.

[64] T. M. Cover and J. A. Thomas. *Elements of Information Theory*. John Wiley and Sons, New York, NY, 1991.

[65] N. Cristianini and J. Shawe-Taylor. *Support Vector Machines and Other Kernel Based Learning Methods*. Cambridge University Press, Cambridge, UK, 2000.

[66] I. Csiszár. Information-type measures of divergence of probability distributions and indirect observations. *Studia Sci. Math. Hung.*, 2:299–318, 1967.

[67] I. Csiszár and J. Korner. *Information Theory: Coding Theorems for Discrete Memoryless Systems*. Academic Press, Orlando FL, 1981.

[68] M. H. DeGroot. *Optimal Statistical Decisions*. McGraw Hill, 1970.

[69] C. Dellacherie and P. A. Meyer. *Probabilities and Potential B: Theory of Martingales*. North-Holland, Amsterdam, 1982.

[70] E. V. Denardo. *Dynamic Programming Models and Applications*. Prentice-Hall, Englewood Cliffs, NJ, 1982.

[71] D. Donoho. Compressed sensing. *IEEE Trans. on Information Theory*, 52(4):1289–1306, April 2006.

[72] A. Doucet. On sequential Monte Carlo methods for Bayesian filtering. Uk. tech. rep., Dept. Eng. Univ. Cambridge, 1998.

[73] A. Doucet, N. de Freitas, and N. Gordon. *Sequential Monte Carlo Methods in Practice*. Springer Publishing, New York, NY, 2001.

[74] A. Doucet, B.-N. Vo, C. Andrieu, and M. Davy. Particle filtering for multi-target tracking and sensor management. In *International Conference on Information Fusion*, 2002.

[75] N. Ehsan and M. Liu. Optimal bandwidth allocation in a delay channel. submitted to *JSAC*.

[76] N. Ehsan and M. Liu. Optimal channel allocation for uplink transmission in satellite communications. submitted to IEEE Transactions on Vehicular Technology.

[77] N. Ehsan and M. Liu. Server allocation with delayed state observation: sufficient conditions for the optimality an index policy. submitted to PEIS.

[78] N. Ehsan and M. Liu. On the optimal index policy for bandwidth allocation with delayed state observation and differentiated services. In *IEEE*

Annual Conference on Computer Communications, volume 3, pages 1974–1983, Hong Kong, April 2004.

[79] N. Ehsan and M. Liu. Properties of optimal resource sharing in delay channels. In *IEEE Conference on Decision and Control*, volume 3, pages 3277–3282, Paradise Island, Bahamas, 2004.

[80] N. El Karoui and I. Karatzas. Dynamic allocation problems in continuous time. *Annals of Applied Probability*, 4(2):255–286, 1994.

[81] V. V. Federov. *Theory of optimal experiments*. Academic Press, Orlando, 1972.

[82] R. A. Fisher. *The design of experiments*. Oliver and Boyd, Edinburgh, 1935.

[83] T. E. Fortmann, Y. Bar-Shalom, M. Scheffé, and S. Gelfand. Detection thresholds for tracking in clutter — A connection between estimation and signal processing. *IEEE Transactions on Automatic Control*, 30(3):221–229, March 1985.

[84] Y. Freund, H. S. Seung, E. Shamir, and N. Tishby. Selective sampling using the query by committee algorithm. *Machine Learning*, 28(2-3):133–168, August 1997.

[85] E. Frostig and G. Weiss. Four proofs of Gittins' multi-armed bandit theorem. Technical report, The University of Haifa, Mount Carmel, 31905, Israel, November 1999.

[86] N. Geng, C. E. Baum, and L. Carin. On the low-frequency natural response of conducting and permeable targets. *IEEE Transactions on Geoscience and Remote Sensing*, 37:347–359, January 1999.

[87] J. C. Gittins. Bandit processes and dynamic allocation indices. *Journal of the Royal Statistical Society: Series B (Methodological)*, 41(2):148–177, 1979.

[88] J. C. Gittins. *Multi-Armed Bandit Allocation Indices*. John Wiley and Sons, New York, NY, 1989.

[89] J. C. Gittins and D. M. Jones. A dynamic allocation index for sequential design of experiments. *Progress in Statistics, Euro. Meet. Statis.*, 1:241–266, 1972.

[90] K. D. Glazebrook, J. Niño Mora, and P. S. Ansell. Index policies for a class of discounted restless bandits. *Advances in Applied Probability*, 34(4):754–774, 2002.

[91] K. D. Glazebrook and D. Ruiz-Hernandez. A restless bandit approach to stochastic scheduling problems with switching costs. Preprint, March 2005.

[92] G. Golubev and B. Levit. Sequential recovery of analytic periodic edges in the binary image models. *Mathematical Methods of Statistics*, 12:95–115, 2003.

[93] N. J. Gordon, D. J. Salmond, and A. F. M. Smith. A novel approach to non-linear and non-Gaussian Bayesian state estimation. *IEE Proceedings on Radar and Signal Processing*, 140:107–113, 1993.

[94] E. Gottlieb and R. Harrigan. The umbra simulation framework. Sand2001-1533 (unlimited release), Sandia National Laboratory, 2001.

[95] C. H. Gowda and R. Viswanatha. Performance of distributed CFAR test under various clutter amplitudes. *IEEE Transactions on Aerospace and Electronic Systems*, 35:1410–1419, 1999.

[96] R. M. Gray. Vector quantization. *IEEE ASSP Magazine*, pages 4–29, Apr. 1984.

[97] J. A. Gubner. *Probability and Random Processes for Electrical and Computer Engineers*. Cambridge University Press, New York, NY, 2006.

[98] P. Hall and I. Molchanov. Sequential methods for design-adaptive estimation of discontinuities in regression curves and surfaces. *Annals of Statistics*, 31(3):921–941, 2003.

[99] P. Hanselman, C. Lawrence, E. Fortunato, B. Tenney, and E. Blasch. Dynamic tactical targeting. In *Conference on Battlefield Digitization and Network-Centric Systems IV*, volume SPIE 5441, pages 36–47, 2004.

[100] J. P. Hardwick and Q. F. Stout. Flexible algorithms for creating and analyzing adaptive sampling procedures. In N. Flournoy, W. F. Rosenberger, and W. K. Wong, editors, *New Developments and Applications in Experimental Design*, volume 34 of *Lecture Notes - Monograph Series*, pages 91–105. Institute of Mathematical Statistics, 1998.

[101] T. Hastie, R. Tibshirani, and J. H. Friedman. *The Elements of Statistical Learning: Data Mining, Inference, and Prediction*. Springer Series in Statistics, Basel, CH, 2001.

[102] J. Havrda and F. Chárvat. Quantification method of classification processes. *Kiberbetika Cislo*, 1(3):30–34, 1967.

[103] Y. He and E. K. P. Chong. Sensor scheduling for target tracking in sensor networks. In *IEEE Conference on Decision and Control*, pages 743–748, 2004.

[104] Y. He and E. K. P. Chong. Sensor scheduling for target tracking: A Monte Carlo sampling approach. *Digital Signal Processing*, 16(5):533–545, September 2006.

[105] M. L. Hernandez, T. Kirubarajan, and Y. Bar-Shalom. Multisensor resource deployment using posterior Cramér-Rao bounds. *IEEE Transactions on Aerospace and Electronic Systems*, 40(2):399–416, April 2004.

[106] A. O. Hero, B. Ma, O. Michel, and J. Gorman. Applications of entropic spanning graphs. *IEEE Signal Processing Magazine*, 19(2):85–95, 2002.

[107] A. O. Hero, B. Ma, O. Michel, and J. D. Gorman. Alpha divergence for classification, indexing and retrieval. Technical Report Technical Report 328, Comm. and Sig. Proc. Lab. (CSPL), Dept. EECS, The University of Michigan, 2001.

[108] K. J. Hintz. A measure of the information gain attributable to cueing. *IEEE Transactions on Systems, Man and Cybernetics*, 21(2):237–244, 1991.

[109] K. J. Hintz and E. S. McVey. Multi-process constrained estimation. *IEEE Transactions on Systems, Man and Cybernetics*, 21(1):434–442, January/February 1991.

[110] M. Horstein. Sequential decoding using noiseless feedback. *IEEE Transactions on Information Theory*, 9(3):136–143, 1963.

[111] R. Howard. *Dynamic Programming and Markov Processes*. John Wiley and Sons, New York, NY, 1960.

[112] C. Hue, J.-P. Le Cadre, and P. Perez. Sequential Monte Carlo methods for multiple target tracking and data fusion. *IEEE Transactions on Signal Processing*, 50:309–325, 2002.

[113] C. Hue, J.-P. Le Cadre, and P. Perez. Tracking multiple objects with particle filtering. *IEEE Transactions on Aerospace and Electronic Systems*, 38:791–812, 2002.

[114] M. Isard and J. MacCormick. BraMBLe: A Bayesian multiple-blob tracker. In *International Conference on Computer Vision*, 2001.

[115] T. Ishikida. *Informational Aspects of Decentralized Resource Allocation.* PhD thesis, University of California, Berkeley, 1992.

[116] T. Ishikida and P. Varaiya. Multi-armed bandit problem revisited. *Journal of Optimization Theory and Applications*, 83:113–154, 1994.

[117] J. Jacod and P. Protter. *Probability Essentials.* Springer-Verlag, 2003.

[118] A. H. Jazwinski. *Stochastic Processes and Filtering Theory.* Academic Press, New York, NY, 1970.

[119] S. Ji, R. Parr, and L. Carin. Non-myopic multi-aspect sensing with partially observable Markov decision processes. *IEEE Transactions on Signal Processing*, 55(6):2720–2730, 2007.

[120] S. Julier and J. Uhlmann. Unscented filtering and non-linear estimation. *Proceedings of the IEEE*, 92:401–422, 2004.

[121] L. P. Kaelbling, M. L. Littman, and A. R. Cassandra. Planning and acting in partially observable stochastic domains. *Artificial Intelligence*, 101:99–134, 1998.

[122] R. Karlsson and F. Gustafsson. Monte Carlo data association for multiple target tracking. In *IEE Workshop on Target Tracking: Algorithms and Applications*, 2001.

[123] H. Kaspi and A. Mandelbaum. Multi-armed bandits in discrete and continuous time. *Annals of Applied Probability*, 8:1270–1290, 1998.

[124] K. Kastella. Discrimination gain for sensor management in multitarget detection and tracking. In *IEEE-SMC and IMACS Multiconference*, volume 1, pages 167–172, 1996.

[125] K. Kastella. Discrimination gain to optimize classification. *IEEE Transactions on Systems, Man and Cybernetics–Part A: Systems and Humans*, 27(1), January 1997.

[126] M. N. Katehakis and U. G. Rothblum. Finite state multi-armed bandit problems: Sensitive-discount, average-reward and average-overtaking optimality. *Annals of Applied Probability*, 6:1024–1034, 1996.

[127] M. N. Katehakis and A. F. Veinott, Jr. The multi-armed bandit problem: Decomposition and computation. *Mathematics of Operations Research*, 12:262–268, 1987.

[128] M. J. Kearns, Y. Mansour, and A. Y. Ng. A sparse sampling algorithm for near-optimal planning in large Markov decision processes. In *International Joint Conference on Artificial Intelligence*, pages 1324–1331, 1999.

[129] F. P. Kelly. Multi-armed bandits with discount factor near one: The Bernoulli case. *Annals of Statistics*, 9:987–1001, 1981.

[130] D. J. Kershaw and R. J. Evans. Optimal waveform selection for tracking systems. *IEEE Transactions on Information Theory*, 40(5):1536–50, September 1994.

[131] D. J. Kershaw and R. J. Evans. Waveform selective probabilistic data association. *IEEE Transactions on Aerospace and Electronic Systems*, 33(4):1180–88, October 1997.

[132] G. P. Klimov. Time sharing service systems I. *Theory of Probability and its Applications (in Russian: Teoriya Veroyatnostei i ee Primeneniya)*, 19:532–551, 1974.

[133] G. P. Klimov. Time sharing service systems II. *Theory of Probability and its Applications (in Russian: Teoriya Veroyatnostei i ee Primeneniya)*, 23:314–321, 1978.

[134] E. D. Kolaczyk and R. D. Nowak. Multiscale likelihood analysis and complexity penalized estimation. *Annals of Statistics*, 32(2):500–527, 2004.

[135] A. Korostelev and J.-C. Kim. Rates of convergence for the sup-norm risk in image models under sequential designs. *Statistics and Probability Letters*, 46:391–399, 2000.

[136] A. P. Korostelev. On minimax rates of convergence in image models under sequential design. *Statistics and Probability Letters*, 43:369–375, 1999.

[137] A. P. Korostelev and A. B. Tsybakov. *Minimax Theory of Image Reconstruction*. Springer Lecture Notes in Statistics, 1993.

[138] C. Kreucher, D. Blatt, A. Hero, and K. Kastella. Adaptive multi-modality sensor scheduling for detection and tracking of smart targets. *Digital Signal Processing*, 16(5):546–567, 2005.

[139] C. Kreucher, A. Hero, K. Kastella, and D. Chang. Efficient methods of non-myopic sensor management for multitarget tracking. In *IEEE Conference on Decision and Control*, 2004.

[140] C. Kreucher, A. O. Hero, and K. Kastella. Multiple model particle filtering for multi-target tracking. In *Workshop on Adaptive Sensor Array Processing*, 2004.

[141] C. Kreucher, K. Kastella, and A. Hero. Multi-target sensor management using alpha divergence measures. In *International Conference on Information Processing in Sensor Networks*, 2003.

[142] C. M. Kreucher, A. O. Hero, and K. Kastella. A comparison of task driven and information driven sensor management for target tracking. In *IEEE Conference on Decision and Control*, 2005.

[143] C. M. Kreucher, A. O. Hero, K. D. Kastella, and M. R. Morelande. An information-based approach to sensor management in large dynamic networks. *Proceedings of the IEEE*, 95(5):978–999, May 2007.

[144] C. M. Kreucher, K. Kastella, and A. O. Hero. Information based sensor management for multitarget tracking. In *SPIE Conference on Signal and Data Processing of Small Targets*, 2003.

[145] C. M. Kreucher, K. Kastella, and A. O. Hero. Multitarget tracking using the joint multitarget probability density. *IEEE Transactions on Aerospace and Electronic Systems*, 39(4):1396–1414, 2005.

[146] C. M. Kreucher, K. Kastella, and A. O. Hero. Sensor management using an active sensing approach. *Signal Processing*, 85(3):607–624, 2005.

[147] V. Krishnamurthy. Algorithms for optimal scheduling and management of hidden Markov model sensors. *IEEE Transactions on Signal Processing*, 50(6):1382–1397, 2002.

[148] V. Krishnamurthy and R. J. Evans. Hidden Markov model multiarmed bandits: A methodology for beam scheduling in multitarget tracking. *IEEE Transactions on Signal Processing*, 49(12):2893–2908, 2001.

[149] V. Krishnamurthy and R. J. Evans. Correction to hidden Markov model multi-arm bandits: A methodology for beam scheduling in multi-target tracking. *IEEE Transactions on Signal Processing*, 51(6):1662–1663, 2003.

[150] A. Krogh and J. Vedelsby. Neural network ensembles, cross validation, and active learning. *Advances in Neural Information Processing Systems*, 7:231–238, 1995.

[151] W. S. Kuklinski. Adaptive sensor tasking and control. In *MITRE 2005 Technology Symposium*. MITRE Corporation, 2005.

[152] S. Kullback. *Information Theory and Statistics*. Dover, 1978.

[153] P. R. Kumar and P. Varaiya. *Stochastic Systems: Estimation, Identification, and Adaptive Control*. Prentice Hall, 1986.

[154] H. Kushner. *Introduction to Stochastic Control*. Holt, Rinehart and Winston, New York, NY, 1971.

[155] B. F. La Scala, B. Moran, and R. Evans. Optimal scheduling for target detection with agile beam radars. In *NATO SET-059 Symposium on Target Tracking and Sensor Data Fusion for Military Observation Systems*, 2003.

[156] T. Lai and H. Robbins. Asymptotically efficient adaptive allocation rules. *Advances in Applied Mathematics*, 6:4–22, 1985.

[157] R. E. Larson and J. L. Casti. *Principles of Dynamic Programming, Parts 1-2*. Marcel Dekker, New York, NY, 1982.

[158] X. Liao, H. Li, and B. Krishnapuram. An m-ary KMP classifier for multi-aspect target classification. In *IEEE International Conference on Acoustics, Speech, and Signal Processing*, volume 2, pages 61–64, 2004.

[159] M. L. Littman. The witness algorithm: Solving partially observable Markov decision processes. Technical Report CS-94-40, Brown University, 1994.

[160] J. Liu and R. Chen. Sequential Monte Carlo methods for dynamic systems. *Journal of the American Statistical Association*, 1998.

[161] C. Lott and D. Teneketzis. On the optimality of an index rule in multi-channel allocation for single-hop mobile networks with multiple service classes. *Probability in the Engineering and Informational Sciences*, 14:259–297, 2000.

[162] W. S. Lovejoy. A survey of algorithmic methods for partially observed Markov decision processes. *Annals of Operations Research*, 28(1):47–65, 1991.

[163] D. MacKay. Information-based objective functions for active data selection. *Neural Computation*, 4:590–604, 1992.

[164] D. MacKay. *Information Theory, Inference and Learning Algorithms*. Cambridge University Press, 2004.

[165] R. Mahler. Global optimal sensor allocation. In *National Symposium on Sensor Fusion*, volume 1, pages 167–172, 1996.

[166] A. Mandelbaum. Discrete multiarmed bandits and multiparameter processes. *Probability Theory and Related Fields*, 71:129–147, 1986.

[167] A. Mandelbaum. Continuous multi-armed bandits and multiparameter processes. *Annals of Probability*, 15:1527–1556, 1987.

[168] S. Maskell, M. Rollason, N. Gordon, and D. Salmond. Efficient particle filtering for multiple target tracking with application to tracking in structured images. In *SPIE Conference on Signal and Data Processing of Small Targets*, 2002.

[169] M. McClure and L. Carin. Matched pursuits with a wave-based dictionary. *IEEE Transactions on Signal Processing*, 45:2912–2927, December 1997.

[170] S. P. Meyn and R. L. Tweedie. *Markov Chains and Stochastic Stability*. Springer-Verlag, London, 1993.

[171] J. Mickova. Stochastic scheduling with multi-armed bandits. Master's thesis, University of Melbourne, Australia, 2000.

[172] M. I. Miller, A. Srivastava, and U. Grenander. Conditional mean estimation via jump-diffusion processes in multiple target tracking/recognition. *IEEE Transactions on Signal Processing*, 43:2678–2690, 1995.

[173] G. E. Monahan. A survey of partially observable Markov decision processes: Theory, models and algorithms. *Management Science*, 28(1):1–16, 1982.

[174] M. Morelande, C. M. Kreucher, and K. Kastella. A Bayesian approach to multiple target detection and tracking. *IEEE Transactions on Signal Processing*, 55(5):1589–1604, 2007.

[175] S. Musick and R. Malhotra. Chasing the elusive sensor manager. In *IEEE National Aerospace and Electronics Conference*, volume 1, pages 606–613, 1994.

[176] D. Mušicki, S. Challa, and S. Suvorova. Multi target tracking of ground targets in clutter with LMIPDA-IMM. In *International Conference on Information Fusion*, Stockholm, Sweden, July 2004.

[177] D. Mušicki and R. Evans. Clutter map information for data association and track initialization. *IEEE Transactions on Aerospace and Electronic Systems*, 40(4):387–398, April 2001.

[178] D. Mušicki, R. Evans, and S. Stankovic. Integrated probabilistic data association. *IEEE Transactions on Automatic Control*, 39(6):1237–1240, June 1994.

[179] P. Nash. *Optimal Allocation of Resources Between Research Projects*. PhD thesis, Cambridge University, 1973.

[180] F. Nathanson. *Radar Design Principles*. McGraw Hill, New York, 1969.

[181] A. Nedic and M. K. Schneider. Index rule-based management of a sensor for searching, tracking, and identifying. In *Tri-Service Radar Symposium*, Boulder Colorado, June 2003.

[182] A. Nehorai and A. Dogandžić. Cramér-Rao bounds for estimating range, velocity and direction with an active array. *IEEE Transactions on Signal Processing*, 49(6):1122–1137, June 2001.

[183] J. Niño-Mora. Restless bandits, partial conservation laws, and indexability. *Advances in Applied Probability*, 33:76–98, 2001.

[184] J. Niño-Mora. Dynamic allocation indices for restless projects and queuing admission control: a polyhedral approach. *Mathematical Programming, Series A*, 93:361–413, 2002.

[185] R. Niu, P. Willett, and Y. Bar-Shalom. From the waveform through the resolution cell to the tracker. In *IEEE Aerospace Conference*, March 1999.

[186] R. Nowak, U. Mitra, and R. Willett. Estimating inhomogeneous fields using wireless sensor networks. *IEEE Journal on Selected Areas in Communications*, 22(6):999–1006, 2004.

[187] M. Orton and W. Fitzgerald. A Bayesian approach to tracking multiple targets using sensor arrays and particle filters. *IEEE Transactions on Signal Processing*, 50:216–223, 2002.

[188] D. G. Pandelis and D. Teneketzis. On the optimality of the Gittins index rule in multi-armed bandits with multiple plays. *Mathematical Methods of Operations Research*, 50:449–461, 1999.

[189] J. Pineau, G. Gordon, and S. Thrun. Point-based value iteration: An anytime algorithm for POMDPs. In *International Joint Conference on Artificial Intelligence*, August 2003.

[190] M. K. Pitt and N. Shephard. Filtering via simulation: Auxiliary particle filters. *Journal of the American Statistical Association*, 94:590–599, 1999.

[191] F. Pukelsheim. *Optimal Design of Experiments*. John Wiley and Sons, New York, NY, 1993.

[192] M. L. Puterman, editor. *Dynamic Programming and its Applications*. Academic Press, New York, NY, 1978.

[193] M. L. Puterman. *Markov Decision Problems: Discrete Stochastic Dynamic Programming*. John Wiley and Sons, New York, NY, 1994.

[194] R. Raich, J. Costa, and A. O. Hero. On dimensionality reduction for classification and its application. In *IEEE International Conference on Acoustics, Speech, and Signal Processing*, Toulouse, May 2006.

[195] A. Rényi. On measures of entropy and information. In *Berkeley Symposium on Mathematics, Statistics and Probability*, volume 1, pages 547–561, 1961.

[196] R. Rifkin and A. Klautau. In defense of one-vs-all classification. *Journal of Machine Learning Research*, 5:101–141, January 2004.

[197] S. M. Ross. *Applied Probability Models with Optimization Applications*. Dover Publications, New York, NY, 1970.

[198] S. M. Ross. *Introduction to Stochastic Dynamic Programming*. Academic Press, New York, NY, 1983.

[199] N. Roy, G. Gordon, and S. Thrun. Finding approximate POMDP solutions through belief compression. *Journal of Artificial Intelligence Research*, 23:1–40, 2005.

[200] P. Runkle, P. Bharadwaj, and L. Carin. Hidden Markov model multi-aspect target classification. *IEEE Transactions on Signal Processing*, 47:2035–2040, July 1999.

[201] P. Runkle, L. Carin, L. Couchman, T. Yoder, and J. Bucaro. Multi-aspect identification of submerged elastic targets via wave-based matching pursuits and hidden Markov models. *J. Acoustical Soc. Am.*, 106:605–616, August 1999.

[202] J. Rust. Chapter 14: Numerical dynamic programming in economics. In H. Amman, D. Kendrick, and J. Rust, editors, *Handbook of Computational Economics*. Elsevier, North Holland, 1996.

[203] J. Rust. Using randomization to break the curse of dimensionality. *Econometrica*, 65:487–516, 1997.

[204] W. Schmaedeke and K. Kastella. Event-averaged maximum likelihood estimation and information-based sensor management. *Proceedings of SPIE*, 2232:91–96, June 1994.

[205] M. K. Schneider, G. L. Mealy, and F. M. Pait. Closing the loop in sensor fusion systems: Stochastic dynamic programming approaches. In *American Control Conference*, 2004.

[206] D. Schulz, D. Fox, and J. Hightower. People tracking with anonymous and ID-sensors using Rao-Blackwellised particle filter. In *International Joint Conference on Artificial Intelligence*, 2003.

[207] N. Secomandi. A rollout policy for the vehicle routing problem with stochastic demands. *Operations Research*, 49:796–802, 2001.

[208] C. E. Shannon. A mathematical theory of communication. *Bell System Technical Journal*, 27:379–423, 1948.

[209] A. N. Shireyaev. *Optimal Stopping Rules*. Springer-Verlag, 1978.

[210] A. N. Shireyaev. *Probability*. Springer-Verlag, 1995.

[211] A. Singh, R. Nowak, and P. Ramanathan. Active learning for adaptive mobile sensing networks. In *International Conference on Information Processing in Sensor Networks*, Nashville, TN, April 2006.

[212] D. Sinno. *Attentive Management of Configurable Sensor Systems*. PhD thesis, Arizona State University, 2000.

[213] D. Sinno and D. Cochran. Dynamic estimation with selectable linear measurements. In *IEEE International Conference on Acoustics, Speech, and Signal Processing*, pages 2193–2196, Seattle, WA, 1998.

[214] D. Sinno, D. Cochran, and D. Morrell. Multi-mode detection with Markov target motion. In *International Conference on Information Fusion*, volume WeD1, pages 26–31, Paris, France, 2000.

[215] S. P. Sira, D. Cochran, A. Papandreou-Suppappola, D. Morrell, W. Moran, S. Howard, and R. Calderbank. Adaptive waveform design for improved detection of low RCS targets in heavy sea clutter. *IEEE Journal on Selected Areas in Signal Processing*, 1(1):55–66, June 2007.

[216] S. P. Sira, A. Papandreou-Suppappola, and D. Morrell. Time-varying waveform selection and configuration for agile sensors in tracking applications. In *IEEE International Conference on Acoustics, Speech, and Signal Processing*, volume 5, pages 881–884, March 2005.

[217] M. I. Skolnik. *Introduction to Radar Systems*. McGraw-Hill, 3rd edition, 2001.

[218] R. D. Smallwood and E. J. Sondik. The optimal control of partially observable Markov processes over a finite horizon. *Operations Research*, 21:1071–1088, 1973.

[219] E. J. Sondik. *The Optimal Control of Partially Observable Markov Processes*. PhD thesis, Stanford University, 1971.

[220] E. J. Sondik. The optimal control of partially observable Markov processes over the infinite horizon: Discounted costs. *Operations Research*, 26(2):282–304, 1978.

[221] N. O. Song and D. Teneketzis. Discrete search with multiple sensors. *Mathematical Methods of Operations Research*, 60:1–14, 2004.

[222] Statlog. Landsat MSS data.

[223] L. D. Stone, C. A. Barlow, and T. L. Corwin. *Bayesian Multiple Target Tracking*. Artech House, Boston, MA, 1999.

[224] M. Stone. Cross-validatory choice and assessment of statistical predictions. *Journal of the Royal Statistical Society, Series B*, 36:111–147, 1974.

[225] C. Striebel. Sufficient statistics in the optimum control of stochastic systems. *Journal of Mathematical Analysis and Applications*, 12:576–592, 1965.

[226] K. Sung and P. Niyogi. Active learning for function approximation. *Proc. Advances in Neural Information Processing Systems*, 7, 1995.

[227] R. Sutton and A. G. Barto. *Reinforcement Learning: An Introduction*. MIT Press, Cambridge, MA, 1998.

[228] S. Suvorova, S. D. Howard, W. Moran, and R. J. Evans. Waveform libraries for radar tracking applications: Maneuvering targets. In *Defence Applications of Signal Processing*, 2004.

[229] I. J. Taneja. New developments in generalized information measures. *Advances in Imaging and Electron Physics*, 91:37–135, 1995.

[230] G. Tesauro. Temporal difference learning and TD-gammon. *Communications of the ACM*, 38(3), March 1995.

[231] S. Tong and D. Koller. Support vector machine active learning with applications to text classification. *International Conference on Machine Learning*, pages 999–1006, 2000.

[232] J. N. Tsitsiklis. A lemma on the multiarmed bandit problem. *IEEE Transactions on Automatic Control*, 31:576–577, 1986.

[233] B. E. Tullsson. Monopulse tracking of Rayleigh targets: A simple approach. *IEEE Transactions on Aerospace and Electronic Systems*, 27:520–531, 1991.

[234] M. Van Oyen, D. Pandelis, and D. Teneketzis. Optimality of index policies for stochastic scheduling with switching penalties. *Journal of Applied Probability*, 29:957–966, 1992.

[235] M. P. Van Oyen and D. Teneketzis. Optimal stochastic scheduling of forest networks with switching penalties. *Advances in Applied Probability*, 26:474–479, 1994.

[236] H. L. van Trees. *Detection, Estimation, and Modulation Theory: Part I*. John Wiley and Sons, New York, NY, 1968.

[237] H. L. van Trees. *Detection, Estimation and Modulation Theory, Part III*. John Wiley and Sons, New York, NY, 1971.

[238] V. N. Vapnik. *Statistical Learning Theory*. John Wiley and Sons, New York, NY, 1998.

[239] V. N. Vapnik. An overview of statistical learning theory. *IEEE Transactions on Neural Networks*, 10(5):988–999, 1999.

[240] P. P. Varaiya, J. C. Walrand, and C. Buyukkov. Extensions of the multiarmed bandit problem: The discounted case. *IEEE Transactions on Automatic Control*, 30:426–439, 1985.

[241] M. Veth, J. Busque, D. Heesch, T. Burgess, F. Douglas, and B. Kish. Affordable moving surface target engagement. In *IEEE Aerospace Conference*, volume 5, pages 2545–2551, 2002.

[242] P. Vincent and Y. Bengio. Kernel matching pursuit. *Machine Learning*, 48:165–187, 2002.

[243] A. Wald. *Sequential Analysis*. John Wiley and Sons, New York, NY, 1947.

[244] J. Wang, A. Dogandžić, and A. Nehorai. Maximum likelihood estimation of compound-Gaussian clutter and target parameters. *IEEE Transactions on Signal Processing*, 54:3884–3898, October 2006.

[245] R. B. Washburn, M. K. Schneider, and J. J. Fox. Stochastic dynamic programming based approaches to sensor resource management. In *International Conference on Information Fusion*, volume 1, pages 608–615, 2002.

[246] R. R. Weber. On Gittins index for multiarmed bandits. *Annals of Probability*, 2:1024–1033, 1992.

[247] R. R. Weber and G. Weiss. On an index policy for restless bandits. *Journal of Applied Probability*, 27:637–648, 1990.

[248] C. C. White III. Partially observed Markov decision processes: A survey. *Annals of Operations Research*, 32, 1991.

[249] P. Whittle. Multi-armed bandits and Gittins index. *Journal of the Royal Statistical Society: Series B (Methodological)*, 42:143–149, 1980.

[250] P. Whittle. Arm-acquiring bandits. *Annals of Probability*, 9:284–292, 1981.

[251] P. Whittle. *Optimization Over Time: Dynamic Programming and Stochastic Control*. John Wiley and Sons, New York, NY, 1983.

[252] P. Whittle. Restless bandits: Activity allocation in a changing world. *Journal of Applied Probability*, 25A:287–298, 1988.

[253] P. Whittle. Tax problems in the undiscounted case. *Journal of Applied Probability*, 42(3):754–765, 2005.

[254] R. Willett, A. Martin, and R. Nowak. Backcasting: Adaptive sampling for sensor networks. In *Information Processing in Sensor Networks*, 26-27 April, Berkeley, CA, USA, 2004.

[255] I. J. Won, D. A. Keiswetter, and D. R. Hanson. GEM-3: A monostatic broadband electromagnetic induction sensor. *J. Environ. Eng. Geophys.*, 2:53–64, March 1997.

[256] G. Wu, E. K. P. Chong, and R. L. Givan. Burst-level congestion control using hindsight optimization. *IEEE Transactions on Automatic Control*, 47:979–991, 2002.

[257] R. W. Yeung. *A First Course in Information Theory*. Springer, 2002.

[258] H. Yu and D. P. Bertsekas. Discretized approximations for pomdp with average cost. In *Conference on Uncertainty in Artificial Intelligence*, pages 619–627, 2004.

[259] Y. Zhang, L. M. Collins, H. Yu, C. E. Baum, and L. Carin. Sensing of unexploded ordnance with magnetometer and induction data: Theory and signal processing. *IEEE Transactions on Geoscience and Remote Sensing*, 41:1005–1015, May 2003.

[260] Y. Zhang, X. Liao, and L. Carin. Detection of buried targets via active selection of labeled data: application to sensing subsurface uxo. *IEEE Transactions on Geoscience and Remote Sensing*, 42(11):2535–2543, 2004.

[261] F. Zhao, J. Shin, and J. Reich. Information-driven dynamic sensor collaboration. *IEEE Signal Processing Magazine*, pages 61–72, March 2002.

Index

Acoustic underwater sensing, 27
Action, 98
Action selector, 102, 103
Action-sequence approximations, 109
Active learning rate, 193
Adaptive partition, 77
Adaptive sampling, 177, 181
ADP, 26
Airborne laser scanning sensor, 178
Airborne sensor, 97
alpha divergence, 37
alpha entropy, 36
Ambiguity function, 243
Ambiguity function, 224, 242
Approximate Dynamic Programming, 26
Approximate dynamic programming, 108
Array, 97
Average reward MDP, 13
Azimuthal ambiguity, 228

Base policy, 111
Battery status, 102
Bayes rule, 23, 49, 103, 183, 247
Bayes update, 23
Bayes' rule, 277
Bayesian CRB, 40
Bayesian filtering, 64
Beam scheduling, 245
Beamshaping, 226
Belief state, 41, 98, 276
Belief-state approximation, 114
Belief-state feedback, 98
Belief-state simplification, 114
Belief-state space, 98, 101
Bellman equation, 12
Bellman's principle, 99
Blind range, 230
Bound, 105, 109
Boundary fragment class, 192
Box-counting dimension, 190
Brownian motion, 65
Burnashev-Zigangirov algorithm, 183
BZ algorithm, 183

Carrier, 222
Cartesian product, 102
CFAR, 71
Chapman-Kolmogorov equation, 275
Chernoff bound, 39, 187
Chernoff exponent, 39
Chirp waveform, 225
Chirp waveform library, 253
Classification, 203
Classification reduction of optimal policy search, 43
Closed loop, 102
Combined innovation, 237
Communication resource, 101
Completely observable rollout, 114
Complexity regularized estimator, 192
Conditional entropy, 270
Constant false alarm rate, 71
Continuation set, 131
Control architecture, 111
Control, receding horizon, 100
Controller, 102
Coupled partitions, 75
Covariance update, 247
CRB, 211
CROPS, 43
Cross ambiguity function, 224
Cross-cue, 260
Cusp-free boundary set, 196

D-optimal experimental design, 210
DAI, 124
DARPA, 261
Data fusion, 262
Discount factor, 276
Discounted reward MDP, 13
Divergence, 96
    Alpha-divergence, 96, 107
    Renyi-divergence, 96, 107
Domain knowledge, 106, 118
Dominating control law, 137
Dominating machine, 136, 137

Doppler aliasing, 232
Doppler processing, 230, 232, 233
DTT, 261
DTT:TEST, 265
Dynamic allocation index, 124
Dynamic programming, 122

Electrically scanned array, 97
Electromagnetic induction sensor, 213
EMI, 213
Empirical loss, 201
Expected information gain, 42
Expected loss, 201
Expected value-to-go, 106

f-divergence, 37
Feature, 108
Feedback, 98
Filter, 102
    Auto-ambiguity function, 224
    Bayesian filter, 64
    Cross-ambiguity function, 224
    Extended Kalman filter, 114
    IPDA tracker, 235
    Kalman filter, 103, 114
    Matched filter, 224, 231
    Measurement filter, 102
    Multi-target particle filter, 73
    Non-linear filter, 65
    Particle filter, 71, 72, 103, 111, 114, 118
    PDA tracker, 235, 241
    SIR particle filter, 71
    Unscented Kalman filter, 114
Fisher information, 40, 234
Fisher information matrix, 233
Fokker-Planck equation, 65
Foresight optimization, 109
Forward induction, 129
Forward prediction, 247
FPE, 65
Function approximator, 107

Gauss-Markov model, 235
Gaussian, 103, 114
Gittins index, 124, 125, 129, 131–133, 138, 141, 146, 280
GMTI, 60, 260
Gridlock, 261
Ground surveillance, 97

Heterogeneous sensors, 96
Heuristics, 106
Hidden Markov model, 276
Hindsight optimization, 109
Hitting time, 280
Homogeneous Markov process, 274
Horizon, 96, 99
Hyperspectral imaging, 54, 55, 258

ID, 258
IMM, 245
Independent partitions, 75
Index-type allocation policies, 122
Indexability, 143
Information divergence, 37
Information gain, 42, 96, 107, 115
Information gain sandwich bound, 45
Information state, 41, 98, 276
Innovation, 236
Innovation covariance matrix, 236
IPDA tracker, 235
ISAR, 60
ISR, 60, 257
Itô equation, 65

JMPD, 59
Joint multi-target probability density, 59
Joint probabilistic data association, 61
JPDA, 61
JPG, 213
JSTARS, 48, 257

Kalman filter, 103, 114, 278
Kalman gain, 247
Kalman update, 247
Kalman update equations, 237, 247
Kernel matching pursuits, 215
Keyhole spacecraft, 258
Kinematic prior, 64
KL divergence, 271
Klimov's problem, 147
KMP, 215
KP, 64

Landmine sensing, 220
Landsat radar, 54
LFM, 225, 243, 253
Likelihood ratio test, 39
LMIPDA, 245
Lookahead, 100

MAB, 121
Magnetometer sensor, 213
MAP detector, 39
Marginalized information gain, 46
Markov chain, 238, 239, 246, 273
Markov decision process, 10, 13
Markov process, 10
Markov property, 273
Markov reward processes, 275
Matched filter, 231
MDP, 10, 13
Measure of effectiveness, 252
Measurement, 102
Measurement filter, 102

INDEX

MHT, 61
MI, 42
MIG, 46
Mobile sensor, 96
Model sensitivity, 51
Monostatic radar, 230
Monte Carlo, 103, 104
Moving target, 96
Moyal's Identity, 232
Moyal's identity, 224
MTI radar, 48
Multi-armed bandit, 121, 122
    Arm-acquiring bandit, 137
    Bandit process, 123
    Classical multi-armed bandit, 123
    Multiple Plays, 140
    Restless bandits, 142
    Superprocess, 134
    Switching Penalties, 138
Multi-target tracking, 48
Multiple hypothesis tracking, 61
Mutual information, 42, 251, 252, 254
Myopic policy, 125

Narrow band approximation, 223
National asset, 258
Neurodynamic programming, 108
Neyman-Pearson detector, 243

Objective function, 99
Obscuration, 97
Observable, 102
Observation, 102
Observation law, 103, 276
Observation space, 276
OID, 72
Optimal design of experiments, 210
Optimal policy, 98

Parallel rollout, 111
Parametric approximation, 107
Partial conservation laws, 144
Partially observable Markov decision process, 19
Partially observable Markov processes, 276
Partially Observed Markov Decision Problems, 19
Particle filter, 103, 111, 114, 118
Passive learning rate, 193
PDA tracker, 235, 237
Phased array antennas, 226
PMHT, 61
Policy, 11, 99, 124
    Adaptive policy, 30
    Admissible policy, MDP, 11
    Admissible policy, POMDP, 20
    Base policy, 110, 111, 114
    Completely observable rollout, 114
    CROPS, 43
    EVTG approximation, 117
    Forward induction policy, 126
    Index-type policy, 124
    Information gain, 47
    Information gain policy, 117
    Iteration operator, 16
    Markov policy, 11, 13
    MDP policy, 12
    Multistep lookahead policy, 125
    Myopic, 95
    Myopic policy, 41, 117, 125
    Non-myopic, 95
    Optimal policy, 12
    Parallel rollout, 111
    Policy iteration, 19, 110
    POMDP policy, 20
    Random policy, 117
    Rollout, 110
    Search, 42
    Single stage policy, 13
    Stationary, 100
    Stationary policy, 13
Policy improvement, 110
Policy iteration, 19
Policy, optimal, 98
POMDP, 19, 41, 95, 278
POMDP approximation, 95
popup threat, 259
Predator, 262
PRI, 221
Principal components, 208
Principle of Optimality, 12
Probabilistic bisection algorithm, 182
Probabilistic multiple hypothesis tracker, 61
Probability of decision error, 39
Probability space, 278
Proxy for performance, 45
Pulse compression, 225
Pulse-Doppler radar, 225

Q-function, 103
Q-learning, 108
Q-value, 100
Q-value approximation, 98, 104

Rényi divergence, 37
Rényi entropy, 36
Radar
    FOPEN, 258
    Hyperspectral imaging, 258
    Laser, 258
    Pulse-Doppler radar, 225
Radar system, 221, 257
Radar:Beam scheduling, 245
Range aliasing, 232
Range ambiguity, 228

Ranking, 101, 104
RDP, 191
Receding horizon, 100
Receiver operating characteristic, 215
Recursive dyadic partitions, 191
Reduction to classification, 43, 115
Regret, 106
Reinforcement learning, 108
Relaxation, 105
Resource management, 95
Revisit time, 245–249
Reward, 99, 275
Reward surrogation, 115
Riccati equations, 236
ROC, 215
Rollout, 110

Sampling importance resampling, 71
Scanned array, 97
Scheduling policy, 124
SDP, 124
Sensor motion, 96
Sensor scheduling, 8, 31, 34, 53, 60, 96, 110, 154, 157, 163, 221, 232, 234, 259
Sensor trajectory, 97
Sensor usage cost, 113
Sequential data selection, 210
Shannon entropy, 270
Shannon entropy policy, 42
Shannon mutual information, 272
Shannon, Claude, 35
Sigma-field, 278
SIR, 71
SNCR, 60
State, 98
State space, 98
State-transition law, 274
State-transition matrix, 274
State-transition probability, 274
Stationary, 100
Stationary MDP, 13
Stationary policy, 100
Stochastic matrix, 274
Stopping set, 131

Stopping time, 126, 279
Sufficient statistic, 22, 131
Surrogate reward, 115
Surveillance, 97
Switching index, 139, 140

T-step-look-ahead policy, 125
Tactical asset, 258
Target identification, 46
Target motion, 96
Target tracking, 46, 96, 102, 106, 110
Tax problem, 147
Terrain classification, 54
Terrain elevation, 97
Theater asset, 258
Time-homogeneous Markov process, 274
Topographical map, 97
Total reward MDP, 13
Track existence, 239
Track existence, 238, 247
Track life, 264
Tracking, 96, 102, 106, 110, 163
Tracking error, 113
Training, 108
Transition law, 103, 274
Transition matrix, 274
Transition probability, 274
Twenty questions game, 178

U-2, 259
UAV, 258
UGS, 258
Uncertainty reduction measures, 41
Unobservable states, 102
UXO, 203, 213

Validation gate, 235, 238, 242
Value iteration, 18
Value-to-go, 106

Waveform libraries, 250
Waveform library utility, 251
Waveform scheduling, 234
Waveform selection, 55

SIGNALS AND COMMUNICATION TECHNOLOGY

(continued from page ii)

Digital Interactive TV and Metadata
Future Broadcast Multimedia
A. Lugmayr, S. Niiranen, and S. Kalli
ISBN 3-387-20843-7

Adaptive Antenna Arrays
Trends and Applications
S. Chandran (Ed.)
ISBN 3-540-20199-8

**Digital Signal Processing
with Field Programmable Gate Arrays**
U. Meyer-Baese
ISBN 3-540-21119-5

**Neuro-Fuzzy and Fuzzy Neural Applications
in Telecommunications**
P. Stavroulakis (Ed.)   ISBN 3-540-40759-6

SDMA for Multipath Wireless Channels
Limiting Characteristics
and Stochastic Models
I.P. Kovalyov   ISBN 3-540-40225-X

Digital Television
A Practical Guide for Engineers
W. Fischer   ISBN 3-540-01155-2

Speech Enhancement
J. Benesty (Ed.)
ISBN 3-540-24039-X

Multimedia Communication Technology
Representation, Transmission
and Identification of Multimedia Signals
J.R. Ohm   ISBN 3-540-01249-4

Information Measures
Information and its Description in Science
and Engineering
C. Arndt   ISBN 3-540-40855-X

Processing of SAR Data
Fundamentals, Signal Processing,
Interferometry
A. Hein   ISBN 3-540-05043-4

Chaos-Based Digital Communication Systems
Operating Principles, Analysis Methods, and
Performance Evalutation
F.C.M. Lau and C.K. Tse
ISBN 3-540-00602-8

Adaptive Signal Processing
Application to Real-World Problems
J. Benesty and Y. Huang (Eds.)
ISBN 3-540-00051-8

**Multimedia Information Retrieval and
Management Technological**
Fundamentals and Applications D. Feng, W.C.
Siu, and H.J. Zhang (Eds.)
ISBN 3-540-00244-8

Structured Cable Systems
A.B. Semenov, S.K. Strizhakov, and I.R.
Suncheley
ISBN 3-540-43000-8

UMTS
The Physical Layer of the Universal Mobile
Telecommunications System
A. Springer and R. Weigel
ISBN 3-540-42162-9

Advanced Theory of Signal Detection
Weak Signal Detection in Generalized
Obeservations
I. Song, J. Bae, and S.Y. Kim
ISBN 3-540-43064-4

Wireless Internet Access over GSM and UMTS
M. Taferner and E. Bonek
ISBN 3-540-42551-9

Printed in the United States of America